A Glimpse into Future Research on Microalgae Diversity, Ecology and Biotechnology

A Glimpse into Future Research on Microalgae Diversity, Ecology and Biotechnology

Editors

Carmela Caroppo
Patrizia Pagliara

MDPI • Basel • Beijing • Wuhan • Barcelona • Belgrade • Manchester • Tokyo • Cluj • Tianjin

Editors

Carmela Caroppo
National Research Council
Water Research Institute
Taranto
Italy

Patrizia Pagliara
Department of Biological and Environmental Sciences and Technologies
University of Salento
Lecce
Italy

Editorial Office
MDPI
St. Alban-Anlage 66
4052 Basel, Switzerland

This is a reprint of articles from the Special Issue published online in the open access journal *Microorganisms* (ISSN 2076-2607) (available at: www.mdpi.com/journal/microorganisms/special_issues/research_microalgae).

For citation purposes, cite each article independently as indicated on the article page online and as indicated below:

LastName, A.A.; LastName, B.B.; LastName, C.C. Article Title. *Journal Name* **Year**, *Volume Number*, Page Range.

ISBN 978-3-0365-6995-6 (Hbk)
ISBN 978-3-0365-6994-9 (PDF)

Contents

About the Editors vii

Preface to "A Glimpse into Future Research on Microalgae Diversity, Ecology and Biotechnology" ix

Carmela Caroppo and Patrizia Pagliara
Microalgae: A Promising Future
Reprinted from: *Microorganisms* **2022**, *10*, 1488, doi:10.3390/microorganisms10081488 1

Veronika Dashkova, Dmitry V. Malashenkov, Assel Baishulakova, Thomas A. Davidson, Ivan A. Vorobjev and Erik Jeppesen et al.
Changes in Phytoplankton Community Composition and Phytoplankton Cell Size in Response to Nitrogen Availability Depend on Temperature
Reprinted from: *Microorganisms* **2022**, *10*, 1322, doi:10.3390/microorganisms10071322 5

Joana Barcelos e Ramos, Susana Chaves Ribeiro, Kai George Schulz, Francisco José Riso Da Costa Coelho, Vanessa Oliveira and Angela Cunha et al.
Emiliania huxleyi—Bacteria Interactions under Increasing CO_2 Concentrations
Reprinted from: *Microorganisms* **2022**, *10*, 2461, doi:10.3390/microorganisms10122461 27

Alexander Okhapkin, Ekaterina Sharagina, Pavel Kulizin, Natalja Startseva and Ekaterina Vodeneeva
Phytoplankton Community Structure in Highly-Mineralized Small Gypsum Karst Lake (Russia)
Reprinted from: *Microorganisms* **2022**, *10*, 386, doi:10.3390/microorganisms10020386 47

Loredana Stabili, Margherita Licciano, Adriana Giangrande and Carmela Caroppo
Filtration of the Microalga *Amphidinium carterae* by the Polychaetes *Sabella spallanzanii* and *Branchiomma luctuosum*: A New Tool for the Control of Harmful Algal Blooms?
Reprinted from: *Microorganisms* **2022**, *10*, 156, doi:10.3390/microorganisms10010156 67

Ewa Żymańczyk-Duda, Sunday Ocholi Samson, Małgorzata Brzezińska-Rodak and Magdalena Klimek-Ochab
Versatile Applications of Cyanobacteria in Biotechnology
Reprinted from: *Microorganisms* **2022**, *10*, 2318, doi:10.3390/microorganisms10122318 81

Patrizia Pagliara, Giuseppe Egidio De Benedetto, Matteo Francavilla, Amilcare Barca and Carmela Caroppo
Bioactive Potential of Two Marine Picocyanobacteria Belonging to *Cyanobium* and *Synechococcus* Genera
Reprinted from: *Microorganisms* **2021**, *9*, 2048, doi:10.3390/microorganisms9102048 101

Dante Matteo Nisticò, Amalia Piro, Daniela Oliva, Vincenzo Osso, Silvia Mazzuca and Francesco Antonio Fagà et al.
A Combination of Aqueous Extraction and Ultrafiltration for the Purification of Phycocyanin from *Arthrospira maxima*
Reprinted from: *Microorganisms* **2022**, *10*, 308, doi:10.3390/microorganisms10020308 117

Amalia Piro, Dante Matteo Nisticò, Daniela Oliva, Francesco Antonio Fagà and Silvia Mazzuca
Physiological and Metabolic Response of *Arthrospira maxima* to Organophosphates
Reprinted from: *Microorganisms* **2022**, *10*, 1063, doi:10.3390/microorganisms10051063 131

Alessia Bani, Katia Parati, Anna Pozzi, Cristina Previtali, Graziella Bongioni and Andrea Pizzera et al.
Comparison of the Performance and Microbial Community Structure of Two Outdoor Pilot-Scale Photobioreactors Treating Digestate
Reprinted from: *Microorganisms* **2020**, *8*, 1754, doi:10.3390/microorganisms8111754 . . . . **149**

Diogo Fleury Azevedo Costa, Joaquín Miguel Castro-Montoya, Karen Harper, Leigh Trevaskis, Emma L. Jackson and Simon Quigley
Algae as Feedstuff for Ruminants: A Focus on Single-Cell Species, Opportunistic Use of Algal By-Products and On-Site Production
Reprinted from: *Microorganisms* **2022**, *10*, 2313, doi:10.3390/microorganisms10122313 . . . . **173**

About the Editors

Carmela Caroppo

Dr. Carmela Caroppo received her Master's degree in Biological Sciences (1988) and specialization in Marine Biochemistry (1994) from the University of Bari (Italy). Currently, she has been working as a researcher at the Water Research Institute of the National Research Council. Her major research interest in the ecology of phytoplankton in various marine ecosystems, from temperate to polar waters. She also worked on the impact of harmful microalgae on aquaculture and on the isolation and culturing of microalgal and cyanobacterial strains for biotechnological applications. She participated in several national and international projects and is the author of about 100 publications in national and international journals.

Patrizia Pagliara

Dr. Patrizia Pagliara is a researcher at the Department of Biological and Environmental Sciences and Technologies (University of Salento, Italy) since 2001. She received her PhD in Biology and Biotechnologies in 2003. The main areas of interest include the invertebrates immune defense strategies and the identification and characterization of novel bioactive substances from marine organisms. A fruitful collaboration with Dr. Carmela Caroppo from CNR has made it possible to know and study the world of microorganisms, focusing on their bioactive compounds. She is author of more than 60 papers published in national and international journals.

Preface to "A Glimpse into Future Research on Microalgae Diversity, Ecology and Biotechnology"

Microalgae are photosynthetic unicellular microorganisms that represent an extremely important component of the aquatic ecosystem productivity, diversity, and functioning. Moreover, these microorganisms, using a network of signals, interact with all the other organisms present in their environment. Signals are often secondary metabolites that play an important role in competition, defense, attraction, and signaling. These molecules are recognized for having bioactive properties, but some of them are still largely underexplored and underexploited. This Special Issue focuses on studies that aim to improve knowledge on microalgal ecology (diversity and dynamics) in aquatic ecosystems, as well as on their capacity to produce bioactive compounds with potential biotechnological applications.

The value of ecological papers in this Special Issue derives from the increase in knowledge of phytoplankton diversity in specific environments (freshwater and brackish systems), but also from providing data on the effects of the increase in CO_2 concentrations in the era of global changes in a cosmopolitan marine microalgal species. Another interesting aspect concerns the biological remediation of microalgal species that are considered "harmful"due to their ability to produce toxins that negatively impact human health.

Investigations on microalgal bioactive compounds are a very promising area in full development. The growing interest is due to the wide range of applications including the food industry, agrochemicals, cosmetics, and pharmaceutical products. The studies presented in this Special Issue range from the biological activity of two new species to the production of protocol for the recovery, fractionation, and purification of high-value molecules (phycocyanin) and the effects of an herbicide (glyphosate) on a cyanobacterial species cultivated for the production of animal and human food additives.

Carmela Caroppo and Patrizia Pagliara

Editors

microorganisms

Editorial

Microalgae: A Promising Future

Carmela Caroppo [1,*] and Patrizia Pagliara [2,*]

1 Water Research Institute, National Research Council (IRSA-CNR), 74123 Taranto, Italy
2 Department of Biological and Environmental Sciences and Technologies, University of Salento, 73100 Lecce, Italy
* Correspondence: carmela.caroppo@irsa.cnr.it (C.C.); patrizia.pagliara@unisalento.it (P.P.)

Microalgae are photosynthetic unicellular microorganisms that represent an extremely important component of the aquatic ecosystem productivity, diversity, and functioning [1]. Particularly, phytoplankton, adapted to live in suspension in water masses, provide about half (49%) of the global net primary production in marine and freshwater systems [2]. Moreover, microalgae exhibit a high diversity and include species highly distant from an evolutionary point of view [3]. Despite their small size, which ranges between 0.2 and 200 μm, they can contribute to climate change mitigation through carbon fixation [4,5]. Indeed, biological capture and sequestration of carbon using microalgae have been recognized as one of the world's most important and effective carbon sequestration methods [6,7]. The promising technique can allow CO_2 capture and recycle into biomass, which in turn could be useful to produce bioenergy and other value-added products. However, because this becomes efficient, it is important to continue investing widely in the research and development of technology and in the knowledge of the microalgal world.

In our day, climate change represents a serious problem, also because these changes hardly affect trophic structure and dynamics of aquatic ecosystems as well as phenology, and physiological and life-history traits of organisms.

In the Special Issue "A Glimpse into Future Research on Microalgae Diversity, Ecology and Biotechnology", studies aimed to improve knowledge on the effects of climate changes on phytoplankton communities' composition and dynamics in freshwater [8] and brackish [9] ecosystems. Dashkova and co-authors [8], using a eutrophic shallow lake mesocosm as a model, demonstrated how the dynamics of structural and morphological changes of phytoplankton responded to N availability under different temperature conditions. These results could contribute to forecast climate change effects on the world's shallow lake ecosystems. Moreover, the use of recent complementary technology to traditional microscopy such as imaging flow cytometry (IFC) has proved useful for the morphological and structural analysis of phytoplankton and as a useful tool for routine monitoring programs of aquatic ecosystems.

On the other hand, Okhapkin and co-authors [9] carried out their research in an atypical brackish system (the small gypsum karstic Lake Klyuchik, Middle Volga basin), characterized by high values of water mineralization and low temperatures. This unique and complex ecosystem, for its peculiarity, represents an interesting model system for the investigation of phytoplankton diversity, largely unexplored in this kind of environment. The value of this research derives not only from the increase in knowledge of phytoplankton diversity in specific conditions, but also from providing helpful tools to decision makers for the management and protection of environments of high naturalistic value.

Another ecological paper reported in this Special Issue focused on the bioremediation of the microalgal species that are considered "harmful" for their ability to produce toxins that negatively impact human health, and to induce blooms with detrimental effects on aquatic ecosystems, fisheries, aquaculture, and tourism [10]. As control actions are urgently needed to suppress harmful algal blooms, Stabili and co-authors [11] demonstrated the high filter-feeding capacity of two sabellid polychaetes, *Branchiomma luctuosum* and *Sabella*

Citation: Caroppo, C.; Pagliara, P. Microalgae: A Promising Future. *Microorganisms* **2022**, *10*, 1488. https://doi.org/10.3390/microorganisms10081488

Received: 20 July 2022
Accepted: 22 July 2022
Published: 24 July 2022

spallanzanii, on a harmful microalga (the dinoflagellate *Amphidinium carterae*). Even if preliminary, these results could represent a sustainable and environmentally friendly method for the restoration of the aquatic ecosystems and a very advantageous tool for the management of the aquaculture plans.

In the field of bioeconomy, the use of innovative processes is extremely important to produce biomaterials and bioenergy, reducing, in the meantime, the consumption of virgin resources. Microalgae-based wastewater treatment represents a valid contribution to this practice. This activity has recently received attention due to its low energy demand, the robust capacity of microalgae to grow under different environmental conditions, and the possibility to recover and transform wastewater nutrients into highly valuable bioactive compounds [12].

Outdoor open systems (i.e., raceway ponds, bubble columns, flat panels) are considered so far as the most viable method of microalgal cultivation on wastewaters [13], even if complex and dynamic microbial consortia are formed inside [14,15]. A contribution to this field derives from a study, included in this Special Issue, which examined and compared the nutrient removal efficiency, biomass productivity, and microbial community structure of two outdoor pilot-scale photobioreactors (bubble column and raceway pond) [13]. Data obtained, including the characterization of bacterial and eukaryotic communities using a metabarcoding approach and quantitative PCR, highlighted a different behavior and a different composition of the microbial communities, which were subjected to variations of the environmental conditions as well as of the reactors' operational parameters. Knowing the factors that influence the structure and dynamics of the microbial consortia allowed establishing which parameters influenced the performance of the reactors. These data encourage the use of microalgal polycultures for a more stable production of biomass in outdoor cultivation systems.

In recent years, an increased interest aimed at expanding knowledge on microalgae has been recorded as they represent a world still partially unknown, but which has many intriguing application aspects being considered as next-generation resources with the potential to address urgent industrial and agricultural demands. In this framework, we must consider that microalgae use a network of signals to interact with all the other organisms living in their environment. The signals are often secondary metabolites generally not necessary for their daily functioning, but they play an important role in competition, defense, attraction, and signaling. These molecules that include pigments, sterols, and polyunsaturated fatty acids can have antioxidant and anti-inflammatory effects [16]. However, even if they are recognized for having bioactive properties, they are still largely underexplored and underexploited.

Among microalgae, cyanobacteria are one of the most investigated microorganisms for biotechnological purposes, as three reports of the Special Issue demonstrate [17–19].

Arthrospira platensis, known as *Spirulina*, is a cyanobacterium with multiple nutritional and therapeutic properties [20]. These algae are in fact rich in proteins (60%–70% by weight), vitamins (4% by weight), essential amino acids, minerals (zinc, selenium, magnesium), essential fatty acids (Linoleni-Co acid), carotenoids, chlorophylls, and phytosterols. For this reason, their production is primarily destined for dietary supplement markets, and they are commercialized in many countries. *Arthrospira* is also a rich and inexpensive source of the pigment like phycocyanin [21,22]. Phycocyanin is a blue-red fluorescent, water-soluble, and non-toxic biliprotein pigment with recognized therapeutic properties, including antioxidant, anti-inflammatory, immune-modulatory, and anti-cancer activities. A combination of different techniques is generally used to achieve the purification of phycocyanin from crude algae extracts. Therefore, an increasingly addressed research towards product improvement has been observed in recent years. As an example, Nisticò and co-authors [17] developed a protocol to obtain a very high phycocyanin yield from *A. maxima* biomass. It represents a sustainable process for the recovery, fractionation, and purification of phycocyanin from a strain of *A. maxima* cultivated in a farm devoted to producing this molecule with food-grade purity. In addition, the authors adopted a combination of ultrafiltration and diafiltration that allowed the removal of about 91.7% of the DNA from the crude extract. This is of particular interest as it leads to an increase in purity degree in the retentate fraction, thus allowing the production of extracts suitable for other uses including therapeutic and biomedical applications.

Being that *Arthrospira* is a species largely cultivated for animal and human food additives and as a source of phycocyanin, a bioactive antioxidant molecule, it is important to know the effects of herbicides on this species. In this context, Piro and co-authors [18] investigated the effects of glyphosate, a broad-spectrum herbicide at the center of a large debate on potential toxicity. On the subject, there are different positions with different results, and this once again also happens regarding the effects on *Spirulina*. Indeed, the authors [18] evidenced that a deliberate treatment with glyphosate on the selected cultivation of this cyanobacterium negatively affected the biomass and the photosynthetic pigments, and it induced resistance in *A. maxima* survival. These results demonstrated that the resistance of *A. maxima* to glyphosate effects is mediated by a key enzyme until now not identified in other *Spirulina* species.

Investigations on bioactive compounds are a very promising area in full development. The growing interest is due to the wide range of applications including the food industry, agrochemicals, cosmetics, and pharmaceutical products. Although new bioactive molecules from the microbial world continue to be discovered, this however remains a field that is still little explored. With the investigations on two cyanobacterial strains, Pagliara and co-authors [19] intended to contribute to increasing the amount of information on the properties of these microorganisms, expanding the cyanobacteria range from which new compounds with significant bioactivity could be identified. The two strains, belonging to *Cyanobium* and *Synechococcus* genera, have been previously identified after their isolation from a Mediterranean marine sponge [23,24]. Intriguing is the presence in these cyanobacteria of important compound as BMAA, 2,4-DAB and microcystin, here evidenced for the first time in cyanobacteria isolated from a marine sponge. Moreover, the strong cytotoxic activity observed for aqueous and methanolic extracts of these two cyanobacteria laid the foundation to produce bioactive compounds of pharmacological interest.

The editors hope that the data reported in this Special Issue could represent a useful reference for researchers and managers interested in microalgae and their biotechnological applications.

Funding: This research received no external funding.

Conflicts of Interest: The authors declare no conflict of interest.

References

1. Naselli-Flores, L.; Padisák, J. Ecosystem services provided by marine and freshwater phytoplankton. *Hydrobiologia* **2022**. [CrossRef] [PubMed]
2. Friend, A.D.; Geider, R.J.; Behrenfeld, M.J.; Sill, C.J. Photosynthesis in Global Scale Models. In *Photosynthesis in Silico: Understanding Complexity from Molecules to Ecosystems*; Laisk, A., Nedbal, L., Govindjee, G., Eds.; Springer: Dordrecht, Germany, 2009; pp. 465–497.
3. De Vargas, C.; Audic, S.; Henry, N.; Decelle, J.; Mahé, F.; Logares, R.; Lara, E.; Berney, C.; Le Bescot, N.; Probert, I.; et al. Eukaryotic plankton diversity in the sunlit ocean. *Science* **2015**, *348*, 1261605. [CrossRef] [PubMed]
4. Ratnapuram, H.P.; Vutukuru, S.; Yadavalli, R. Mixotrophic transition induced lipid productivity in *Chlorella pyrenoidosa* under stress conditions for biodiesel production. *Heliyon* **2018**, *4*, e00496. [CrossRef] [PubMed]
5. Vu, C.H.T.; Lee, H.-G.; Chang, Y.K.; Oh, H.-M. Axenic cultures for microalgal biotechnology: Establishment, assessment, maintenance, and applications. *Biotechnol. Adv.* **2018**, *36*, 380–396. [CrossRef] [PubMed]
6. Moreira, D.; Pires, J.C. Atmospheric CO_2 capture by algae: Negative carbon dioxide emission path. *Bioresour. Technol.* **2016**, *215*, 371–379. [CrossRef] [PubMed]
7. Alami, A.H.; Alasad, S.; Ali, M.; Alshamsi, M. Investigating algae for CO_2 capture and accumulation and simultaneous production of biomass for biodiesel production. *Sci. Total Environ.* **2020**, *759*, 143529. [CrossRef] [PubMed]
8. Dashkova, V.; Malashenkov, D.V.; Baishulakova, A.; Davidson, T.A.; Vorobjev, I.A.; Jeppesen, E.; Barteneva, N.S. Changes in Phytoplankton Community Composition and Phytoplankton Cell Size in Response to Nitrogen Availability Depend on Temperature. *Microorganisms* **2022**, *10*, 1322. [CrossRef]

9. Okhapkin, A.; Sharagina, E.; Kulizin, P.; Startseva, N.; Vodeneeva, E. Phytoplankton Community Structure in Highly-Mineralized Small Gypsum Karst Lake (Russia). *Microorganisms* **2022**, *10*, 386. [CrossRef]
10. Karlson, B.; Andersen, P.; Arneborg, L.; Cembella, A.; Eikrem, W.; John, U.; West, J.J.; Klemm, K.; Kobos, J.; Lehtinen, S.; et al. Harmful algal blooms and their effects in coastal seas of Northern Europe. *Harmful Algae* **2021**, *102*, 101989. [CrossRef]
11. Stabili, L.; Licciano, M.; Giangrande, A.; Caroppo, C. Filtration of the Microalga *Amphidinium carterae* by the Polychaetes *Sabella spallanzanii* and *Branchiomma luctuosum*: A New Tool for the Control of Harmful Algal Blooms? *Microorganisms* **2022**, *10*, 156. [CrossRef]
12. López-Sánchez, A.; Silva-Gálvez, A.L.; Aguilar-Juárez, Ó.; Senés-Guerrero, C.; Orozco-Nunnelly, D.A.; Carrillo-Nieves, D.; Gradilla-Hernández, M.S. Microalgae-based livestock wastewater treatment (MbWT) as a circular bioeconomy approach: Enhancement of biomass productivity, pollutant removal and high-value compound production. *J. Environ. Manag.* **2022**, *308*, 114612. [CrossRef] [PubMed]
13. Bani, A.; Parati, K.; Pozzi, A.; Previtali, C.; Bongioni, G.; Pizzera, A.; Ficara, E.; Bellucci, M. Comparison of the Performance and Microbial Community Structure of Two Outdoor Pilot-Scale Photobioreactors Treating Digestate. *Microorganisms* **2020**, *8*, 1754. [CrossRef] [PubMed]
14. Ofiţeru, I.D.; Lunn, M.; Curtis, T.P.; Wells, G.F.; Criddle, C.S.; Francis, C.A.; Sloan, W.T. Combined niche and neutral effects in a microbial wastewater treatment community. *Proc. Natl. Acad. Sci. USA* **2010**, *107*, 15345–15350. [CrossRef]
15. Van Der Gast, C.J.; Ager, D.; Lilley, A.K. Temporal scaling of bacterial taxa is influenced by both stochastic and deterministic ecological factors. *Environ. Microbiol.* **2008**, *10*, 1411–1418. [CrossRef]
16. Lauritano, C.; Saide, A. Editorial of Special Issue "Microalgal Molecules and Enzymes". *Int. J. Mol. Sci.* **2021**, *22*, 13450. [CrossRef] [PubMed]
17. Nisticò, D.M.; Piro, A.; Oliva, D.; Osso, V.; Mazzuca, S.; Fagà, F.A.; Morelli, R.; Conidi, C.; Figoli, A.; Cassano, A. A Combination of Aqueous Extraction and Ultrafiltration for the Purification of Phycocyanin from *Arthrospira maxima*. *Microorganisms* **2022**, *10*, 308. [CrossRef] [PubMed]
18. Piro, A.; Nisticò, D.M.; Oliva, D.; Fagà, F.A.; Mazzuca, S. Physiological and Metabolic Response of *Arthrospira maxima* to Organophosphates. *Microorganisms* **2022**, *10*, 1063. [CrossRef] [PubMed]
19. Pagliara, P.; De Benedetto, G.E.; Francavilla, M.; Barca, A.; Caroppo, C. Bioactive Potential of Two Marine Picocyanobacteria Belonging to *Cyanobium* and *Synechococcus* Genera. *Microorganisms* **2021**, *9*, 2048. [CrossRef]
20. Wu, Q.; Liu, L.; Miron, A.; Klímová, B.; Wan, D.; Kuča, K. The antioxidant, immunomodulatory, and anti-inflammatory activities of *Spirulina*: An overview. *Arch. Toxicol.* **2016**, *90*, 1817–1840. [CrossRef]
21. Kato, T. Blue pigment from *Spirulina*. *New Food Indust.* **1994**, *29*, 17–21.
22. McCarty, M.F. Clinical potential of *Spirulina* as a source of phycocyanobilin. *J. Med. Food* **2007**, *10*, 566–570. [CrossRef] [PubMed]
23. Pagliara, P.; Caroppo, C. Cytotoxic and antimitotic activities in aqueous extracts of eight cyanobacterial strains isolated from the marine sponge *Petrosia ficiformis*. *Toxicon* **2011**, *57*, 889–896. [CrossRef] [PubMed]
24. Pagliara, P.; Barca, A.; Verri, T.; Caroppo, C. The Marine Sponge *Petrosia ficiformis* Harbors Different Cyanobacteria Strains with Potential Biotechnological Application. *J. Mar. Sci. Eng.* **2020**, *8*, 638. [CrossRef]

microorganisms

Article

Changes in Phytoplankton Community Composition and Phytoplankton Cell Size in Response to Nitrogen Availability Depend on Temperature

Veronika Dashkova [1,2,*], Dmitry V. Malashenkov [2,3,†], Assel Baishulakova [2], Thomas A. Davidson [4], Ivan A. Vorobjev [2], Erik Jeppesen [4,5,6,7] and Natasha S. Barteneva [2,8,*]

1 School of Engineering and Digital Sciences, Nazarbayev University, Nur-Sultan 00010, Kazakhstan
2 School of Sciences and Humanities, Nazarbayev University, Nur-Sultan 00010, Kazakhstan; dvmalashenkov@gmail.com (D.V.M.); assel.baishulakova@alumni.nu.edu.kz (A.B.); ivan.vorobyev@nu.edu.kz (I.A.V.)
3 National Laboratory Astana, Nur-Sultan 00010, Kazakhstan
4 Department of Ecoscience, Aarhus University Center for Water Technology (WATEC), 8000 Aarhus, Denmark; thd@bios.au.dk (T.A.D.); ej@bios.au.dk (E.J.)
5 Sino-Danish Centre for Education and Research, Beijing 100049, China
6 Limnology Laboratory, Department of Biological Sciences and Centre for Ecosystem Research and Implementation, Middle East Technical University, Ankara 06800, Turkey
7 Institute of Marine Sciences, Middle East Technical University, Erdemli-Mersin 33731, Turkey
8 The Environment & Resource Efficiency Cluster, Nazarbayev University, Nur-Sultan 00010, Kazakhstan
* Correspondence: veronika.dashkova@nu.edu.kz (V.D.); natalie.barteneva@nu.edu.kz (N.S.B.)
† Current address: Department of Hydrobiology, Faculty of Biology, Moscow State University, 119991 Moscow, Russia.

Citation: Dashkova, V.; Malashenkov, D.V.; Baishulakova, A.; Davidson, T.A.; Vorobjev, I.A.; Jeppesen, E.; Barteneva, N.S. Changes in Phytoplankton Community Composition and Phytoplankton Cell Size in Response to Nitrogen Availability Depend on Temperature. *Microorganisms* **2022**, *10*, 1322. https://doi.org/10.3390/microorganisms10071322

Academic Editors: Patrizia Pagliara and Carmela Caroppo

Received: 6 May 2022
Accepted: 15 June 2022
Published: 30 June 2022

Abstract: The climate-driven changes in temperature, in combination with high inputs of nutrients through anthropogenic activities, significantly affect phytoplankton communities in shallow lakes. This study aimed to assess the effect of nutrients on the community composition, size distribution, and diversity of phytoplankton at three contrasting temperature regimes in phosphorus (P)–enriched mesocosms and with different nitrogen (N) availability imitating eutrophic environments. We applied imaging flow cytometry (IFC) to evaluate complex phytoplankton communities changes, particularly size of planktonic cells, biomass, and phytoplankton composition. We found that N enrichment led to the shift in the dominance from the bloom-forming cyanobacteria to the mixed-type blooming by cyanobacteria and green algae. Moreover, the N enrichment stimulated phytoplankton size increase in the high-temperature regime and led to phytoplankton size decrease in lower temperatures. A combination of high temperature and N enrichment resulted in the lowest phytoplankton diversity. Together these findings demonstrate that the net effect of N and P pollution on phytoplankton communities depends on the temperature conditions. These implications are important for forecasting future climate change impacts on the world's shallow lake ecosystems.

Keywords: phytoplankton; biodiversity; biovolume; cell size; eutrophication; mesocosm; temperature; nitrogen pollution; climate change; imaging flow cytometry

1. Introduction

Ongoing climate change strongly impacts freshwater aquatic ecosystems, altering trophic structure and dynamics [1–4], phenology, physiological and life-history traits of organisms, and intensifying the magnitude of eutrophication [2,4]. It involves shifts in many environmental factors (temperature, CO_2 rise, nutrients loading) and significantly affects phytoplankton communities' diversity, composition, and planktonic cell size in shallow lakes [5–8]. The prevalence of potentially toxic cyanobacteria in freshwater lakes is increasing, being favored under warmer and nutrient-rich conditions [9]. The harmful algal blooms (HABs) produced by cyanobacteria deteriorate water quality [10], and accelerate eutrophication by releasing more N and P into the environment [11].

However, the impact of climate change on phytoplankton communities' composition and dynamics of freshwater ecosystems is not fully understood. Studies of phytoplankton communities' composition and the taxonomic diversity of phytoplankton in response to warming and its consequences, e.g., increased nutrient concentration provide contrasting results [12–16]. In the short-term, warmer temperatures were shown to decrease phytoplankton diversity in temperate water bodies [12] and promote changes in the dominance and dynamics of phytoplankton species [13]. Long exposure to warming may result in the adaptation of phytoplankton species to warmer conditions [14], and an increase in functional phytoplankton diversity [15]. Warming-associated increased nutrient loading leads to decreased phytoplankton diversity [16]. It is reported that intensified N pollution in P-enriched eutrophic lakes may lead to changes in the phytoplankton composition, depending on the concentration of N input. Low-to-moderate N concentrations were seen to lead to the dominance of potentially toxic cyanobacteria (such as *Microcystis* spp., *Planktothrix* spp.) [17–21], and high concentrations of N may result in the prevalence of green algae [17,20].

Traditional microscopy is a method of choice in exploring changes in phytoplankton communities and phytoplankton cell size [22–25]; however, it is time-consuming with limited sampling capacity [26]. The phytoplankton communities' structure is highly complex, and the use of other techniques such as spectral fluorometry and conventional flow cytometry is limited by a high degree of morphological heterogeneity and the microscopic size of the planktonic cells. Imaging flow cytometry (IFC) is a recent alternative and/or complementary approach to traditional microscopy. It allows for examining plankton communities, combining the large statistical power of flow cytometry and image-based analysis [27–30]. Large datasets of single-cell morphological and size parameters acquired via IFC enable a comprehensive evaluation of the response of plankton communities to environmental conditions [31–34]. It is a powerful technology for investigating microalgae [29] and enables the effective analysis of phytoplankton in a size range of ca. 10–300 μ [35]. To our knowledge, no research had explored the effect of N variation in different temperature scenarios on the dynamics of phytoplankton communities of shallow lake systems applying IFC.

This study aims to assess the temporal shifts in phytoplankton structure, biomass, size, and diversity in response to different N availability and temperature scenarios using eutrophic shallow lake mesocosms as a model. More specifically, we are interested in these questions: (1) Does N limitation leads to the dominance of phytoplankton taxa adapted to N-limited conditions, including N-fixing cyanobacteria? (2) Does N-limitation lead to decreased cell size and diversity due to the dominance of a few well-adapted species? (3) How does temperature variation impact phytoplankton community, particularly phytoplankton structure, biomass, cell/colony size, and diversity?

2. Materials and Methods

2.1. Experimental Setup

The study was conducted as part of the Lake Mesocosm Warming Experiment (LMWE) at the facility owned by Aarhus University, located at Lemming, Central Jutland, Denmark (56°14′ N, 9°31′ E). The LMWE has been running continuously since 2003 and it is the longest freshwater mesocosm experiment in the world. Overall, the facility includes 24 outdoor freshwater cylindrical steel tanks of 1.9 m diameter and 1.5 m depth with ventilated paddles installed at the bottom mixing the water to maintain uniform conditions. The water in mesocosms is sourced from groundwater with a water retention time of approximately 2.5 months [36]. The temperature regimes include unheated control with ambient water temperature (AMB) and two elevated temperature settings based on the IPCC climate scenarios for the period 2071–2100, IPCC A2 (ca. +3 °C) and IPCC A2 + 50%

(ca. +4.5 °C), with four replications for each regime (AMB tanks: A1, D1, F1, G1; IPCC A2 tanks: A2, D2, F2, G2; IPCC A2 + 50% tanks: A3, D3, F3, G3) (Figure 1).

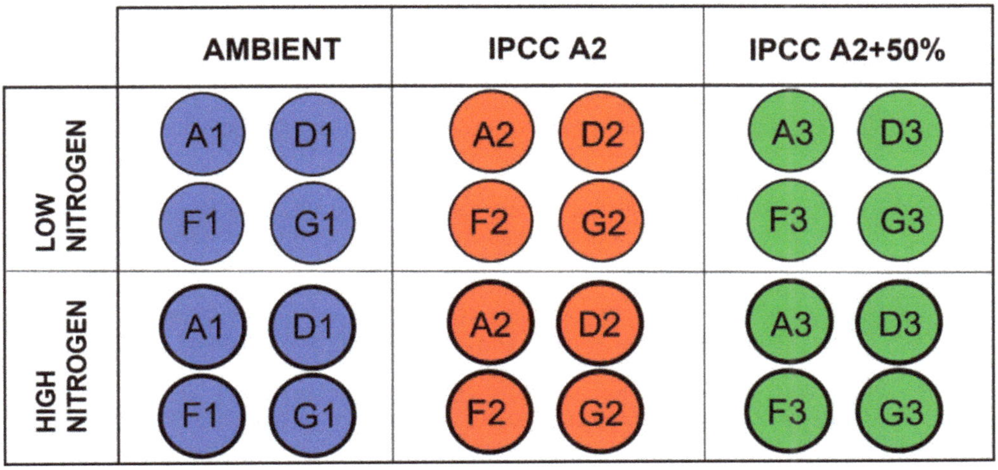

Figure 1. Schematic representation of the experimental design. Regular circles correspond to low nitrogen treatment (N0), and bold circles correspond to high nitrogen treatment (N+). Each temperature treatment has four replicate tanks: ambient temperature (A1, D1, F1, G1), IPCC A2 scenario (A2, D2, F2, G2), IPCC A2 + 50% scenario (A3, D3, F3, G3).

The water temperature is maintained in the heated tanks using an automatic heating system with reference to the ambient air temperature on that daytime throughout the seasons. The current study was run in 12 mesocosms with high nutrient level which have been constantly supplied with 108.6 mg N per m^3 per day and 2.7 mg P per m^3 per day in the form $Ca(NO_3)_2$ and Na_2HPO_4, respectively. In total, each mesocosm was artificially supplied with 2152 mg N and 54 mg P per week in addition to the nutrient inputs from groundwater of approximately 2–5 mg P and 30–63 mg N per week. As a part of the experiment, N input was terminated in nutrient-rich tanks for one year to study the effect of N limitation on the lake community under euthrophic conditions. Resumption of N input the following year allowed us to study the response of the aquatic community to the N addition in terms of composition, biomass, size and diversity changes. In June 2018, the N supply in the high nutrient mesocosms was terminated for one year, maintaining the P input only. During this period, only insignificant quantities of N were supplied with the groundwater. In June 2019, the N additions were resumed along with continuing P addition. The initial average nutrient concentrations were TN 2.40 mg L^{-1} and TP 165 µg L^{-1} for the period 2017–2018 [37]. To study the response of phytoplankton community to N addition in eutrophic shallow lake systems, we compared the biological parameters of the high N period when N addition was resumed with the control low N period when N supply was terminated. The experimental design, sampling period and analysis procedures were identical in both treatments.

2.2. Sample Collection and Measurement of Water Parameters

Integrated water samples were collected using a 1 m long tube water sampler weekly or biweekly from the HN tanks from June to October in 2018 and in 2019. The samples collected for imaging flow cytometry and light microscopy were fixed with 1% glutaraldehyde for preservation before the analyses. Water temperature, turbidity, conductivity, oxygen, and pH were measured using YSI probes (Xylem Inc., College Station, TX, USA). Freshly collected unpreserved samples were analyzed for total nitrogen (TN), total phosphorus (TP), a sum of nitrites and nitrates (NO_2 + NO_3), ammonium (NH_4), orthophosphates

(PO_4), and total Chl a as described in Søndergaard and co-authors [38], and total iron was measured spectrophotometrically (as described in Gibbs [39]).

2.3. FlowCAM Analysis and Microscopy

A benchtop FlowCAM VS-4 imaging particle analyzer (Yokagawa Fluid Imaging Technologies, Scarborough, ME, USA) equipped with 532 nm excitation laser was used for the analyses. In total, 324 phytoplankton samples were collected and processed. The samples were analyzed in laser trigger mode using 10× and 20× objectives as described earlier [40]. The obtained particle images were classified into taxonomic phytoplankton groups based on image and size filters in Visual Spreadsheet software version 4.0 (Yokagawa Fluid Imaging Technologies, Scarborough, ME, USA) and manual inspection. The nanoplankton group based on 5–20 μm (diameter ABD) size filter was excluded from the analysis as it mostly contained cell debris, especially in dense blooming samples. Identifiable cell images under 20 μm were manually transferred to one of the morphological groups. Quantitative parameters of each group, including cell concentration and size parameters, were then obtained from the software, and used for biomass estimation and statistical analysis.

Taxonomic identification of phytoplankton cells and colonies was performed using a Leica DM2500 microscope (Leica Microsystems, Wetzlar, Germany) equipped with differential interference contrast (DIC). Phytoplankton species were identified using 63× and 100× objectives.

2.4. Biovolume and Biomass Estimation

Size parameters such as width, length, diameter (ABD) were used for biovolume estimation of classified phytoplankton groups. Appropriate geometric shapes were assigned to the single cells and colonies of the present phytoplankton groups and biovolume was estimated using the corresponding formulae (according to Olenina [41], and Bergkemper and Weisse [42]. For unidentified species, FlowCAM biovolume estimations were based on ABD [43,44]. A full list of the shapes and formulae used is given in Table S1.

The biovolume estimates for most phytoplankton groups were obtained using size parameters retrieved from FlowCAM and standard microscopy biovolume formulae with some adaptations to two-dimensional image-based and colony-based size measurements (Table S1). The obtained biovolume estimates were converted to wet biomass units assuming that the plasma density equals 1 g cm^{-3}. Mean values of single-cell biomass measurements for each phytoplankton group and date were obtained and converted into mg per L units.

2.5. Statistical Data Analysis

The significance of water temperature variation between the two years was assessed using a non-parametric Kendall's Tau b test available in SPSS software (IBM, Armonk, NY, USA). Multivariate analyses, including analysis of similarities ANOSIM, analysis of similarity percentages SIMPER, and PERMANOVA were performed using Primer-e software v.7 (PRIMER-E Ltd., Auckland, New Zealand). For ANOSIM and SIMPER, square root transformation was applied to raw data with average biomass values for each date from mid-June to September 2018 and 2019 for all phytoplankton groups, and the results were then converted to a resemblance matrix based on Bray–Curtis similarity distances. Two separate one-way (A) ANOSIM analyses based on Spearman correlation were performed to test the similarity of the phytoplankton community between the N0 and N+ treatments and between the temperature treatments. Bootstrap averages (150 bootstrap averages per group) were performed based on the resemblance matrix separately for N and temperature factor and displayed in mMDS plots. Square root transformed data was used to perform one-way SIMPER analysis based on Bray–Curtis similarity separately for N as a factor and temperature. Repeated measures PERMANOVA was performed on the data converted to a resemblance matrix based on binomial deviance with temperature and treatment as fixed factors and a time factor nested in the treatment. Permutation of residuals was performed using a reduced model with 9999 permutations, selecting type III (partial) sum of squares. To test the variation within the factor groups, PERMDISP, a distance-based test for homogeneity of multivariate dispersions, was also performed separately for each factor. The distances are calculated to centroids and p-values are obtained using permutations.

Diversity indices were estimated as Pielou's evenness (J′), Shannon–Wiener diversity index (H′), and Simpson index of diversity (1-D) applying Primer-e software. Diversity estimators were calculated for each tank and date from mid-June to September for the N0 and N+ treatments. Average values for each tank from mid-June to September during the N0 and N+ treatments were then calculated. Average values from the replicate tanks, corresponding to the AMB, IPCC A2, and IPCC A2 + 50% treatments, were then used to calculate the percentage change between the N0 and the N+ treatments. A Wilcoxon Signed Rank Test was used to test if the observed changes differed significantly from 0.

Multivariate ordination analysis RDA was performed to infer relationships between phytoplankton data and environmental parameters using Canoco 5 software (Microcomputer Power, Ithaca, NY, USA). The response data had a gradient of 2.7 SD units for phytoplankton classes and 3.8 SD units for phytoplankton groups, suggesting that the optimum solution is to use a linear RDA method in the first case and a unimodal method CCA in the latter case. However, the two types of analyses may still be applied for both, and RDA was the preferred method. The analysis initially considered the data on 12 species and 11 environmental variables (TN, NO_2 + NO_3, NH_4, TP, PO_4, TFe, Chl a, N:P, temperature, conductivity, turbidity). Before the analysis, the species data were log-transformed and centered by species. The significance of canonical axes was tested using 4999 time series permutations. Relationships between the phytoplankton groups and environmental factors were evaluated based on RDA bi-plots and explanatory response tables containing regression coefficients of phytoplankton group × environmental factor combination pairs.

Single-cell measurements of area-based (ABD) diameter retrieved from VisualSpreadsheet software for each tank from the period mid-June to September were used to estimate the relative frequency distribution as percentages using GraphPad Prism software (Dotmatics, Boston, MA, USA). The average of the single dates and standard deviation (SD) were calculated for each bin class for each tank for the N0 and N+ periods. Averages of the replicate tanks and SD were then calculated and used to plot the size distribution graphs. Descriptive statistics (mean, median, 25% percentile, 75% percentile) of the ABD measurements were obtained for the temperature treatments based on the average values of the replicate tanks and used to calculate the percentage change between the N0 and N+ treatments. The data were tested for normality using Anderson–Darling, D'Agostino–Pearson, Shapiro–Wilk, and Kolmogorov–Smirnov tests. One-sample t-test and Wilcoxon Signed Rank Test were used to test if the observed changes differed significantly from 0.

3. Results

3.1. Environmental Changes

The simulated temperature regimes (IPCC A2 and IPCC A2 + 50% scenarios) differed, as expected, from the ambient temperature during the study period (Figure 2).

Mean water temperature did not differ significantly between the N0 and N+ sampling seasons (Figure 2; Table S2), implying that the observed differences in phytoplankton composition between the two treatment regimes can be attributed to the experimental conditions and not natural year-to-year variations in temperature.

Variations of the main environmental variables, including total nitrogen (TN), inorganic forms of nitrogen (NO_2 + NO_3 and NH_4), total phosphorus (TP), and phosphates (PO_4), in the N0 and N+ treatments are displayed in Figures 3 and S1.

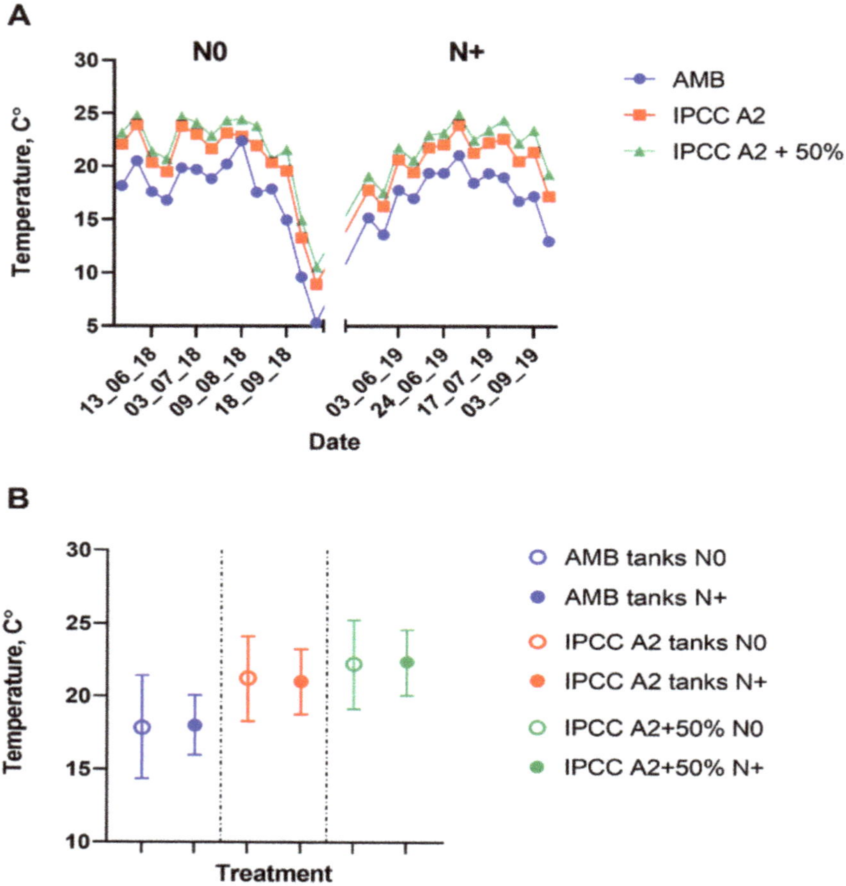

Figure 2. Water temperature measurements in the individual tanks. (**A**) Water temperature in the ambient temperature (AMB), IPCC A2, and IPCC A2 + 50% tanks from mid-June to September during the N0 and N+ treatments. (**B**) Mean water temperature for the mid-June–September period in AMB, IPCC A2 and IPCC A2 + 50% during the N0 and N+ treatments. Bars indicate SD of the mean.

Overall, large variability in nutrient concentrations among the individual replicate tanks was observed (Figure 3). Thus, TN and PO_4 concentrations differed between the temperature treatments. TN was relatively higher in AMB than in IPCC A2 + 50%, whereas PO_4 was the highest in IPCC A2 + 50% (Figure 3A,C). However, no clear effect of N treatment on TN was observed; TN varied between 0–4.3 mg/L and 0.1–5.4 mg/L during the N0 and N+ treatments, respectively. By contrast, NO_2 + NO_3 tended to increase in all temperature treatments after N addition resuming and was significantly higher in AMB and IPCC A2 than in IPCC A2 + 50%. NH_4, TP, and PO_4 concentrations did not respond significantly to the N treatment.

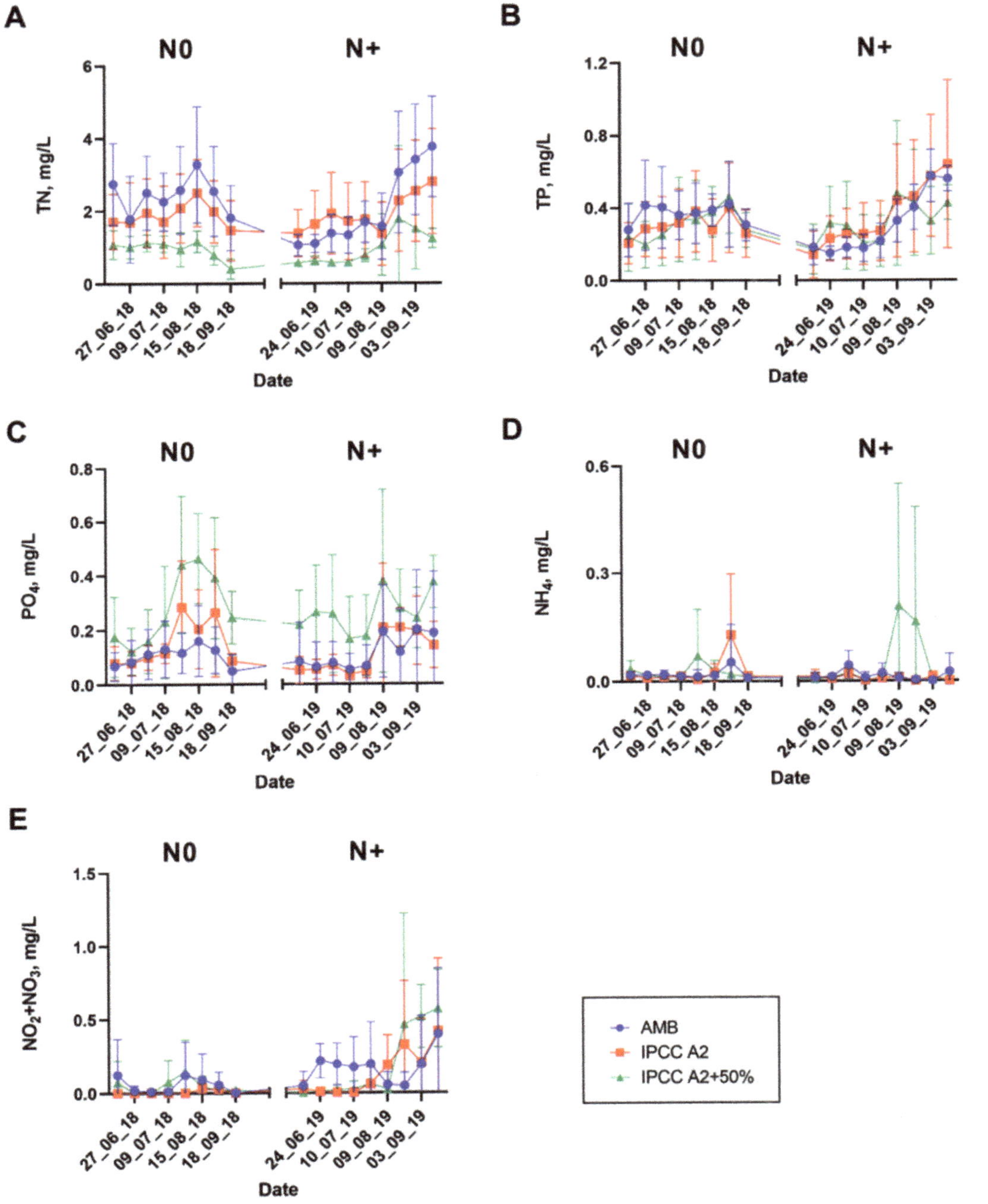

Figure 3. Variations in nutrients from mid-June to September during the N0 and N+ periods: (**A**) TN; (**B**) TP; (**C**) PO_4; (**D**) NH_4; (**E**) $NO_2 + NO_3$. Bars—SD of the mean.

3.2. Total Phytoplankton Biomass and Composition

The phytoplankton community was composed of taxonomic groups belonging to Cyanophyta, Chlorophyta, Cryptophyta, Miozoa, Bacillariophyta, and Euglenozoa phyla (Table S3; Figure 4).

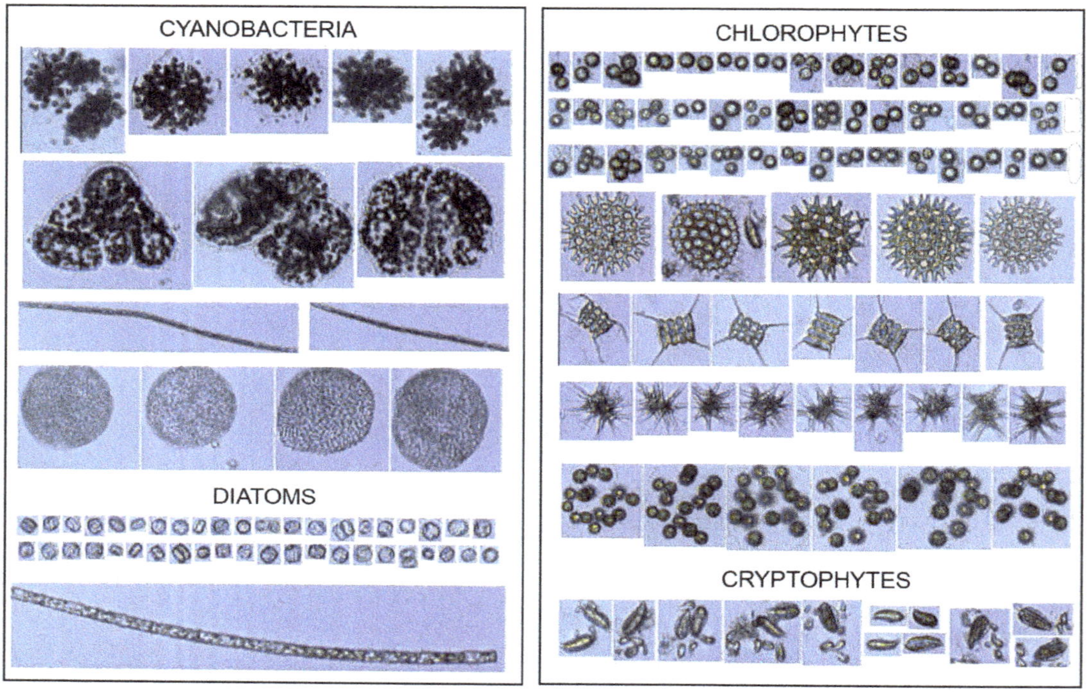

Figure 4. FlowCAM image library showing the different phytoplankton groups (10× objective).

Sixteen major morphologically distinct groups were classified using a FlowCAM imaging flow cytometer (Table S3).

Total phytoplankton biomass varied significantly among the individual tanks. Generally, the lowest total phytoplankton biomass was found in the A tanks (A1, A2, A3) and in the tanks with the highest temperature (IPCC A2 + 50%). A1–A3 tanks had an extremely low cell biomass throughout the experiment and occasional maximum peak values were associated with the development of cyanobacteria (A1 = 290 mg/L) and chlorophytes (A2 = 79 mg/L) (Figure 5A,B).

The community biomass of IPCC A2 + 50% was mainly composed of cryptophytes and chlorophytes. Cryptophytes dominated the total biomass in the N0 treatment and had several blooms; however, when the N supply was resumed a shift in dominance to chlorophytes occurred (Figure 5).

Except for A1–A2 tanks, AMB and IPCC A2 temperature tanks had relatively higher phytoplankton biomasses, reaching 17×10^5 mg/L of total biomass during blooming events in some of the tanks (Figure 5). Although there was no uniform response of total biomass to the N treatment in AMB and IPCC A2, common trends in phytoplankton composition occurred. During the N0 treatment cyanobacteria dominated the summer community in most of the AMB and IPCC A2 tanks, whereas the following summer when N supply was resumed the contribution of chlorophytes to the total biomass increased significantly, except for the G1 tank (Figure 5). Cyanobacteria biomass mostly consisted of filamentous and potential N-fixing *Cuspidothrix* spp., *Pseudanabaena* spp., and non-N-fixing colony-forming *Microcystis* spp. (Figure S2). As expected, N-fixing filamentous cyanobacteria, including *Cuspidothrix* sp., had several major outbreaks in summer and autumn during the N0 treatment in D1 and D2. While the biomass of filamentous cyanobacteria decreased significantly in the N+ period, *Microcystis* spp. did not show a uniform response to resumed N addition in terms of biomass. Despite the fact *Microcystis* spp. still dominated total community biomass in several tanks (D2, G1, G2) after the N supply was resumed, the magnitude of the biomass outbreaks decreased. In the N+ treatment, the proportion of chlorophytes, mainly *Micractinium* spp. and *Pediastrum* spp. (Figure S2), of total community biomass increased. Blooming events of *Micractinium* spp. developed towards the end of summer in most of the tanks.

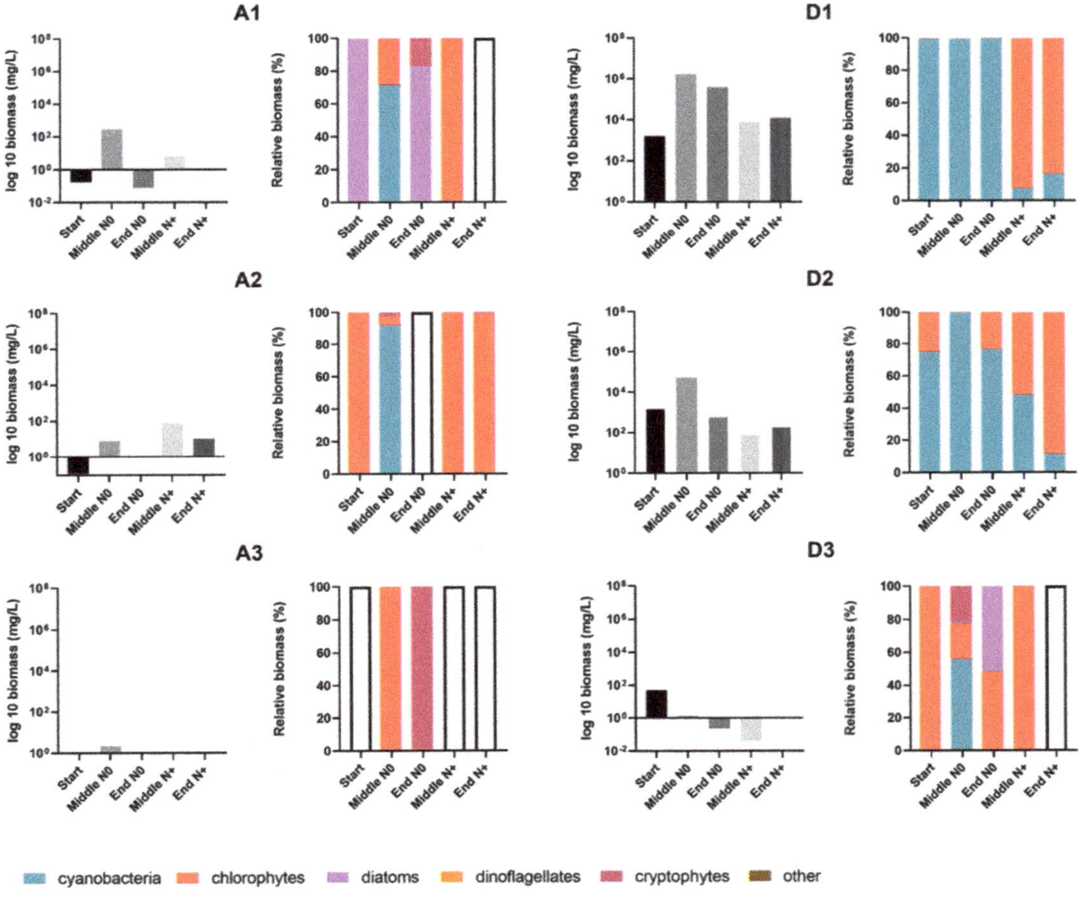

(A)

Figure 5. *Cont.*

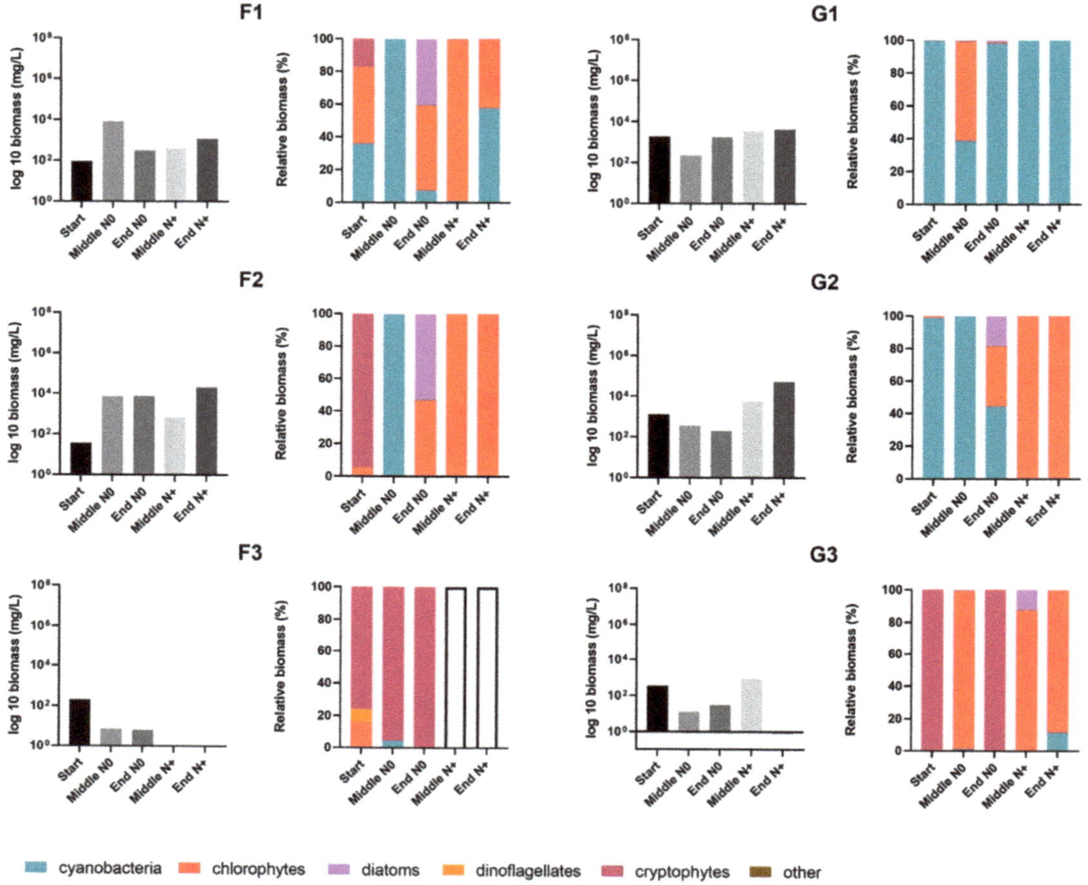

(**B**)

Figure 5. Total phytoplankton biomass and the relative contribution of the different phytoplankton groups to total biomass in the individual tanks at the start, in the middle and at the end of the N0 treatment and in the middle and at the end of the N+ treatment. (**A**) A1–A3, and D1–D3 individual tanks; (**B**) F1–F3, and G1–G3 individual tanks. Empty bars indicate samples with no cell biomass.

3.3. Changes in Phytoplankton Community Structure

According to the ANOSIM analysis of similarities, there were significant differences in the entire phytoplankton community between the N0 and N+ treatments ($R = 0.069$, $p < 0.001$). The phytoplankton community of IPCC A2 + 50% differed significantly different from that of AMB ($R = 0.412$, $p < 0.001$) and A2 ($R = 0.293$, $p < 0.001$), while there was no significant difference between AMB and IPCC A2 ($R = -0.002$, $p = 0.45$). The observed dissimilarities are graphically illustrated in Figure 6 (below) using estimated bootstrap averages based on Bray–Curtis similarity matrix in mMDS space.

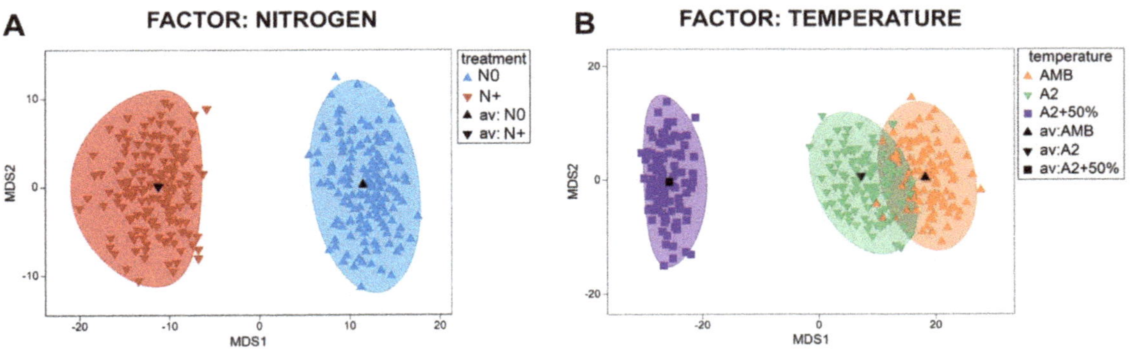

Figure 6. Bootstrap regions for phytoplankton community averages plotted in an mMDS plot with 150 bootstraps per group and 95% coverage. (**A**) Nitrogen treatment factor; (**B**) Temperature treatment factor.

ANOSIM analysis of the individual tanks showed significant differences between the N0 and N+ treatments for majority of tanks (A1: R = 0.068, $p < 0.17$; A2: R = 0.109, $p < 0.11$; A3: 0.168, $p < 0.03$; D1: R = 0.836, $p < 0.001$; D2: R = −0.003, $p < 0.43$; D3: R = −0.086, $p < 0.92$; F1: R = 0.58, $p < 0.002$; F2: R = 0.272, $p < 0.004$; F3: R = 0.177, $p < 0.03$; G1: R = −0.005, $p < 0.4$; G2: R = 0.248, $p < 0.012$; G3: R = 0.285, $p < 0.003$). No significant differences were found between the treatments attributed to A tanks and all IPCC + 50% tanks, which can be explained by low cell biomass in those tanks. Additionally, there is some variability among the replicate tanks differing in the extent of the community dissimilarity between the treatments. Dissimilarities based on bootstrap regions for the phytoplankton communities from each tank are displayed in Figure 7.

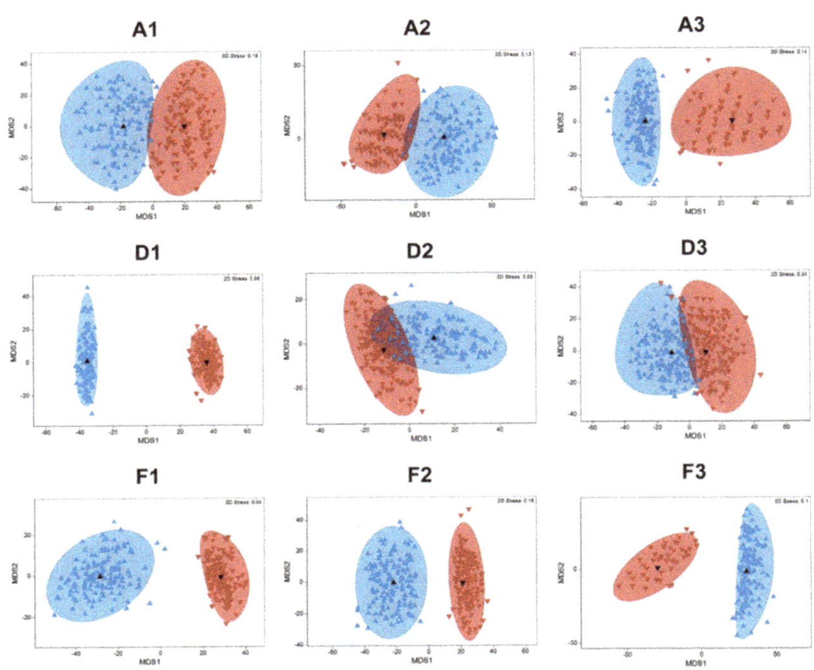

Figure 7. *Cont.*

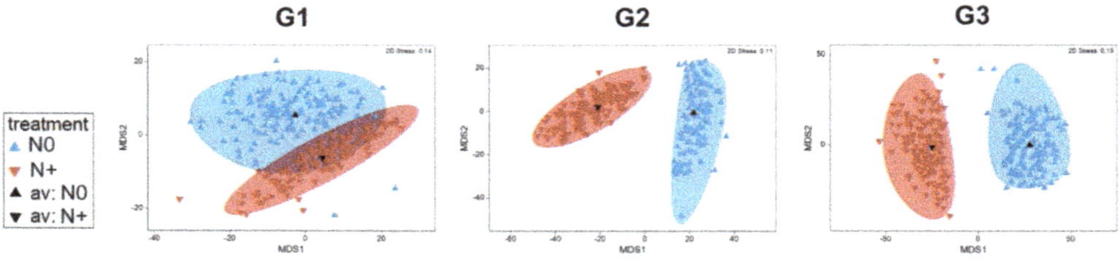

Figure 7. Bootstrap regions for phytoplankton community averages for tanks A1–A3, D1–D3, F1–F3, and G1–G3 plotted in an mMDS plot with 150 bootstraps per group and 95% coverage.

PERMANOVA analysis confirmed the significant effect of singular factors, such as N variation (PERMANOVA, pseudo-F = 6.02, $p < 0.003$) and temperature (PERMANOVA pseudo-F = 4.2, $p < 0.05$), on the phytoplankton taxonomic structure, but also revealed a significant interaction effect of N and temperature variation (PERMANOVA, pseudo-F = 4, $p < 0.05$). Moreover, significant variation in phytoplankton structure was found among the N (PERMDISP, F = 7.6, $p < 0.0001$) and the temperature (PERMDISP, F = 6.7, $p < 0.0001$) treatments.

Analysis of similarity percentages, SIMPER, was performed to identify the phytoplankton groups creating dissimilarity among the treatments. Overall, biomass changes in cyanobacteria, chlorophytes, and cryptophytes contributed mostly to the differences observed between the N treatments and the temperature regimes (Table S4). However, the exact contribution percentage to the dissimilarity varied among the tanks. Thus, cyanobacteria biomass decreased during the N+ treatment compared to N0 treatment, whereas the biomass of chlorophytes increased after N addition was resumed. In IPCC A2 + 50%, cryptophytes were added to the groups accounting for the dissimilarity between treatments; they were almost exclusively found in the tanks with the highest temperature. The differences in community composition between the temperature treatments were also caused by the biomass variations of cyanobacteria and chlorophytes (Table S4).

3.4. Changes in Phytoplankton Size Distribution

Single-cell measurements of area-based diameter (ABD) were used to obtain size frequency distributions for the AMB, IPCC A2, and IPCC A2 + 50% temperature treatments to evaluate changes in phytoplankton community size in response to N addition (Figure 8).

The size medians in AMB and IPCC A2 were similar and showed a decrease from 37 to 26 µm in AMB and from 37 to 28 µm in IPCC A2 in response to the N addition (Figure 8). In contrast, the size median in IPCC A2 + 50% tanks from 22 to 31 µm during the N+ treatment (Figure 8). The same pattern was observed for other descriptive statistics indicators, including ABD diameter mean, and 25% and 75% percentiles.

3.5. Changes in Phytoplankton Diversity

Changes in phytoplankton diversity among the treatments were assessed as a total of Pielou's evenness (J′), Shannon–Wiener diversity index (H′), and Simpson diversity index (1-Lambda (D)) (Figure 9).

In the N0 treatment, the highest diversity was observed in IPCC A2 + 50% (Figure 9) compared to AMB and IPCC A2. Whereas the J′ index value associated with community evenness did not change in the IPCC A2 + 50% treatment, the H′ and 1-D indices decreased almost twice as much when N addition was resumed (Figure 9). Similarly, phytoplankton diversity tended to decline in IPCC A2, while AMB demonstrated a slight diversity increase in the N+ treatment (Figure 9). The highest diversity based on H′ and 1-D indices was observed in AMB in the N+ treatment (Figure 9).

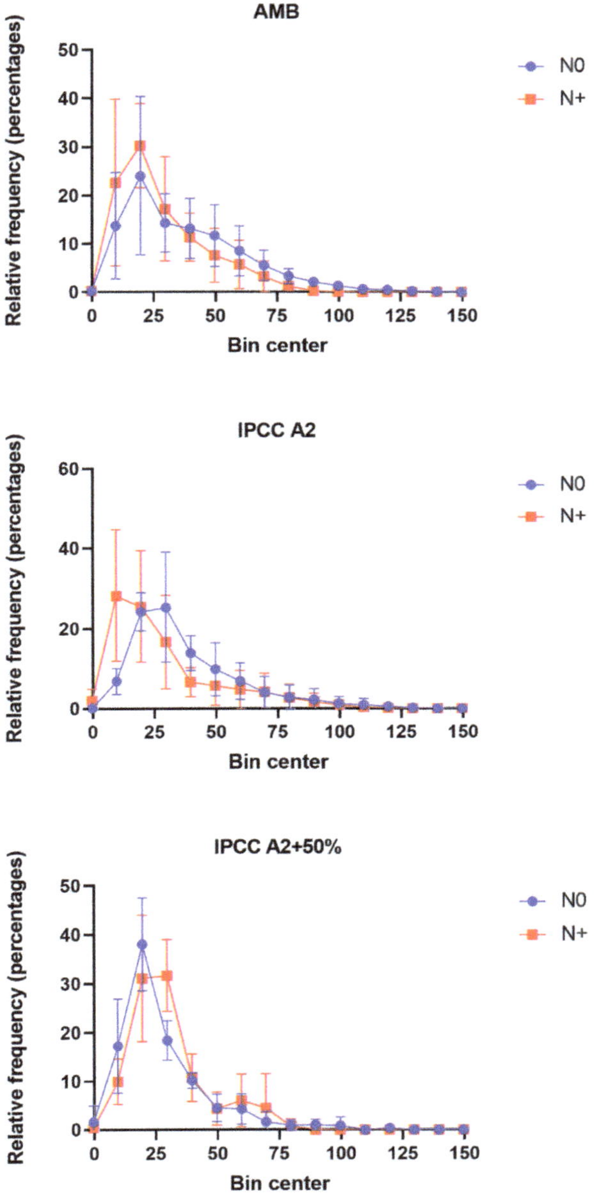

Figure 8. Phytoplankton (cell and colony) size distributions in AMB, IPCC A2, and IPCC A2 + 50% in the N0 and N+ treatments based on the average distributions of area-based diameter (ABD) in the individual tanks for the period mid-June–September. Each bin contains the number of size values in certain range of values. The bin width equals 10. Bars—SD of the mean.

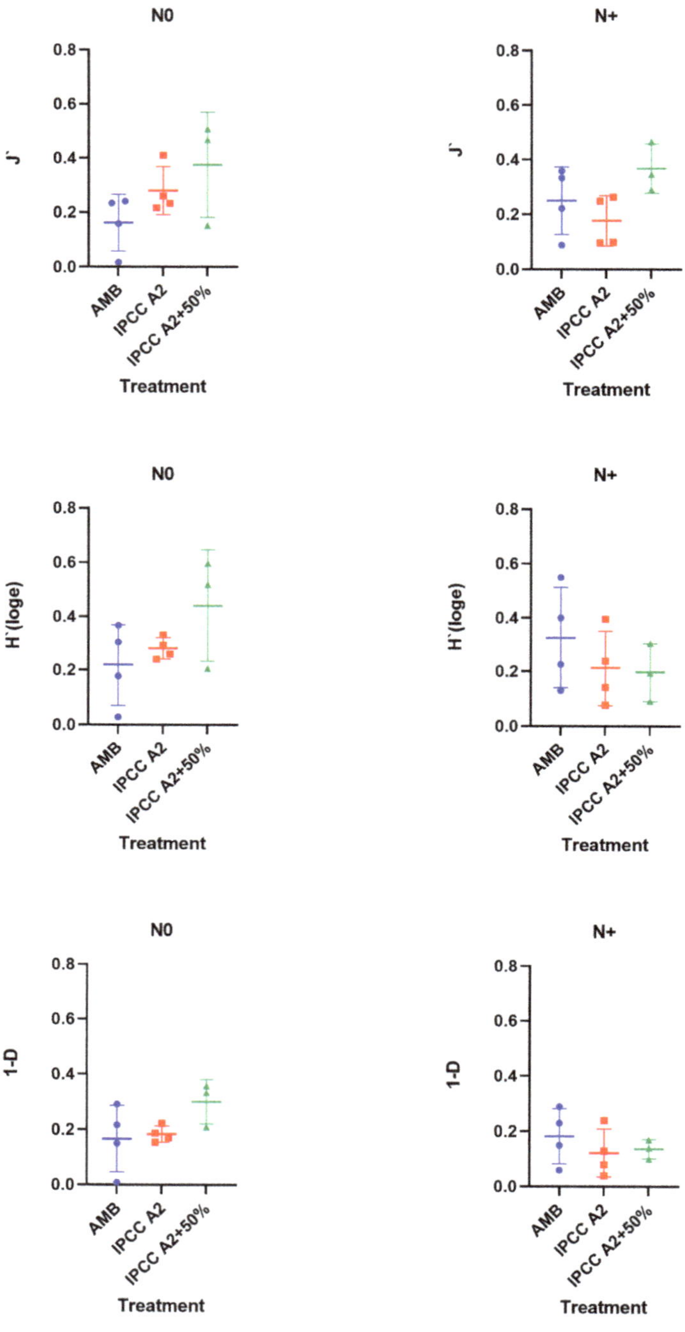

Figure 9. Phytoplankton diversity estimated using Pielou's evenness (J′), Shannon–Wiener diversity index (H′), and Simpson diversity index (1-Lambda (D)) in AMB, A2, and A2 + 50% in the N0 and +N treatments.

3.6. RDA of Phytoplankton Community and Environmental Factors

We performed an RDA analysis to evaluate the relative contribution of different environmental factors to the phytoplankton community variation. Overall, the available environmental variables explained 48.3% and 40.0% of the phytoplankton biomass variation in the RDA performed for phytoplankton classes and groups, respectively (Figure 10).

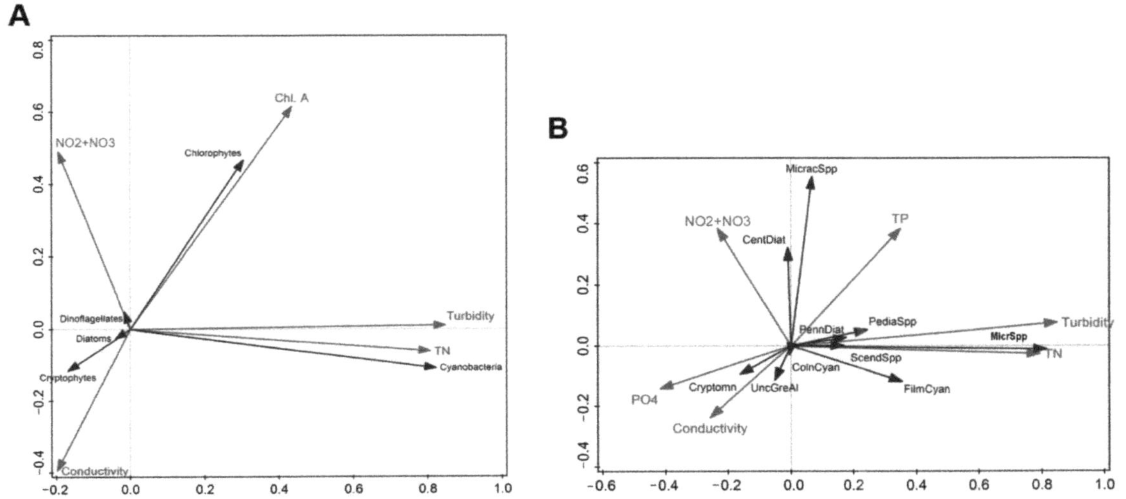

Figure 10. RDA plots for phytoplankton biomass for the whole study period. (**A**) phytoplankton presented as classes; (**B**) phytoplankton presented as groups. Blue arrows—environmental parameters; black arrows—phytoplankton taxa: MicracSpp—*Micractinium* spp., PennDiat—pennate diatoms, PediaSpp—*Pediastrum* spp., MicrSpp—*Microcystis* spp., ScendSpp—*Scenedesmus* spp., FilmCyan—filamentous cyanobacteria, ColnCyan—colonial cyanobacteria, UncGreAl—unicellular green algae, Cryptomn—cryptomonads. Only significant environmental factors ($p < 0.05$) are displayed. Graphical relationships between different parameters reflect the contribution percentage of each parameter accounting for simple effects, which vary from the percentage contribution based on the conditional effects, used for results interpretations.

In both cases, turbidity contributed most to the total community variation (28.5%, $p < 0.004$ and 19.6%, $p < 0.004$, respectively), followed by Chl a, conductivity, TN concentration, and NO_2 + NO_3 concentration, each of which contributed less than 7.4% (Figure 10A,B). Additionally, TP (1.2%, $p < 0.04$) and PO_4 (1.4%, $p < 0.02$) concentrations contributed significantly to the variation among the phytoplankton groups (Figure 10B).

Several significant correlations appeared between the biotic data and environmental variables. Turbidity correlated positively with the biomass of both *Microcystis* spp. and, to a lesser extent, filamentous cyanobacteria. Similarly, the biomass of cyanobacteria, including *Microcystis* spp. and filamentous cyanobacteria, was associated with high TN (Figure 10A, B) ($R^2 \geq 0.9$, $p < 0.01$). There was a negative relationship between NO_2 + NO_3 and the biomass of *Microcystis* spp., whereas the opposite pattern was observed for *Micractinium* spp. However, the strength of these relationships was weak ($R^2 \leq 0.18$, $p < 0.01$). Both *Microcystis* spp. and *Micractinium* spp. correlated positively with TP ($R^2 = 0.39$ and 0.18, respectively, $p < 0.01$), while there was a strong negative relationship between *Microcystis* spp. and PO_4 ($R = -0.66$, $R^2 = 0.43$, $p < 0.01$) and, to a lesser extent, between *Micractinium* spp., filamentous cyanobacteria, and PO4 ($R^2 \leq 0.07$, $p < 0.01$).

4. Discussion

Temperature-related shifts in algal communities have already been demonstrated in a number of natural phytoplankton community studies [23,45,46]. Classically, the increasing dominance of large-size phytoplankton species is highly dependent on the higher input of new nutrients. In the last decade, IFC has been used to assess the composition, abundance, size, and biovolume of phytoplankton and zooplankton in mesocosm experiments [47–52]. The IFC-based analysis allowed us to record frequently phytoplankton population dynamics based on biomass and cell size of thousands of cells in 16 phytoplankton groups in response to N and temperature variation. However, to our knowledge, no studies have explored the dynamics of phytoplankton communities of shallow lake systems in response to the combined treatment of N and temperature variation using IFC technology. We believe that this work may contribute to understanding climate change and nutrient pollution effects on shallow freshwater ecosystems.

4.1. Phytoplankton Cell Size

Phytoplankton cell size is an essential physiological parameter influencing food web structure, the efficiency of energy transfer, and the flux of carbon in aquatic ecosystems (ref. [53]). It was reported before that phytoplankton cell size decreases with warming ([54,55]), whereas increased nutrient availability is predicted to promote bigger cell sizes [56] and greater size diversity.

In the LMWE experiment, we observed a decrease in phytoplankton size in AMB and IPCC A2 after resuming the N addition. The reduced cell size may reflect the shift in dominance from *Microcystis* colonies to *Micractinium* single cells and colonies, with *Micractinium* being smaller in size. Generally, the average cell size in the highest temperature tanks was smaller than in AMB and IPCC A2, and there was a significant difference between AMB and IPCC A2 + 50% in the N0 period. In contrast, the average cell/colony size increased in the tanks with the highest temperature regimes when the N supply was resumed. Assuming that a relatively smaller size is the survival strategy adopted by phytoplankton populations under high-temperature stress and nutrient deficiency [57], the availability of N favored the greater size diversity. However, as in the other temperature treatments, the shift in size distribution may be attributed to the replacement of cryptomonads by relatively larger colonies of *Micractinium* spp. rather than size changes within the phytoplankton classes.

4.2. Phytoplankton Community Composition and Biomass Shifting

We applied IFC to assess phytoplankton composition and diversity, biomass, and size of planktonic cells. Recorded images were classified into 16 distinct groups based on morphological features (Table S3) with accuracy comparable to the measurements obtained by traditional microscopy [27,32,58]. However, due to the estimation algorithm using the area based on the two-dimensional images, instead of true three-dimensional shapes, biovolumes obtained by IFC tend to be misestimated [58]. The alternative is to use linear dimensions derived from taxonomic information on a particular cell to apply a shape-specific biovolume estimation in microscopy [42,58,59]. Yet, in the case of complicated shapes such as those of *Desmodesmus* spp. and/or unidentified species, FlowCAM-derived ABD volumes with the appropriate thresholding of the dark and light pixels of the images were applied.

Different temperature regimes led to shifts in phytoplankton community composition and biomass. The lowest biomass was found at ca. 4.5 °C above the ambient temperature and all A tanks. The latter observation can be explained by the presence of macrophytes in those tanks compared with the other introduced by chance. Phytoplankton community composition at ca. 4.5 °C above the ambient temperature was different from the community at ambient temperature and ca. +3 °C temperature. Phytoplankton community composition in ambient and ca. +3 °C temperature did not differ significantly and responded similarly to the variation in N. Mesocosms at both temperature regimes were dominated by cyanobacteria and green algae, whereas mesocosms with the highest temperature regime were mostly dominated by cryptomonads and green algae. The phytoplankton composition and dynamics in the high-nutrient tanks resembles those in the natural eutrophic lakes [60–62]. Bloom-forming *Microcystis* spp. and *Micractinium* spp. were the most abundant representatives of Cyanobacteria and Chlorophyta, respectively. The resuming of the N supply in P-enriched mesocosms led to a significant increase in the proportion of green algae, including the genus *Micractinium* spp., and a decrease in the proportion of cyanobacteria in AMB and IPCC A2 temperature regimes.

A shift towards dominance by green algae of the phytoplankton community with increasing N has been reported in mesocosm studies simulating shallow lakes [63] and natural field samples in shallow lakes at high N loading [17,64–66]. Despite the well-recognized harm of cyanobacteria blooms to the aquatic ecosystem, the blooming of other phytoplankton groups may also have serious implications for the ecosystem's biodiversity and functioning [67,68]. In their recent paper, Amorim and do Nascimento Moura [68] showed that both one-type and mixed (Cyanobacteria and other algal taxa) blooms could impact water quality, biodiversity, trophic dynamics, and ecosystem functioning. Our results demonstrate that variation in the amount of N in P-enriched systems may shift the dominance between the bloom-forming cyanobacteria and cyanobacteria/green algae. Because of the expected intensification of algal blooms in the future years due to climate change [69], management efforts should be concentrated on the mixed-type blooms as well.

We found a positive correlation between cyanobacteria and TN amount, despite the observed prevalence of cyanobacteria in N-limited conditions, likely reflecting overall low inorganic N concentration, with most of the N being stored in the phytoplankton. Moreover, the relationships between TP, PO_4, and *Microcystis* spp. showed opposite patterns. *Microcystis* spp. correlated positively with TP and negatively with PO_4. Different relationships between TP, PO_4, and cyanobacteria have been shown previously (e.g., [70]), resulting from the fact that TP includes internal stores of phosphorus in phytoplankton, while PO_4 represents soluble reactive phosphorus, which is affected by many processes including sediment–water interactions [71].

4.3. Phytoplankton Diversity

The contrasting patterns observed for the different temperature regimes imply an interaction effect of temperature and N availability on phytoplankton diversity. Phytoplankton diversity tended (though not significantly so) to increase in AMB and decrease in IPCC A2 and IPCC A2 + 50% after resuming the N supply. The combination of high temperature and high N led to the lowest diversity compared with the N0 conditions, where the highest diversity was observed in the IPCC A2 + 50% tanks (Figure 9). Studies of the effect of temperature on diversity have yielded contrasting results. In a mesocosm experiment, Yvon-Durocher and colleagues [72] observed a 67% increase in phytoplankton species richness and evenness in response to 4 °C warming. A field study also found increases in phytoplankton diversity with warming; however, in this case, it was possibly attributable to increased nutrient concentrations [73]. Other studies revealed a reduction in phytoplankton diversity with warming [74,75]. However, it was also shown that the effect of warming on biodiversity might vary with the nutrient levels [23,73,76], as in our study. Based on results, it is evident that temperature increase accompanied by increased nutrient pollution will negatively impact phytoplankton diversity.

5. Conclusions

In conclusion, we were able to capture the response of phytoplankton communities to N availability under contrasting temperature conditions using IFC. Our findings showed that N variation in P-enriched environments profoundly affects the composition, biomass, size distribution, and diversity of phytoplankton communities. Moreover, the net impact of N loading on the phytoplankton community depended on the temperature conditions. Based on the temperature regime, the phytoplankton community had a differential response to N variation in species composition, cell size, and diversity, changing the dominance between the bloom-forming cyanobacteria and cyanobacteria/green algae. The average cell size in the highest temperature tanks was smaller but increased in the tanks with the highest temperature regimes when the N supply was resumed. Therefore, although it adds to the complexity, it is important to consider the interaction of nutrient level with temperature when assessing the impact of nutrient loading on freshwater ecosystems in the context of climate change. Our results demonstrate that the IFC-based approach can be a valuable tool in capturing the dynamics of structural and morphological changes in phytoplankton communities.

Supplementary Materials: The following supporting information can be downloaded at: https://www.mdpi.com/article/10.3390/microorganisms10071322/s1, Table S1: Shapes and formulae used for biovolume estimation; Table S2: Average temperature data for temperature treatment tanks for N0 and N+ periods during June–October and correlation results between the two datasets obtained using Kendall's Tau test; Table S3: Phytoplankton composition in the high-nutrient tanks captured by FlowCAM imaging flow cytometer and microscopy; Table S4: SIMPER analysis results for different temperature tanks; Figure S1: Concentrations of the main nutrients (TN, NH4, NO2+NO3, TP, PO4) in the tanks exposed to different temperature treatments (AMB, IPCC A2, IPCC A2+50%); Figure S2: Relative contribution of the different phytoplankton groups to the total biomass in the individual tanks during the N0 and +N treatments.

Author Contributions: Conceptualization: V.D., I.A.V., E.J. and N.S.B.; data curation: V.D., I.A.V., E.J. and N.S.B.; formal analysis: V.D., D.V.M. and A.B.; funding: N.S.B., I.A.V. and E.J.; investigation: V.D., D.V.M. and A.B.; methodology: V.D., T.A.D., I.A.V., E.J. and N.S.B.; resources: T.A.D. and N.S.B.; supervision: I.A.V., E.J. and N.S.B.; visualization: V.D.; writing—original draft: V.D.; writing—review and editing: V.D., D.V.M., T.A.D., I.A.V., E.J. and N.S.B. All authors have read and agreed to the published version of the manuscript.

Funding: This study was funded by AnaEE Denmark (anaee.dk), the TÜBITAK and program BIDEB 2232 (project 118C250) to E.J., a Transnational Access granted to N.S.B. through AQUACOSM project (#IFCPHYTO and #SCPCRTNY), and by the European Commission EU H2020-INFRAIA-project (No. 731065) to T.A.D., by FDCRG #110119FD4513 and OPCRP2020018 from Nazarbayev University to N.S.B, and Ministry of Sciences, Republic of Kazakhstan grants MES #4350/GF4 and MES #AP05134153/GF4, to I.A.V. and N.S.B.

Institutional Review Board Statement: Not applicable.

Informed Consent Statement: Not applicable.

Data Availability Statement: The data that support the findings of this study are available from the corresponding author, upon reasonable request.

Acknowledgments: We are very thankful to Ann Lene Vigh, Kathrine Tabermann Uhrenholt, Eti Ester Levi, and Ivan Nielsen from Aarhus University (Silkeborg, Denmark) for Assistance with sample collection and measurements of environmental variables and Anne Mette Poulsen for help with manuscript editing. We are grateful to Harry Nelson, Yokagawa Fluid Technologies Inc. (Scarborough, ME, USA) for his help and advice and Vladimir Novokhatsky from Nazarbayev University (Nur-Sultan, Kazakhstan) for his outstanding service support of the FlowCAM instrument.

Conflicts of Interest: The authors declare no conflict of interest. The funders had no role in the design of the study, in the collection, analyses, or interpretation of data, in the writing of the manuscript, or in the decision to publish the results.

References

1. Kosten, S.; Huszar, V.; Bécares, E.; Costa, L.S.; van Donk, E.; Hansson, L.-A.; Jeppesen, E.; Kruk, C.; Lacerot, G.; Mazzeo, N.; et al. Warmer climates boost cyanobacterial dominance in shallow lakes. *Glob. Chang. Biol.* **2012**, *18*, 118–126. [CrossRef]
2. Jeppesen, E.; Kronvang, B.; Meerhoff, M.; Søndergaard, M.; Hansen, K.M.; Andersen, H.E.; Lauridsen, T.L.; Liboriussen, L.; Beklioglu, M.; Özen, A.; et al. Climate change effects on runoff, catchment phosphorus loading and lake ecological state, and potential adaptations. *J. Environ. Qual.* **2009**, *38*, 1930–1941. [CrossRef] [PubMed]
3. Woodward, G.; Perkins, D.M.; Brown, L.E. Climate change and freshwater ecosystems impacts across multiple levels of organization. *Philos. Trans. R. Soc. B* **2010**, *365*, 2093–2106. [CrossRef] [PubMed]

4. Moss, B.; Kosten, S.; Meerhoff, M.; Battarbee, R.W.; Jeppesen, E.; Mazzeo, N.; Havens, K.E.; Lacerot, G.; Liu, Z.W.; De Meester, L.; et al. Allied attack: Climate change and eutrophication. *Inland Waters* **2011**, *1*, 101–105. [CrossRef]
5. Meis, S.; Thackeray, S.; Jones, I. Effects of recent climate change on phytoplankton phenology in a temperate lake. *Freshw. Biol.* **2009**, *54*, 1888–1898. [CrossRef]
6. Lewandowska, A.; Sommer, U. Climate change and the spring bloom: A mesocosm study on the influence of light and temperature on phytoplankton and mesozooplankton. *Mar. Ecol. Prog. Ser.* **2010**, *405*, 101–111. [CrossRef]
7. Winder, M.; Sommer, U. Phytoplankton response to a changing climate. *Hydrobiologia* **2012**, *698*, 5–16. [CrossRef]
8. Jeppesen, E.; Canfield, D.E.; Bachmann, R.W.; Søndergaard, M.; Havens, K.E.; Johansson, L.S.; Lauridsen, T.L.; Sh, T.; Rutter, R.P.; Warren, G.; et al. Toward predicting climate change effects on lakes: A comparison of 1656 shallow lakes from Florida and Denmark reveals substantial differences in nutrient dynamics, metabolism, trophic structure, and top-down control. *Inland Waters* **2020**, *10*, 197–211. [CrossRef]
9. Paerl, H.; Paul, V. Climate change: Links to global expansion of harmful cyanobacteria. *Water Res.* **2012**, *46*, 1349–1363. [CrossRef]
10. Paerl, H.; Huisman, J. Climate change: A catalyst for global expansion of harmful cyanobacterial blooms. *Environ. Microbiol. Rep.* **2009**, *1*, 27–37. [CrossRef]
11. Yan, X.; Xu, X.; Wang, M.; Wang, G.; Wu, S.; Li, Z.; Sun, H.; Shi, A.; Yang, Y. Climate warming and cyanobacteria blooms: Looks at their relationships from a new perspective. *Water Res.* **2017**, *125*, 449–457. [CrossRef] [PubMed]
12. Yang, Y.; Stenger-Kovács, C.; Padisák, J.; Pettersson, K. Effects of winter severity on spring phytoplankton development in a temperate lake (Lake Erken, Sweden). *Hydrobiologia* **2016**, *780*, 47–57. [CrossRef]
13. Machado, K.B.; Vieira, L.C.G.; Nabout, J.C. Predicting the dynamics of taxonomic and functional phytoplankton compositions in different global warming scenarios. *Hydrobiologia* **2019**, *830*, 115–134. [CrossRef]
14. Schaum, C.; Barton, S.; Bestion, E.; Buckling, A.; Garcia-Carreras, B.; Lopez, P.; Lowe, C.; Pawar, S.; Smirnoff, N.; Trimmer, M.; et al. Adaptation of phytoplankton to a decade of experimental warming linked to increased photosynthesis. *Nat. Ecol. Evol.* **2017**, *1*, 94. [CrossRef]
15. Abonyi, A.; Ács, É.; Hidas, A.; Grigorszky, I.; Várbíró, G.; Borics, G.; Kiss, K.T. Functional diversity of phytoplankton highlights long-term gradual regime shift in the middle section of the Danube River due to global warming, human impacts and oligotrophication. *Freshw. Biol.* **2018**, *63*, 456–472. [CrossRef]
16. Fu, H.; Yuan, G.; Ozkan, K.; Sander Johansson, L.; Sondergaard, M.; Lauridsen, T.L.; Jeppesen, E. Seasonal and long-term trends in the spatial heterogeneity of lake phytoplankton communities over two decades of restoration and climate change. *Sci. Total Environ.* **2020**, *748*, 141106. [CrossRef]
17. Barica, J.; Kling, H.; Gibson, J. Experimental manipulation of algal bloom composition by nitrogen addition. *Can. J. Fish. Aquat. Sci.* **1980**, *37*, 1175–1183. [CrossRef]
18. Finlay, K.; Patoine, A.; Donald, D.B.; Bogard, M.J.; Leavitt, P.R. Experimental evidence that pollution with urea can degrade water quality in phosphorus-rich lakes of the Northern Great Plains. *Limnol. Oceanogr.* **2010**, *55*, 1213–1230. [CrossRef]
19. Donald, D.B.; Bogard, M.J.; Finlay, K.; Bunting, L.; Leavitt, P.R. Phytoplankton-specific response to enrichment of phosphorus-rich surface waters with ammonium, nitrate, and urea. *PLoS ONE* **2013**, *8*, e53277. [CrossRef]
20. Bogard, M.; Vogt, R.; Hayes, N.; Leavitt, P. Unabated nitrogen pollution favors growth of toxic cyanobacteria over Chlorophytes in most hypereutrophic lakes. *Environ. Sci. Technol.* **2020**, *54*, 3219–3227. [CrossRef]
21. Swabrick, V.J.; Simpson, G.L.; Glibert, P.M.; Leavitt, P.R. Differential stimulation and suppression of phytoplankton growth by ammonium enrichment in eutrophic hardwater lakes over 16 years. *Limnol. Oceanogr.* **2019**, *64*, S130–S149.
22. McKee, D.; Atkinson, D.; Collings, S.; Eaton, J.; Gill, A.; Harvey, I.; Hatton, K.; Heyes, T.; Wilson, D.; Moss, B. Response of freshwater microcosm communities to nutrients, fish, and elevated temperature during winter and summer. *Limnol. Oceanogr.* **2003**, *48*, 707–722. [CrossRef]
23. Verbeek, L.; Gall, A.; Hillebrand, H.; Striebel, M. Warming and oligotrophication cause shifts in freshwater phytoplankton communities. *Glob. Chang. Biol.* **2018**, *24*, 4532–4543. [CrossRef] [PubMed]
24. Richardson, J.; Feuchtmayr, H.; Miller, C.; Hunter, P.; Maberly, S.; Carvalho, L. Response of cyanobacteria and phytoplankton abundance to warming, extreme rainfall events and nutrient enrichment. *Glob. Chang. Biol.* **2019**, *25*, 3365–3380. [CrossRef] [PubMed]
25. Filiz, N.; Işkın, U.; Beklioğlu, M.; Öğlü, B.; Cao, Y.; Davidson, T.A.; Søndergaard, M.; Lauridsen, T.L.; Jeppesen, E. Phytoplankton community response to nutrients, temperatures, and a heat wave in shallow lakes: An experimental approach. *Water* **2020**, *12*, 3394. [CrossRef]
26. Haraguchi, L.; Jakobsen, H.H.; Lundholm, N.; Carstensen, J. Monitoring natural phytoplankton communities: A comparison between traditional methods and pulse-shape recording flow cytometry. *Aquat. Microb. Ecol.* **2017**, *80*, 77–92. [CrossRef]
27. Sieracki, C.; Sieracki, M.; Yentsch, C. An imaging-in-flow system for automated analysis of marine microplankton. *Mar. Ecol. Prog. Ser.* **1998**, *168*, 285–296. [CrossRef]
28. Olson, R.; Sosik, H. A submersible imaging-in-flow instrument to analyze nano-and microplankton: Imaging FlowCytobot. *Limnol. Oceanogr. Methods* **2007**, *5*, 195–203. [CrossRef]
29. Dashkova, V.; Malashenkov, D.; Poulton, N.; Vorobjev, I.; Barteneva, N. Imaging flow cytometry for phytoplankton analysis. *Methods* **2017**, *112*, 188–200. [CrossRef]

30. Dashkova, V.; Segev, E.; Malashenkov, D.; Kolter, R.; Vorobjev, I.; Barteneva, N. Microalgal cytometric analysis in the presence of endogenous autofluorescent pigments. *Algal Res.* **2016**, *19*, 370–380. [CrossRef]
31. Zarauz, L.; Irigoien, X.; Fernandes, J. Changes in plankton size structure and composition, during the generation of a phytoplankton bloom, in the central Cantabrian Sea. *J. Plankton Res.* **2008**, *31*, 193–207. [CrossRef]
32. Álvarez, E.; Moyano, M.; López-Urrutia, A.; Nogueira, E.; Scharek, R. Routine determination of plankton community composition and size structure: A comparison between FlowCAM and light microscopy. *J. Plankton Res.* **2014**, *36*, 170–184. [CrossRef]
33. Laney, S.; Sosik, H. Phytoplankton assemblage structure in and around a massive under-ice bloom in the Chukchi Sea. *Deep Sea Res. Part II Top. Stud. Oceanogr.* **2014**, *105*, 30–41. [CrossRef]
34. Reul, A.; Muñoz, M.; Bautista, B.; Neale, P.J.; Sobrino, C.; Mercado, J.M.; Segovia, M.; Salles, S.; Kulk, G.; León, P.; et al. Effect of CO_2, nutrients and light on coastal plankton. III. Trophic cascade, size structure and composition. *Aquat. Biol.* **2014**, *22*, 59–76. [CrossRef]
35. Poulton, N.J.; Martin, J.L. Imaging flow cytometry for quantitative phytoplankton analysis-FlowCAM. In *Microscopic and Molecular Methods for Quantitative Phytoplankton Analysis*; UNESCO: Paris, France, 2010; p. 47.
36. Liboriussen, L.; Landkildehus, F.; Meerhoff, M.; Bramm, M.E.; Søndergaard, M.; Christoffersen, K.; Richardson, K.; Søndergaard, M.; Lauridsen, T.L.; Jeppesen, E. Global warming: Design of a flow-through shallow lake mesocosm climate experiment. *Limnol. Oceanogr. Methods* **2005**, *3*, 1–9. [CrossRef]
37. Hao, B.; Wu, H.; Zhen, W.; Jo, H.; Cai, Y.; Jeppesen, E.; Li, W. Warming effects in periphoton community and abundance in different seasons are influenced by nutrient state and plant type: Shallow lake mesocosm. *Front. Plant Sci.* **2020**, *11*, 404. [CrossRef]
38. Søndergaard, M.; Jeppesen, E.; Mortensen, E.; Dall, E.; Kristensen, P.; Sortkjær, O. Phytoplankton biomass reduction after planktivorous fish reduction in a shallow, eutrophic lake: A combined effect of reduced internal P-loading and increased zooplankton grazing. *Hydrobiologia* **1990**, *200*, 229–240. [CrossRef]
39. Gibbs, M. A simple method for the rapid determination of iron in natural waters. *Water Res.* **1979**, *13*, 295–297. [CrossRef]
40. Malashenkov, D.; Dashkova, V.; Zhakupova, K.; Vorobjev, I.; Barteneva, N. Comparative analysis of freshwater phytoplankton communities in two lakes of Burabay National Park using morphological and molecular approaches. *Sci. Rep.* **2021**, *11*, 16130. [CrossRef]
41. Olenina, I. Biovolumes and size-classes of phytoplankton in the Baltic Sea. *Balt. Sea Environ. Proc.* **2006**, *106*, 6–22.
42. Bergkemper, V.; Weisse, T. Do current European lake monitoring programmes reliably estimate phytoplankton community changes? *Hydrobiologia* **2017**, *824*, 143–162. [CrossRef]
43. Jakobsen, H.; Carstensen, J. FlowCAM: Sizing cells and understanding the impact of size distributions on biovolume of planktonic community structure. *Aquat. Microb. Ecol.* **2011**, *65*, 75–87. [CrossRef]
44. Kydd, J.; Rajakaruna, H.; Briski, E.; Bailey, S. Examination of a high resolution laser optical plankton counter and FlowCAM for measuring plankton concentration and size. *J. Sea Res.* **2018**, *133*, 2–10. [CrossRef]
45. Hsieh, C.H.; Ishikawa, K.; Sakai, Y.; Ishikawa, T.; Ichise, S.; Yamamoto, Y.; Kuo, T.C.; Park, H.D.; Yamamura, N.; Kumagai, M. Phytoplankton community reorganization driven by eutrophication and warming in Lake Biwa. *Aquat. Sci.* **2010**, *72*, 467–483. [CrossRef]
46. De Senerpont Domis, L.N.; Van de Waal, D.B.; Helmsing, N.R.; Van Donk, E.; Mooij, W.M. Community stoichiometry in a changing world: Combined effects of warming and eutrophication on phytoplankton dynamics. *Ecology* **2014**, *95*, 1485–1495. [CrossRef]
47. Nielsen, L.; Jakobsen, H.; Hansen, P. High resilience of two coastal plankton communities to twenty-first century seawater acidification: Evidence from microcosm studies. *Mar. Biol. Res.* **2010**, *6*, 542–555. [CrossRef]
48. Jónasdóttir, S.; Dutz, J.; Koski, M.; Yebra, L.; Jakobsen, H.; Vidoudez, C.; Pohnert, G.; Nejstgaard, J. Extensive cross-disciplinary analysis of biological and chemical control of *Calanus finmarchicus* reproduction during an aldehyde forming diatom bloom in mesocosms. *Mar. Biol.* **2011**, *158*, 1943–1963. [CrossRef]
49. Calbet, A.; Sazhin, A.F.; Nejstgaard, J.C.; Berger, S.A.; Tait, Z.S.; Olmos, L.; Sousoni, D.; Isari, S.; Martinez, R.A.; Bouquet, J.M.; et al. Future climate scenarios for a coastal productive planktonic food web resulting in microplankton phenology changes and decreased trophic transfer efficiency. *PLoS ONE* **2014**, *9*, e94388.
50. Whitten, A.; Marin Jarrin, J.; McNaught, A. A mesocosm investigation of the effects of quagga mussels (Dreissena rostriformis bugensis) on Lake Michigan zooplankton assemblages. *J. Great Lakes Res.* **2018**, *44*, 105–113. [CrossRef]
51. Detmer, T.; Broadway, K.; Potter, C.; Collins, S.; Parkos, J.; Wahl, D. Comparison of microscopy to a semi-automated method (FlowCAM) for characterization of individual-, population-, and community-level measurements of zooplankton. *Hydrobiologia* **2019**, *838*, 99–110. [CrossRef]
52. Mirasbekov, Y.; Zhumakhanova, A.; Zhantuyakova, A.; Sarkytbayev, K.; Malashenkov, D.; Baishulakova, A.; Dashkova, V.; Davidson, T.; Vorobjev, I.A.; Jeppesen, E.; et al. Semi-automated classification of colonial *Microcystis* by FlowCAM imaging flow cytometry in mesocosm experiment reveals high heterogeneity during seasonal bloom. *Sci. Rep.* **2021**, *11*, 9377. [CrossRef] [PubMed]
53. Hillebrand, H.; Acevedo-Trejos, E.; Moorthi, S.D.; Ryabov, A.; Striebel, M.; Thomas, P.K.; Schneider, M.L. Cell size as driver and sentinel of phytoplankton community structure and functioning. *Funct. Ecol.* **2022**, *36*, 276–293. [CrossRef]
54. Yvon-Durocher, G.; Montoya, J.; Trimmer, M.; Woodward, G. Warming alters the size spectrum and shifts the distribution of biomass in freshwater ecosystems. *Glob. Chang. Biol.* **2011**, *17*, 1681–1694. [CrossRef]

55. Zohary, T.; Flaim, G.; Sommer, U. Temperature and the size of freshwater phytoplankton. *Hydrobiologia* **2020**, *848*, 143–155. [CrossRef]
56. Irwin, A.; Finkel, Z.; Schofield, O.; Falkowski, P. Scaling-up from nutrient physiology to the size-structure of phytoplankton communities. *J. Plankton Res.* **2006**, *28*, 459–471. [CrossRef]
57. Peter, K.; Sommer, U. Phytoplankton Cell Size Reduction in Response to Warming Mediated by Nutrient Limitation. *PLoS ONE* **2013**, *8*, e71528. [CrossRef]
58. Álvarez, E.; López-Urrutia, A.; Nogueira, E. Improvement of plankton biovolume estimates derived from image-based automatic sampling devices: Application to FlowCAM. *J. Plankton Res.* **2012**, *34*, 454–469. [CrossRef]
59. Hrycik, A.; Shambaugh, A.; Stockwell, J. Comparison of FlowCAM and microscope biovolume measurements for a diverse freshwater phytoplankton community. *J. Plankton Res.* **2019**, *41*, 849–864. [CrossRef]
60. Moss, B.; Mckee, D.; Atkinson, D.; Collings, S.; Eaton, J.; Gill, A.; Harvey, I.; Hatton, K.; Heyes, T.; Wilson, D. How important is climate? Effects of warming, nutrient addition and fish on phytoplankton in shallow lake microcosms. *J. Appl. Ecol.* **2004**, *40*, 782–792. [CrossRef]
61. Lv, J.; Wu, H.; Chen, M. Effects of nitrogen and phosphorus on phytoplankton composition and biomass in 15 subtropical, urban shallow lakes in Wuhan, China. *Limnologica* **2011**, *41*, 48–56. [CrossRef]
62. Moraska Lafrancois, B.; Nydick, K.R.; Caruso, B. Influence of nitrogen on phytoplankton biomass and community composition in fifteen Snowy Range Lakes (Wyoming, USA). *Arct. Antarct. Alp. Res.* **2003**, *35*, 499–508. [CrossRef]
63. Beklioğlu, M.; Bucak, T.; Levi, E.E.; Erdoğan, Ş.; Özen, A.; Filiz, N.; Bezirci, G.; Çakıroğlu, A.İ.; Tavşanoğlu, Ü.N.; Gökçe, D.; et al. Influences of climate and nutrient enrichment on the multiple trophic levels of Turkish shallow lakes. *Inland Waters* **2020**, *10*, 173–185. [CrossRef]
64. Ahoutou, M.K.; Yao, E.K.; Djeha, R.Y.; Kone, M.; Tambosco, K.; Duval, C.; Hamlaoui, S.; Bernard, C.; Bouvy, M.; Marie, B.; et al. Impacts of nutrient loading and fish grazing on the phytoplankton community and cyanotoxin production in a shallow tropical lake: Results from mesocosm experiments. *Microbiol. Open* **2022**, *11*, e1278. [CrossRef] [PubMed]
65. Jensen, J.; Jeppesen, E.; Olrik, K.; Kristensen, P. Impact of Nutrients and Physical Factors on the Shift from Cyanobacterial to Chlorophyte Dominance in Shallow Danish Lakes. *Can. J. Fish. Aquat. Sci.* **1994**, *51*, 1692–1699. [CrossRef]
66. Ma, J.; Qin, B.; Paerl, H.; Brookes, J.; Wu, P.; Zhou, J.; Deng, J.; Guo, J.; Li, Z. Green algal over cyanobacterial dominance promoted with nitrogen and phosphorus additions in a mesocosm study at Lake Taihu, China. *Environ. Sci. Pol. Res.* **2014**, *22*, 5041–5049. [CrossRef]
67. Escalas, A.; Catherine, A.; Maloufi, S.; Cellamare, M.; Hamlaoui, S.; Yéprémian, C.; Louvard, C.; Troussellier, M.; Bernard, C. Drivers and ecological consequences of dominance in periurban phytoplankton communities using networks approaches. *Water Res.* **2019**, *163*, 114893. [CrossRef]
68. Amorim, C.A.; do Nascimento Moura, A. Ecological impacts of freshwater algal blooms on water quality, plankton biodiversity, structure, and ecosystem functioning. *Sci. Total Environ.* **2021**, *758*, 143605. [CrossRef]
69. Griffith, A.; Gobler, C. Harmful algal blooms: A climate change co-stressor in marine and freshwater ecosystems. *Harmful Algae* **2020**, *91*, 101590. [CrossRef]
70. Li, D.; Wu, N.; Tang, S.; Su, G.; Li, X.; Zhang, Y.; Wang, G.; Zhang, J.; Liu, H.; Hecker, M.; et al. Factors associated with blooms of cyanobacteria in a large shallow lake, China. *Environ. Sci. Eur.* **2018**, *30*, 27. [CrossRef]
71. Søndergaard, M.; Jensen, J.P.; Jeppesen, E. Role of sediment and internal loading of phosphorus in shallow lakes. *Hydrobiologia* **2003**, *506*, 135–145. [CrossRef]
72. Yvon-Durocher, G.; Allen, A.P.; Cellamare, M.; Dossena, M.; Gaston, K.J.; Leitao, M.; Montoya, J.M.; Reuman, D.C.; Woodward, G.; Trimmer, M. Five years of experimental warming increases the biodiversity and productivity of phytoplankton. *PLoS Biol.* **2015**, *13*, e1002324. [CrossRef] [PubMed]
73. Schindler, D.W.; Beaty, K.G.; Fee, E.J.; Cruikshank, D.R.; DeBruyn, E.R.; Findlay, D.L.; Linsey, G.A.; Shearer, J.A.; Stainton, M.P.; Turner, M.A. Effects of climatic warming on lakes of the central boreal forest. *Science* **1990**, *250*, 967–970. [CrossRef] [PubMed]
74. Rasconi, S.; Winter, K.; Kainz, M. Temperature increase and fluctuation induce phytoplankton biodiversity loss-Evidence from a multi-seasonal mesocosm experiment. *Ecol. Evol.* **2017**, *7*, 2936–2946. [CrossRef] [PubMed]
75. Urrutia-Cordero, P.; Ekvall, M.; Ratcovich, J.; Soares, M.; Wilken, S.; Zhang, H.; Hansson, L.-A. Phytoplankton diversity loss along a gradient of future warming and brownification in freshwater mesocosms. *Freshw. Biol.* **2017**, *62*, 1869–1878. [CrossRef]
76. Rosset, V.; Lehmann, A.; Oertli, B. Warmer and richer? Predicting the impact of climate warming on species richness in small temperate waterbodies. *Glob. Chang. Biol.* **2010**, *16*, 2376–2387. [CrossRef]

microorganisms

Article

Emiliania huxleyi—Bacteria Interactions under Increasing CO_2 Concentrations

Joana Barcelos e Ramos [1,*], Susana Chaves Ribeiro [1], Kai George Schulz [2], Francisco José Riso Da Costa Coelho [3], Vanessa Oliveira [3], Angela Cunha [3], Newton Carlos Marcial Gomes [3], Colin Brownlee [4], Uta Passow [5] and Eduardo Brito de Azevedo [1]

1 Group of Climate, Meteorology and Global Change, IITAA, University of the Azores, Rua Capitão d'Ávila, São Pedro, 9700-042 Angra do Heroísmo, Portugal
2 Centre for Coastal Biogeochemistry, School of Environmental Science and Management, Southern Cross University, P.O. Box 157, Lismore, NSW 2480, Australia
3 CESAM—Centre for Environmental and Marine Studies, Department of Biology, University of Aveiro, 3810-193 Aveiro, Portugal
4 The Marine Biological Association of the United Kingdom, The Laboratory Citadel Hill, Plymouth PL1 2PB, UK
5 Ocean Science, Faculty of Sciences, Memorial University of Newfoundland, St. John's, NL A1C 5S7, Canada
* Correspondence: joana.b.ramos@uac.pt

Abstract: The interactions established between marine microbes, namely phytoplankton–bacteria, are key to the balance of organic matter export to depth and recycling in the surface ocean. Still, their role in the response of phytoplankton to rising CO_2 concentrations is poorly understood. Here, we show that the response of the cosmopolitan *Emiliania huxleyi* (*E. huxleyi*) to increasing CO_2 is affected by the coexistence with bacteria. Specifically, decreased growth rate of *E. huxleyi* at enhanced CO_2 concentrations was amplified in the bloom phase (potentially also related to nutrient concentrations) and with the coexistence with *Idiomarina abyssalis* (*I. abyssalis*) and *Brachybacterium* sp. In addition, enhanced CO_2 concentrations also affected *E. huxleyi*'s cellular content estimates, increasing organic and decreasing inorganic carbon, in the presence of *I. abyssalis*, but not *Brachybacterium* sp. At the same time, the bacterial isolates only survived in coexistence with *E. huxleyi*, but exclusively *I. abyssalis* at present CO_2 concentrations. Bacterial species or group-specific responses to the projected CO_2 rise, together with the concomitant effect on *E. huxleyi*, might impact the balance between the microbial loop and the export of organic matter, with consequences for atmospheric carbon dioxide.

Keywords: *Emiliania huxleyi*; CO_2; coccolithophores; *Idiomarina abyssalis*; *Brachybacterium* sp.; phytoplankton–bacteria interactions; changing ocean; functional profiling of marine bacteria

Citation: Barcelos e Ramos, J.; Ribeiro, S.C.; Schulz, K.G.; Coelho, F.J.R.D.C.; Oliveira, V.; Cunha, A.; Gomes, N.C.M.; Brownlee, C.; Passow, U.; de Azevedo, E.B. *Emiliania huxleyi*—Bacteria Interactions under Increasing CO_2 Concentrations. *Microorganisms* **2022**, *10*, 2461. https://doi.org/10.3390/microorganisms10122461

Academic Editors: Carmela Caroppo and Patrizia Pagliara

Received: 3 November 2022
Accepted: 28 November 2022
Published: 13 December 2022

Publisher's Note: MDPI stays neutral with regard to jurisdictional claims in published maps and institutional affiliations.

1. Introduction

Earth's climate and biosphere are strongly interlinked. The interactions established in the upper ocean between eukaryotic phytoplankton, bacteria and viruses play an important role in the pelagic energy flow and nutrient cycling [1,2] with consequences for biogeochemical cycles and feedback to climate.

The interactions established between marine phytoplankton and bacteria vary in complexity, from simply sharing the same environment, competing for the same resources and tightly relying on each other (mutualism) to living from the other in a host–parasite relationship. Phytoplankton growth and production rates affect organic matter characteristics, influencing bacteria community composition, abundance and production rates [3]. In parallel, bacteria community composition affects production rates and potentially functional redundancy and plasticity to changing environmental conditions [4]. As a result, relative abundances of some bacteria families correlate with phytoplankton biomass indicators during bloom events, while others remain unaltered [3]. CO_2 concentrations are projected to reach about 700 μatm by the year 2100 according to the 'intermediate scenario,

RCP 6.0′ [5], with a concomitant decrease in seawater pH. This has been seen to affect the important players of marine elemental cycling, phytoplankton and bacteria, differently. In spite of the urgency of increasing knowledge on how these planktonic communities respond to global change, little is known about specific phytoplankton–bacteria interactions.

Coccolithophores, and particularly the cosmopolitan *Emiliania huxleyi* (*E. huxleyi*), have been shown to be sensitive to increasing CO_2 concentrations (e.g., see review by [6,7]). How their response is affected by interactions with bacteria is poorly understood. Moreover, the response of the heterotrophic bacteria to increasing CO_2 concentrations is also not well-known (e.g., [8]). Organic matter degradation can be significantly affected by changes in seawater pH due to changes in the efficiency of hydrolytic enzymes [9]. However, ocean acidification experiments often consider bacteria within entire planktonic communities in mesocosms (e.g., [10]) or shipboard incubations [11], complicating the disentanglement of specific bacterial responses and interactions. Relatively recent studies addressed bacteria associated with phytoplankton cultures, demonstrating the establishment of a core microbiome [12] in which the *Roseobacter* clade is often well represented [13] and that the interactions vary with environmental conditions, time and partner species. Examples of this are the known frequent associations of Marinobacter and Marivita [14], the studied symbiotic/commensal relationships [13,15] and the *E. huxleyi* relationships with the symbiotic and pathogenic α-proteobacteria *Phaeobacter inhibens* and *Phaeobacter gallaeciensis* (*Roseobacter clade*) (references in [16]). However, there is still scarce information about single bacterial isolates or groups and virtually none considering the bacteria-specific phytoplankton interactions under increasing CO_2 concentrations.

This study aims to address the role of interactions established between a coccolithophore and two bacteria with unknown responses to increasing CO_2 concentrations. For this, we focused on *E. huxleyi* and two bacteria strains previously isolated from offshore Terceira island: (1) the γ-proteobacteria *Idiomarina abyssalis* (*I. abyssalis*), found in association with phytoplankton and sinking particles and with higher average relative abundance in the epipelagic [17]; and (2) the Actinobacterium *Brachybacterium* sp., potentially associated with planktonic communities and in the wake of sinking particles [18]. It is known that *E. huxleyi* produces higher concentrations of carbohydrates under stressful conditions, such as enhanced CO_2 concentrations [19]. This could lead to an increase in the relative number of heterotrophic bacteria in the proximities of the phytoplankton cells. Still, it remains unclear how these interactions could be differently affected and affect increasing CO_2 concentrations. Here, we hypothesised that: (1) the response of *E. huxleyi* to increasing CO_2 concentrations is affected by coexisting bacteria; (2) the response of the chosen bacteria to increasing CO_2 concentrations is affected by coexisting *E. huxleyi*; (3) the response of *E. huxleyi* to increasing CO_2 concentrations is bacteria species-specific.

2. Methods

2.1. Experimental Setup

Monospecific cultures of the cosmopolitan coccolithophore *E. huxleyi* (371, CCMP, axenic) were grown semi-continuously until the experiment for a minimum of 10 generations, under two CO_2 concentrations (average pCO_2: ~475 and 1056 μatm, $pH_{total\ scale}$ of ~7.99 and 7.69, respectively; Tables S1 and S2). Cultures previous to and during the experiment were grown in 0.1 μm sterile filtered North Atlantic seawater (salinity of 36) enriched with phosphate and nitrate, reaching 4.7 μmol L^{-1} and 83 μmol L^{-1}, respectively (Table S3), and with trace metals and vitamins following f/8 [20], at 20 °C, a photon flux density of 185 (+/− 10) μmol m^{-2} s^{-1} (supplied from OSRAM L 18W/840, Lumilux, coolwhite) and a 14/10 h light/dark cycle. The absence of bacteria in the *E. huxleyi* cultures, as well as contamination in the cultures with bacteria, was confirmed every ~4 days in the pre-cultures and during the experiments by means of microscopy and agar plating. Considering that *Brachybacterium* sp. forms yellow colonies and *I. abyssalis* transparent, it was easier to check agar plates for contamination throughout the experiment and in all treatments.

The two bacteria (*Idiomarina abyssalis* PhyBa_CO2_1 and *Brachybacterium* sp. PhyBa_CO2_2), chosen from our culture collection, were originally isolated by direct plating in Marine Agar (Difco, Le Pont de Claix, France). The isolates were analysed for their potential for carbohydrate fermentation (Table S4) to assure functional diversity and then selected for the experiment. Specifically, *I. abyssalis* was isolated in April 2018 from an *E. huxleyi* strain isolated from surface water off Biscoitos (Terceira island, 38°80′05″ N, 27°25′93″ W) and belongs to the Proteobacteria phylum. Related strains were *Idiomarina abyssalis* strain KMM 227 and *Idiomarina loihiensis* strain L2TR (100% and 99% 16S rRNA sequence similarity, respectively). *Brachybacterium* sp. was isolated in May 2019 from surface samples off the coast of Terceira (38°38′00″ N 27°09′00″ W) and belongs to the Actinobacteria phylum. Its most closely related strains identified with 16S were *Brachybacterium paraconglomeratum* strain LMG 19861 and *Brachybacterium conglomeratum* strain J 1015 (both with 100% 16S rRNA sequence similarity). However, based on Average Nucleotide Identity (ANI) information, the description of a new species is being prepared.

The isolates were kept at −20 °C in 30% (*v/v*) glycerol and revitalised in Marine Broth prior to the experiments at 20 °C under the same light conditions as *E. huxleyi*. Bacteria cultures were reactivated in Marine Broth for ~12 generations (division cycles) before the start of the experiment. On the day of the experiment, bacteria strains were centrifuged for 10 min at 14,000 rcf, washed twice with sterile seawater and resuspended with the corresponding medium (present and future) to a concentration of approximately 10^7 colony-forming units (CFU) mL^{-1} (determined with the Track-Dilution method) and left in the incubation chamber overnight at 20 °C, a photon flux density of 185 (+/− 10) $\mu mol\ m^{-2}\ s^{-1}$ (supplied from OSRAM L 18W/840, Lumilux, coolwhite) and a 14/10 h light/dark cycle. The culture media were always acclimated to the temperature of the experiment before inoculation of the bacteria and *E. huxleyi*.

At the start of the experiment, the axenic *E. huxleyi* was inoculated first (final concentration 2300 to 5600 cells mL^{-1}), followed by the intended bacteria (final concentration *I. abyssalis* ~2×10^5 CFU ml^{-1} and *Brachybacterium* sp. ~9×10^4 CFU mL^{-1}). During the exponential phase, *E. huxleyi* cell abundances ranged from ~2 and 6×10^4 cell mL^{-1} and *I. abyssalis* and *Brachybacterium* sp. were ~1×10^5 CFU mL^{-1}. The low cell concentrations minimised changes in seawater carbonate chemistry (average DIC drawdown of 2.6%). All cultures were manually vertically rotated (15 times, gently) one hour after the beginning of the light phase to avoid aggregation, sedimentation and self-shading. Sampling occurred at the beginning and after 4 and 13/14 days of incubation. At the end of the experiment (bloom, day 13/14), *E. huxleyi* reached a maximum of ~1×10^5 cell mL^{-1} while *I. abyssalis* reached ~7×10^6 CFU mL^{-1} and *Brachybacterium* sp. ~3×10^2 (present CO_2) and ~9×10^6 (high CO_2) CFU mL^{-1}.

2.2. Carbonate System

Total alkalinity was measured with an open cell potentiometric titration following Dickson et al. [21], using a Metrohm Titrino Plus 848 equipped with an 869 Compact Sample Changer and corrected with certified reference material supplied by A. Dickson (batch 128). The pH_t was determined via two methods: (1) through a glass electrode (WTW, pH 340i), which was calibrated with a TRIS seawater buffer, supplied by A. Dickson and; (2) colorimetrically by adding known amounts of m-cresol purple [22,23].

The carbonate system was manipulated by the addition of specific amounts of $NaHCO_3$ and HCl in a closed system following [24]. All carbonate chemistry parameters were calculated from measured salinity, temperature, phosphate concentrations and pH and TA using CO2sys [25], with the equilibrium constants determined by Mehrbach et al. [26] as refitted by Dickson and Millero [27] (Tables S1 and S2).

2.3. Nutrients

Samples were taken for the determination of dissolved inorganic nutrients at the beginning and the exponential phase. They were filtered through a polyethersulfone

(PES) 0.2 μm syringe filter and stored at −20 °C until analysis. Concentrations of nitrate and phosphate were measured following Hansen and Koroleff [28], using a Varian Cary 50 spectrophotometer.

2.4. Cell Numbers and Growth Rates

E. huxleyi abundances were determined with an inverted microscope (Nikon Eclipse TS 100) at 200× magnification from samples fixed with buffered Lugol (2% final concentrations) shortly after sample collection and counting on average ~1102 cells per sample +/− 88 SE. Bacteria were quantified after 48 h by counting viable colony counts (colony forming units, CFU) from agar plates after dilution steps, as well as total numbers of cells through flow cytometry on day 4 (exponential). More specifically, on day 4, it was important to complement the counts of viable cells (CFU) that grow and are responsible for bacterial rates with total bacteria (flow cytometry), essential to calculating cellular quotas. Hence, for the cellular quota data, flow cytometry was chosen. This strengthens the quality of the data and, therefore, assures comparison with previous works. Flow cytometry was conducted on frozen, glutaraldehyde (0.6% final concentration) preserved samples. The suspensions were stained with Syto BC in DMSO (Invitrogen) and incubated in the dark for 5 min at room temperature, beads were added at known concentration and reading duration was controlled to optimise the measurements. Samples were analysed with a FACSCalibur on the following gain settings: FL1 = 650; FL2 = 650; FL3 = 650; SS = 450 (event rate always below 1000 events per second).

Cell division rate (μ) was calculated as:

$$\mu = (\ln Ce - \ln Ci)/\Delta t \quad (1)$$

where Ce and Ci refer to end and initial numbers of cells, respectively, and Δt to the duration of the incubation period in days.

2.5. Particulate Organic Matter and Cellular Element Quotas

Samples for cellular particulate total carbon (TPC), organic carbon (POC) and nitrogen (PON) were gently filtered (200 mbar) through pre-combusted GF/F filters (6 h, 450 °C) and stored at −20 °C until analysis. TPC and PON samples were directly dried (4 h, 60 °C) while POC filters were firstly exposed to an acidified environment inside an exicator with 1 cm HCl (35%) for 2 h and then dried (4 h, 60 °C). All filters were then packed in tin boats and analysed in a gas chromatograph (EURO EA Elemental Analyser, EUROVECTOR equipped with a thermal conductivity detector and an element analyser) following Sharp [29]. Particulate inorganic carbon (PIC) was calculated by subtracting POC from TPC. Samples were taken during the exponential phase (fourth day of incubation) to allow comparison with previous studies.

In monocultures, cellular quotas for each species were calculated for day 4 (exponential) from POC, PON and PIC and the respective cell numbers of *E. huxleyi*, *Bacteriastrum* sp. and *I. abyssalis*. In the co-cultures, cellular POC, PON and PIC quota of each species was estimated by assuming no change as opposed to the mono-cultures in two ways:

$$E.\ huxleyi_{\mathrm{quotas}} = (C/N_{\mathrm{total}} - ((A_b \times Q_b))/A_{\mathrm{Ehuxleyi}} \quad (2)$$

$$Bacteria_{\mathrm{quotas}} = (C/N_{\mathrm{total}} - ((A_{\mathrm{Ehux}} \times Q_{\mathrm{Ehux}}))/A_b \quad (3)$$

where C/N_{total} corresponds to the total particulate matter in the filter (μg L^{-1}) carbon (C) or nitrogen (N), from which the portion of the other organism in the co-culture is subtracted. This was achieved by multiplying their A (abundance) in the co-culture with the cellular quotas (Q) determined for the monoclonal cultures of the same CO_2 concentration. The result was then divided by the abundance (A) using microscopy for *E. huxleyi*/flow cytometry in the bacteria) of the species intended to be estimated. These calculations assumed that the cell quota of one of the two organisms is not affected by co-existence.

Comparing the unaffected with the potentially affected quota provides the full range of the potential response. It is worth noting that measured POC and PON can also include exudates such as transparent exopolymer particles (TEPs). The percentage (estimated from $A_{Ehux} \times Q_{Ehux}$) of particulate organic matter corresponding to *E. huxleyi* in the co-cultures was ~80% in all conditions, except in coexistence with *I. abyssalis* under enhanced CO_2 concentrations (~90%). Therefore, changes in bacteria quotas would have less impact on *E. huxleyi* than the inverse (Table S5).

Finally, bacterial cellular quotas of the mono-cultures were always calculated from untreated TPC filters since they do not have relevant particulate inorganic carbon. In the co-cultures, bacteria cellular quotas were determined from *E. huxleyi* total and organic carbon quotas.

2.6. Coccolith Morphology

Samples collected for scanning electron microscopy (SEM) were filtered through 0.45 µm nitrate cellulose membranes and dried in an exsiccator. Filters were glued to supports, coated with charcoal and processed with SEM. Resulting images were then analysed with image J. All photographs were taken at the same magnification and the measured lengths and widths were calibrated using the scale bar present in all pictures. An average of ~130 (with the exception of E + AU under high CO_2 concentrations where only 6 coccoliths were measured) coccoliths were manually measured per condition, namely distal shield length (DSL) and width (DSW), and central area length (CAL) and width (CAW). From these measurements, the corresponding areas (distal shield area, DSA, and central area area, CAA) were calculated assuming an elliptical shape for the coccolith (e.g., [30]) as:

$$DSA/CAA = \pi \times ((\text{length} \times \text{width})/4) \quad (4)$$

2.7. Extracellular Enzyme Activity

Extracellular enzyme activity was determined in triplicate in 96-well black plates for each bottle by means of 4-methyl-coumarinyl-7-amide (MCA) and 4-methylumelliferone (MUF) fluorogenic analogues using a high throughput plate reader approach (following [31,32]). The activities of protease, β-glucosidase, α-glucosidase, phosphatase, lipase and chitinase were measured using L-leucine-7-amino-4-methylcoumarin, 4-methylumbelliferyl β-D-glucopyranoside, 4-methylumbelliferyl α-D-glucopyranoside, 4-methylumbelliferyl phosphate, 4-methylumbelliferyl 4-oleate and 4-methylumbelliferyl N-acetyl-β-D-glucosaminide, respectively. Substrates were added to the samples at a 39 µM final concentration. Samples were taken at time 0 and after 3h of incubation under the experimental conditions (20 °C and 185 µmol m^{-2} s^{-1}) and with minimum headspace by using a transparent, plastic, tight cover. After the incubation sample, fluorescence was measured at 365/460 nm (excitation/emission) for MUF and 380/440 nm (excitation/emission) for MCA with Fluostar Omega directly or after the addition of a 'STOP' buffer for the β- and α-glucosidase. Both final concentration and incubation duration were previously determined.

2.8. Utilisation of Carbohydrates

Fermentation of carbohydrates of the two bacterial isolates was determined using a colorimetric method prior to the experiment. Bacterial cells were suspended in culture medium to turbidity of 2 McFarland into a suspension medium with bromocresol purple as pH indicator. Thereafter, 100 µL of each sugar (Glucose, Frutose, Maltose, Trehalose, Ribose, Xylose, Dextrin, Starch, Lactose, Galactose, Arabinose, Rhamnose, Sucrose, Raffinose, Mannitol, Sorbitol and Inulin) was added at a concentration of 1% (*w/v*) to 400 µL of bacterial suspension. After incubation at 25 °C for 48 h, the color change was monitored. Results were expressed as clear positive (++), positive (+) or negative (−) according to color change (blue to yellow).

2.9. Bacteria Identification and Genome Analyses

2.9.1. DNA Extraction and Sequencing

Idiomarina abyssalis PhyBa_CO2_1 and *Brachybacterium* sp. PhyBa_CO2_2 were cultivated in Marine Broth at 20 °C, in 2 consecutive cultures of 48 h each. After cultivation, genomic DNA for identification (Tables S6 and S7, and Figures S1 and S2) and sequencing was extracted by means of a commercial bacterial DNA isolation kit (PureLinkTM Microbiome, Carlsbad, CA, USA).

16S ribosomal RNA (rRNA) gene amplicons were generated with PCR using primers 27F (5′-AGAGTTTGGATCMTGGCTCAG-3′) and 1492R (5′-CGGTTACCTTGTTACGACTT-3′). The PCR reaction mixture contained 10 µL 5X GoTaq reaction buffer (Promega), 1 µL dNTPs (10 mM), 2.5 µL primer 27F (10 µM), 2.5 µL primer 1492R (10 µM), 0.5 µL GoTaq Polymerase (5 U/µL) (Promega) and 1 µL of the extracted DNA. Nuclease-free water (Promega) was added to reach a total reaction volume of 50 µL. The following conditions were used for the bacterial 16S rRNA gene amplification: initial denaturation at 98 °C for 10 min followed by 35 cycles of denaturation at 98 °C for 20 s, annealing at 52 °C for 20 s, elongation at 72 °C for 45 s and a final extension step at 72 °C for 5 min. PCR products were purified using the Gene-JET PCR purification kit (Thermo Fisher Scientific, Waltham, MA, USA) and quantified using a Nanodrop 2000c spectrophotometer (Thermo Fisher Scientific, Waltham, MA, USA). The purified PCR products were sent for Sanger sequencing with primers 27F and 1492R (GATC Biotech, Cologne, Germany; now part of Eurofins Genomics Germany GmbH). Trimming (99% good bases, quality value > 20, 25-base window) and contig assembly were conducted with DNA Baser (version 3.5.4.2). Genome sequencing of strains was performed using the in Illumina Novaseq platform (150 bp paired-end reads). The genomes were sequenced at STAB VIDA (Lisbon, Portugal).

2.9.2. Genome Assembly and Quality Control

The quality of the reads was assessed with FASTQC 0.11.5 (Andrews S. FASTQC: a quality control tool for high throughput sequence data; 2010; available online at: http://www.bioinformatics.babraham.ac.uk/projects/fastqc, accessed on 27 November 2022). Trimmomatic 0.38 was used to remove adapters and quality filtering with the parameters: Leading: 8; Trailing: 8; Slidingwindow: 4:15; and Minlen: 100 [33]. Genome sequences generated were de novo assembled with the SPAdes 3.15.2 [34]. The quality of the draft assemblies was evaluated using QUAST 5.0.2 [35]. Completeness and contamination of analysed genomes were estimated using CheckM 1.1.3 with the default set of marker genes [36]. Only genomes that were at least 95% complete and had no more than 5% contamination were used. Prediction of CDSs of the assembled genome was performed with the RAST 2.0 server using the 'classic RAST' algorithm [37].

2.9.3. Phylogenetic and Phylogenomic Analyses

The phylogenetic analyses were conducted on sequences of the 16S rRNA genes of both strains and representatives of closely related described strains were retrieved from NCBI GenBank and included in the analysis. A phylogenetic tree was constructed using the web server Phylogeny.fr (http://www.phylogeny.fr/, accessed on 27 November 2022): sequences are aligned using Muscle, the alignment is curated using G-blocks and the phylogeny is established using PhyML-aLRT. Finally, the tree was drawn using TreeDyn [38].

For genome-wide assessments of phylogeny, we compute whole genome average nucleotide identity (ANI) for each pair of genomes using FastANI 1.32 [39]. Furthermore, whole-genome sequences of strains and related genomes were compared by using ANI-BLAST (ANIb) and ANI-MUMmer (ANIm) algorithms within the JSpeciesWS web service [40,41].

2.9.4. Genome Annotation

Functional annotations based on clusters of orthologous groups of proteins (COGs) and protein families (Pfams 24.0) were performed with the webserver WebMGA (evalue = 0.001) [42] using the amino acid fast file obtained from RAST. Carbohydrate-active enzymes (CAZymes) were annotated based on HMMER searches (HMMER 3.0b) [43] against the dbCAN database release 9.0 [44,45].

2.10. Statistical Analysis

Statistical significance of the data was evaluated with the parametric t-test, Welch's test (significance determined as 95%, $p < 0.05$), using the program R. The sample size, n, varied between 3 and 9.

3. Results

3.1. Comparison between Monospecific Cultures and Co-Cultures under Present CO_2 Concentrations

Under present CO_2 concentrations, *E. huxleyi* cell numbers (Figure 1) and growth rate (Figure 2) were not significantly affected by the presence of the bacteria tested. In contrast and under the same conditions, *I. abyssalis* abundances increased (~10^6 CFU mL^{-1}) significantly, but only after the longer incubation period (here called bloom phase) of coexistence with *E. huxleyi* (initial 4 days ~10^5 CFU mL^{-1}, Figure 3), while *Brachybacterium* sp. decreased to ~10^2 CFU mL^{-1} during the same time frame. Accordingly, within the first nutrient replete (Table S3) 4 days under present CO_2 concentrations, estimated cellular element quotas for *E. huxleyi* (Figures 4 and 5) and bacteria (Figure 6) did not vary significantly, apart from decreasing nitrogen ($p_{\text{Brachybacterium}} = 0.03$, Figure 4) and organic carbon quotas of *E. huxleyi* in coexistence with *Brachybacterium* sp. in relation to the axenic culture ($p_{\text{Brachybacterium}} = 0.04$, Figure 5). Although no difference in cellular PIC quotas was detected after the initial days as a result of coexistence with *I. abyssalis*. At the end of the incubation, coccolith distal shield width was significantly larger than other cultures (Table 1). Finally, at the end of the incubation period under present conditions, extracellular enzyme activities were low, but leucine aminopeptidase and β-glucosidase (only E + B) activities were higher in coexistence than in the single cultures (Table 2).

Table 1. Significance (p value) of t-test comparing coccolith measures and communities analysed. *E. huxleyi* (E) and *E. huxleyi* co-cultured with *Brachybacterium* sp. (E + B) and with *I. abyssalis* (E + I). Coccolith measures are distal shield length (DSL) and width (DSW), central area length (CAL) and width (CAW) and calculated distal shield area (DSA) and central area area (CAA).

	Present CO_2						High CO_2					
	DSL	DSW	DSA	CAL	CAW	CAA	DSL	DSW	DSA	CAL	CAW	CAA
E vs. E + B	0.26	0.9	0.6	0.6	0.3	0.3	0.3	0.7	0.6	0.6	0.6	0.5
E vs. E + I	0.4	0.02	0.07	0.07	0.6	0.6	0.07	0.02	0.02	0.01	0.04	0.01
E+B vs. E + I	0.73	0.02	0.2	0.2	0.1	0.1	0.9	0.44	0.6	0.6	0.5	0.5

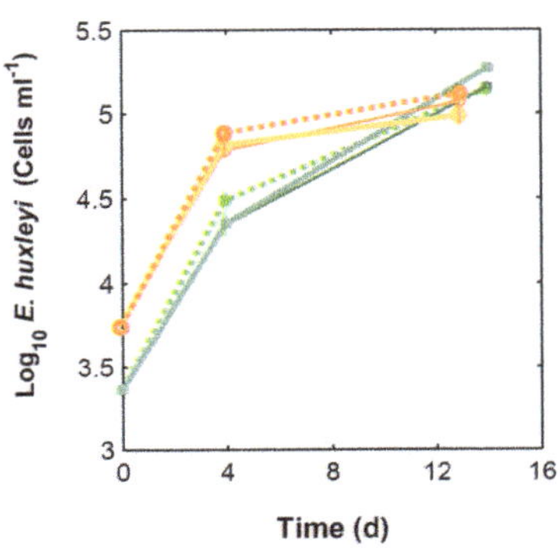

Figure 1. *E. huxleyi* abundances through time. Solid lines correspond to co-cultures and dashed lines to monocultures, under 475 (green) and 1056 μatm (orange). Thin solid lines correspond to *E. huxleyi* + *Brachybacterium* sp.; thicker solid lines to *E. huxleyi* + *I. abyssalis*.

Table 2. Enzymatic activities at the start (0) and after 4 and 13/14 days of incubation, under present and future CO_2 concentrations.

		Leucine (µmol L⁻¹ h⁻¹)			Alkaline Phosphatase (µmol L⁻¹ h⁻¹)			α-Glucosidase (nmol L⁻¹ h⁻¹)			β-Glucosidase (nmol L⁻¹ h⁻¹)		
		0	**4**	**13/14**	**0**	**4**	**13/14**	**0**	**4**	**13/14**	**0**	**4**	**13/14**
p	*E. huxleyi*	0.02 (+/−0.001)	0.03 (+/−0.009)	0.14 (+/−0.01)	0	0	0	0.79 (+/−0.39)	0	0	0	0	0.42 (+/−0.28)
f	*E. huxleyi*	0.05 (+/−0.0006)	0.12 (+/−0.01)	0.42 (+/−0.02)	0.3 (+/−0.03)	0.18 (+/−0.07)	0.43 (+/−0.08)	0	0	0	0	0.57 (+/−0.57)	7.21 (+/−3.30)
p	*I. abyssalis*	7.89 (+/−0.1)	0.01 (+/−0.002)	0	157.64 (+/−2.82)	0.18 (+/−0.08)	0.14 (+/−0.11)	0	0	0	0	0	0
f	*I. abyssalis*	6.70 (+/−0.08)	0.006 (+/−0.001)	0	47.25 (+/−0.63)	0.48 (+/−0.2)	0.94 (+/−0.03)	0	0	0	0	0	4.52
p	*Brachybacterium* sp.	1.56 (+/−0.02)	0	0	5.5 (+/−0.03)	0	0.02 (+/−0.01)	878.52 (+/−3.81)	0	0	26.47 (0.82)	0	0
f	*Brachybacterium* sp.	1.24 (+/−0.01)	0	0	6.43 (+/−0.04)	0.62 (+/−0.18)	0.81 (+/−0.05)	649.25 (+/−7.12)	0	0	297.76 (+/−39.84)	0	4.84 (+/−2.27)
p	*E + I*	0	0.06 (+/−0.005)	0.71 (+/−0.03)	-	0.19 (+/−0.03)	0.29 (+/−0.15)	-	0	0	-	0	0
f	*E + I*	0	0.12 (+/−0.02)	0.77 (+/−0.03)		0.33 (+/−0.11)	0.56 (+/−0.07)		0	0		0	1.89 (+/−1.95)
p	*E + B*	0	0.03 (+/−0.003)	0.28 (+/−0.05)	-	0	0	-	0	0	-	0	3.66 (3.28)
f	*E + B*	0	0.12 (+/−0.02)	0.51 (+/−0.03)		0.3 (+/−0.12)	0.38 (+/−0.10)		0	0.02 (+/−0.02)		0	4.24 (+/−2.73

Chitinase and Lipase below detection limits, Alkaline phosphatase is underestimated, since no buffer was added to improve fluorescence.

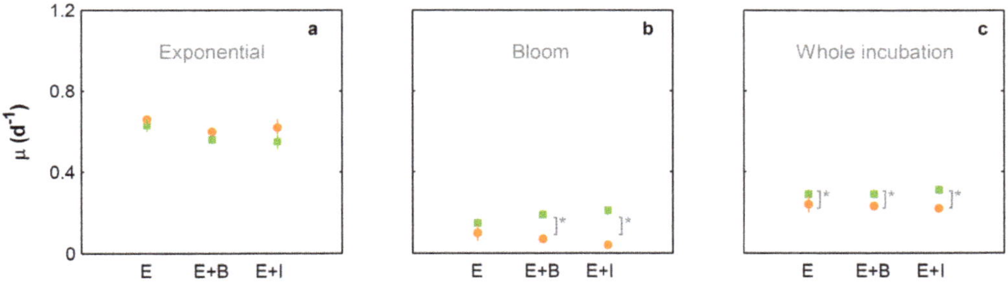

Figure 2. Growth rates based on cell counts of monocultures of *E. huxleyi* and *E. huxleyi* co-cultured with *Brachybacterium* sp. (E + B) and with *I. abyssalis* (E + I), under 475 (green) and 1056 μatm (orange). Cell division rates were calculated during the exponential (from 0 to 4 days, **a**), bloom phase (from 4 to 13/14 days, **b**) and whole incubation (**c**). * Denotes significant differences.

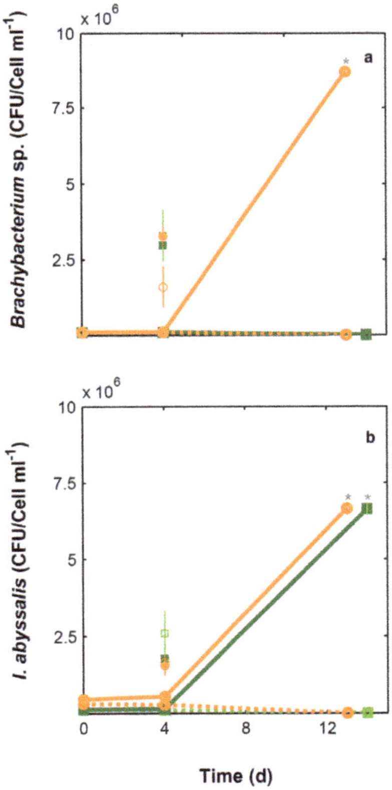

Figure 3. *Brachybacterium* sp. (**a**) and *I. abyssalis* (**b**) abundances through time. Lines refer to CFU data and isolated markers to counts obtained by means of flow cytometry. Solid lines correspond to co-cultures and dashed lines to monocultures of each bacteria isolate, under 475 (green) and 1056 μatm (orange). * Denotes significant differences.

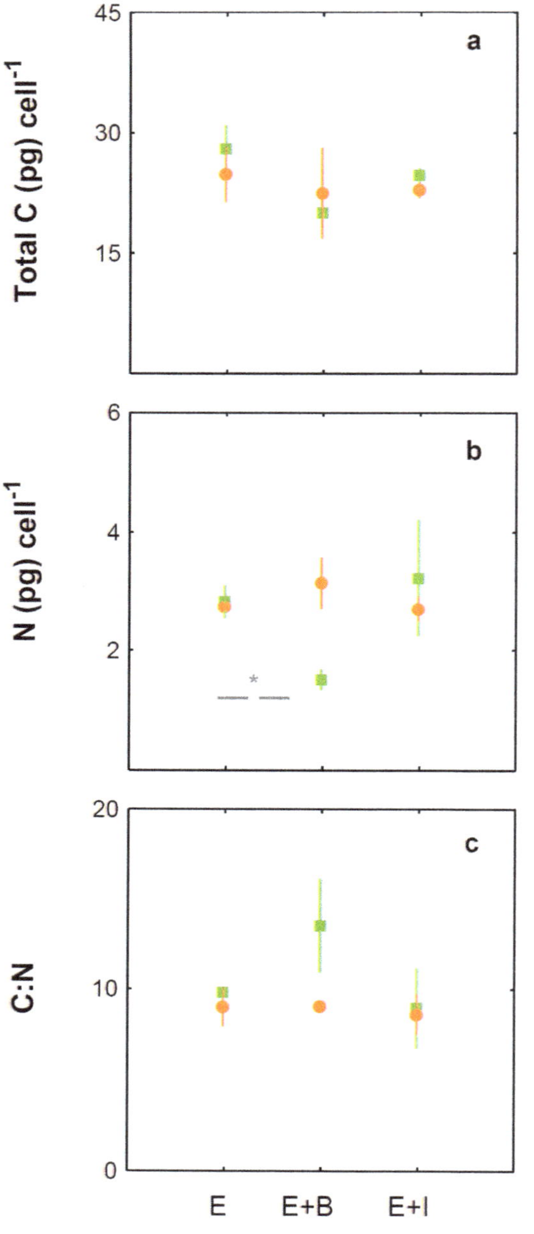

Figure 4. Cellular element quotas of *E. huxleyi* monocultures and *E. huxleyi* co-cultured with *Brachybacterium* sp. (E + B) and *I. abyssalis* (E + I), under 475 (green) and 1056 µatm (orange). Total particulate carbon (**a**), nitrogen (**b**) and carbon to nitrogen ratio (**c**). * Denotes significant differences.

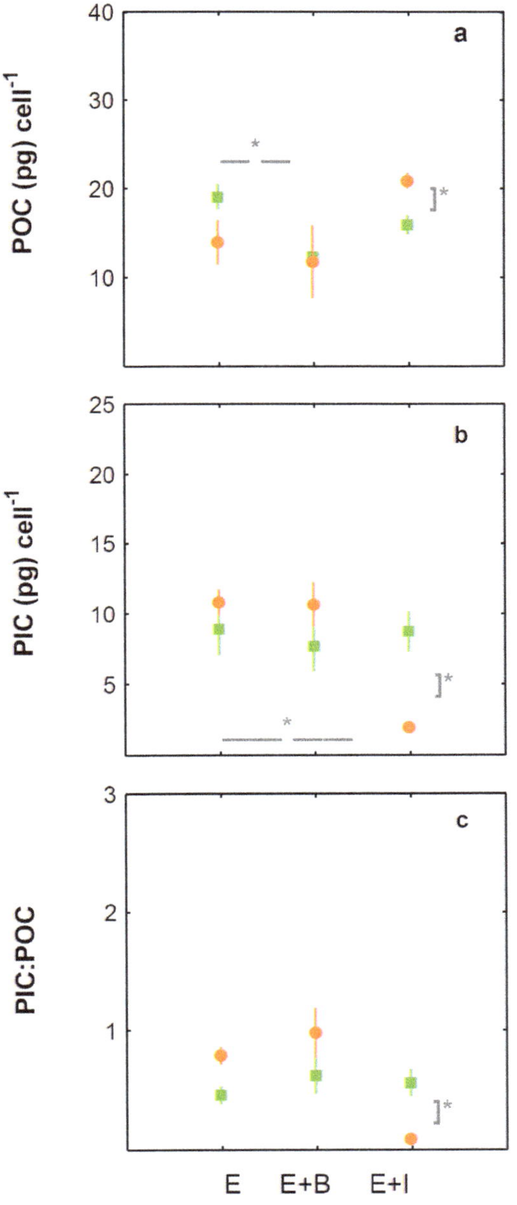

Figure 5. Cellular elemental quotas of *E. huxleyi* monocultures and *E. huxleyi* co-cultured with *Brachybacterium* sp. (E + B) and with *I. abyssalis* (E + I), under 475 (green) and 1056 μatm (orange). Particulate organic carbon (**a**), particulate inorganic carbon (**b**) and particulate inorganic to organic carbon ratio (**c**). * Denotes significant differences.

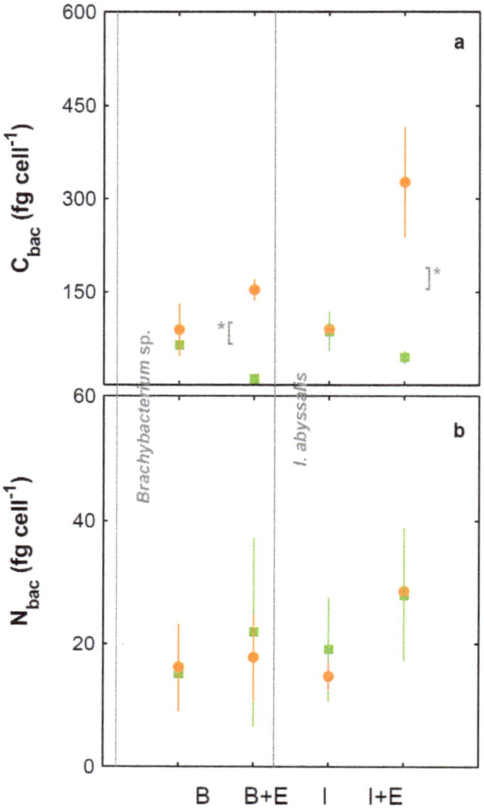

Figure 6. Cellular element quotas (based on total and organic particulate) of *Brachybacterium* sp. (B) and *I. abyssalis* (I) monocultures and those bacteria co-cultured with *E. huxleyi* (*Brachybacterium* sp., B + E; *I. abissalis*, I + E), under 475 (green) and 1056 μatm (orange). Total particulate carbon (**a**) and particulate nitrogen (**b**). * Denotes significant differences.

3.2. Comparison between Monospecific Cultures and Co-Cultures under High CO_2 Treatments

Under high CO_2 concentration, abundances (Figure 1) and growth rates (Figure 2) of axenic cultures of *E. huxleyi* were not significantly different from co-cultures (Figure 2). However, the two bacteria benefited from the presence of *E. huxleyi*, dramatically increasing in abundance under enhanced CO_2 concentrations (from ~10^2 CFU mL^{-1} to ~10^6 CFU mL^{-1}, Figure 3) after an initial lag phase (Figure 3). This longer coexistence and enhanced bacterial abundance corresponded to decreased growth rates of *E. huxleyi* in relation to the exponential phase (Figure 2). Within the initial 4 days (exponential period), cellular contents of *E. huxleyi* did not show significant differences in estimated particulate total nitrogen and carbon (Figure 4), decreasing only particulate inorganic carbon when *E. huxleyi* was co-cultured with *I. abyssalis* at enhanced CO_2 concentrations ($p = 0.057$), resulting in the lowest average PIC/POC (Figure 5) and, at the end of the incubation, significant increase in the coccoliths' distal shield width ($p = 0.02$), distal shield area ($p = 0.02$), central area length ($p = 0.01$), central area width ($p = 0.04$) and central area area ($p = 0.01$) in relation to the axenic culture (Table 1). At the same time, bacteria carbon quotas at enhanced CO_2 concentrations were higher in co-existence with *E. huxleyi* (Figure 6).

3.3. Responses to Increasing CO_2 Concentrations

E. huxleyi abundances increased similarly under either CO_2 treatment in the first 4 days of incubation (exponential, Figure 1) and consequently with no significant difference in growth rate (Figure 2) or in the cellular quotas of axenic *E. huxleyi* treatments (Figure 4). However, the co-existence of *E. huxleyi* with *I. abyssalis* significantly increased the coccolithophore's organic carbon ($p = 0.002$) and decreased inorganic carbon quotas ($p = 0.04$), resulting in a PIC/POC decrease ($p = 0.06$) in relation to the axenic culture at increasing CO_2 concentrations (Figure 5). At the end of the incubation, almost all measures of coccolith size, with the exception of coccolith central area width, were larger under present CO_2 concentrations in the axenic *E. huxleyi*, while only coccolith central area length was significantly affected in the co-cultures (Tables 1 and 3).

Table 3. Significance (*p* value) of *t*-test comparing coccolith measures and carbon dioxide conditions. *E. huxleyi* (E) and *E. huxleyi* co-cultured with *Brachybacterium* sp. (E + B) and with *I. abyssalis* (E + I).

	CO_2 Effect					
	DSL	**DSW**	**DSA**	**CAL**	**CAW**	**CAA**
E	0.01	0.003	0.002	0.002	0.1	0.04
E + B	0.47	0.45	0.4	0.002	0.4	0.4
E + I	0.21	0.09	0.1	0.01	0.4	0.3

Although *E. huxleyi's* cellular contents only varied significantly in the presence of *I. abyssalis*, estimated bacterial carbon quotas increased significantly ($p_{TPC} = 0.05$) with increasing CO_2 concentrations in all conditions and for both bacteria (assuming that CO_2-specific *E. huxleyi* cellular quotas did not vary with co-existence) (Figure 6). These changes in cellular quotas occurred despite virtually no variation in bacterial numbers within the shorter incubation (Figure 3).

Prolonging the incubation period (bloom phase), resulted in the absence of both bacteria cultured alone and of *Brachybacterium* sp. in all treatments except when cultured with *E. huxleyi* at enhanced CO_2 concentrations. Specifically, the growth rate of the two bacteria in co-culture from day 4 to 13/14 was 0.5 for *Brachybacterium* sp. and 0.27 for *I. abyssalis* at increased CO_2 concentrations and 0.38 in the latter under present CO_2. The observed differences in growth rate might be related to the two bacteria distinct functional profiling (Figure 7 and Table S5) according to the cluster of orthologous genes of proteins (COG) and protein family (Pfam)-based annotations (Figure 7 and Tables S8–S10). However, genome information does not directly reflect expression, here we can only consider the potential of the two species. In the *Brachybacterium* sp. genome, a high number of CDSs were assigned to the COG class 'Carbohydrate transport and metabolism' (Class G, Figure 7), while in the *I. abyssalis* genome a high number of CDSs were assigned to the COG classes 'Cell motility' and 'Signal transduction mechanisms' (Classes N and T, respectively, Table S5). Unlike *Brachybacterium* sp., the *I. abyssalis* genome possessed complete metabolic pathways for flagellum biosynthesis (flgBCDEFGHIKM and fliDEGHJKLMNOPLS) and motility (cheABRWZ and motABY). Moreover, only in the *I. abyssalis* genome were several protein domains identified found in TonB-dependent receptors (PF00593, PF03544, PF07660 and PF07715), involved in the transport of siderophores, as well as vitamin B12, nickel complexes and carbohydrates (Figure 7, *Idiomarina abyssalis*). Finally, the observed differences showed concomitant higher β-glucosidase activities under enhanced CO_2 concentrations and also leucine aminopeptidase in the cultures with *E. huxleyi* and *I. abyssalis* (Table S3).

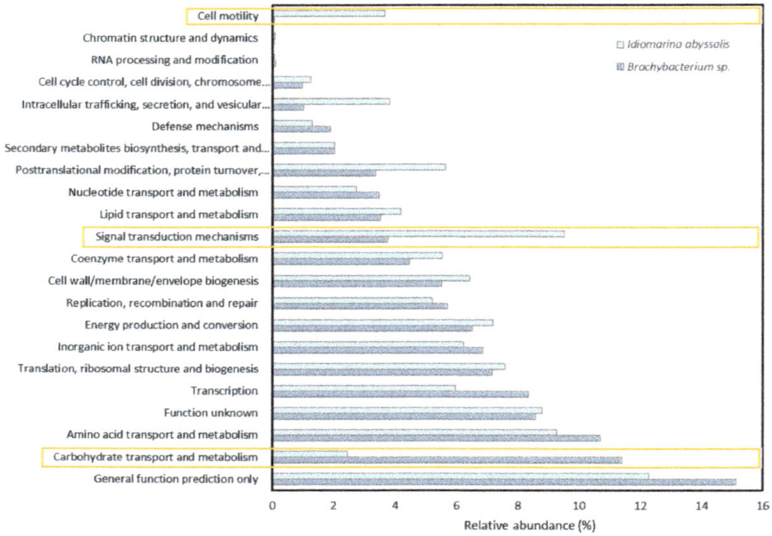

Figure 7. Percentage of genes assigned to specific functional activities of *Brachybacterium* sp. and *I. abyssalis*.

4. Discussion

4.1. Co-Existence in the Present Ocean

Phytoplankton–bacteria interactions play a crucial role in the balance between organic matter production, recycling and transport into the deep. How specific relationships influence biogeochemical cycling under present conditions is still not fully understood. Testing the relationships between a cosmopolitan coccolithophore and two bacteria with distinct functional profiles could hint at the complexity of the relationships occurring at a given timepoint. *Idiomarina* (γ-proteobacteria) and *Brachybacterium* (Actinobacteria) are frequently found in natural communities (e.g., [18,46]). γ-proteobacteria are often abundant in free living communities (e.g., [47]) and are well represented in nonaxenic cultures [15]), while *Brachybacterium* is relatively less abundant (Southern Atlantic, [48]). Contrary to Roseobacter, the presence of the bacteria tested in this study at abundances of ~10^5 CFU mL^{-1} did not affect the average growth rate of *E. huxleyi* in relation to its axenic cultures under present CO_2 concentrations. Additionally, no considerable changes were observed in most enzymatic activities and cellular quotas of the coccolithophore or the bacteria in the present ocean, with the exception of particulate nitrogen of *E. huxleyi* in co-culture with *Brachybacterium* sp. on day 4, potentially as a result of the decay of *Brachybacterium* sp. This is supported by the observation that *Brachybacterium* sp. is not able to grow alone under present CO_2 conditions for a longer period of time, potentially due to being restricted to inorganic nutrients and insufficient dissolved organic matter from the medium and *E. huxleyi* exudates. *I. abyssalis* was isolated from a fresh (months) *E. huxleyi* culture (isolated from offshore Terceira in 2016), potentially indicating a long-term commensal relation between the two organisms. Although *E. huxleyi* did not appear to benefit, *I. abyssalis* increased its growth rate in the presence of the coccolithophore. This only occurred after the initial days of exponential growth, when *E. huxleyi* abundances were higher, increasing the probability of phytoplankton–bacteria encounter rate and the provision of carbon sources from the coccolithophore. This species-specific effect [12] agrees with the response found here and could have consequences for carbon recycling.

4.2. Co-Existence-Driven Changes on Organisms' Response to the Future Ocean

Our results showed that the effects of rising CO_2 concentrations on the physiology of *E. huxleyi* was affected by the presence of bacteria, which in turn were differently impacted by it. A significant decrease of the growth rate of *E. huxleyi* was found after the longer incubation (higher abundances) at higher CO_2 concentration. More importantly, this previously described decrease (e.g., [6,49]) was augmented by the presence of bacteria. The growth rates of the tested bacteria were lower than those observed when the isolates were incubated in the rich Marine Broth (μ ~2), but similar to other isolates grown with 0.5 mL Zobell Medium in 1 l filtered seawater [50]. The growth of heterotrophic bacteria has been found to benefit from enhanced exudation of organic matter (e.g., carbohydrates) by phytoplankton and resulting aggregated polysaccharides in the form of TEP that occur under nutrient limitation at the end of phytoplankton blooms [10,51] or associated with the exposure to increased CO_2 concentrations [9]. Enhanced exudation of *E. huxleyi* as a response to higher CO_2/lower pH might, indeed, provide more substrate, but bacteria might also increase their organic carbon demand to function as was seen here and in previous studies [52]. Whether this is species-dependent is unclear, but it might be related to the observed dependence of survival of bacteria isolates on the stress response of *E. huxleyi*. Moreover, the two bacteria tested have distinct capacities for carbohydrate utilisation, potentially related to differences observed in their responses.

In the natural environment, senescent phytoplankton and aggregates are quickly colonised by bacteria (e.g., [53,54]) further degrading particulate organic matter into dissolved organic matter [55], which is then reutilised. However, bacteria also has the potential to stimulate aggregation of phytoplankton cells between species [56]. Bacterial production has been found to increase with increasing phytoplankton biomass (e.g., Chlorophyl *a*) and with concomitantly enhanced organic matter at increased CO_2 concentrations [56,57]. While both bacteria in the present study appear to benefit from the coccolithophore's exudations without negative effects on the growth of *E. huxleyi's* (commensalism) under future CO_2 concentrations, under present conditions, *Brachybacterium* sp. was unable to survive in association with *E. huxleyi*. Dissolved organic carbon is mostly comprised of total dissolved carbohydrates and protein components [57,58]. Hence, the difference between the response of the two bacteria might be related to the fact that *Brachybacterium* sp. was not able to be as responsive to the exuded dissolved organic matter as *I. abyssalis*. *I abyssalis* was seen to be a common associate of our isolates of *E. huxleyi*, but also sinking particles [18]. The reasoning could also be related to the presence of the complement of TONB-dependent transporters in the *I. abyssalis* genome, which is known to be involved with the uptake of several compounds such as large protein fragments [58,59]. In fact, *Idiomarina* was first isolated from seawater of the deep sea and characterised as having a relevant protein metabolism for its carbon source [60]. How much of the exudates stay in the water column as TEP will impact the retention versus remineralisation of particulate organic matter [61] and, therefore, the strength of the biological pump.

No differences between axenic and co-cultures were found for most analysed extracellular enzymatic activities after the first four days, potentially due to low bacteria abundances and reduced quorum sensing. In fact, previous studies found maximum Vmax of leu-aminopeptidase and β-glucosidase at the end of a mesocosm experiment with a natural community [56,57]. The strongest response was found for proteases (leucine aminopeptidase) in the co-cultures after the longer incubation, corroborating the difference in the TONB-dependent transporters discussed above. The activity of these hydrolytic enzymes would potentially enable an efficient utilisation of polymeric organic material exuded by *E. huxleyi*. After the longer incubation, monospecific cultures showed higher enzymatic activities under enhanced CO_2 concentrations, in contrast to previous studies [62]. Higher *E. huxleyi* and bacteria abundances would increase the probability of cells encountering each other and also potentially stimulating quorum sensing signals and affecting their responses.

More broadly, contradictory results have been found for the effect of elevated CO_2 concentrations/reduced pH on marine heterotrophic bacteria, from morphological modifications and a temporary inhibition of growth in the case of *Vibrio* sp. [62,63], to lower bacterial abundances due to higher viral lysis rates [63–65]. Studies considering CO_2 levels relevant to projected concentrations (190 to 1050 μatm) showed varying bacteria total abundance during phytoplankton blooms with small direct effects of CO_2 concentrations [64,66,67]. Moreover, γ-proteobacteria and a few rare taxa (increased abundance) were significantly affected by rising CO_2, after nutrient addition [66,68] Despite differences in relative abundances, the activity of free-living bacteria has been found to remain unaltered in a mesocosm study [66,67] while a reduction of 0.5 pH units resulted in a twofold enhancement of the β-glucosidase and leu-aminopeptidase rates [56,57]. Many enzymatic processes involved in the bacterial utilisation of organic substrates were shown to be pH sensitive in previous studies. Indeed, the efficiency of leucine aminopeptidase has been seen to triple at the highest CO_2 concentration (280 to 3000 μatm), both in terms of total activity and per cell [51]. In line with Yamada and Suzumura [69], protease activity rates expressed by *I. abyssalis* and *Brachybacterium* sp. were mostly undetectable during the incubation. The highest activity rates were observed at increased CO_2 concentrations. β-glucosidase rates were similar per hour to previous results published by Grossart, [62], but showed different results concerning CO_2 concentrations. Increased enzymatic hydrolysis [9,51,68,70] together with enhanced presence of gel particles that can be utilised as food and surface for growth [51] at enhanced CO_2 concentrations stimulates carbon and nutrient cycling as organic matter is degraded.

5. Summary

The expected changes in a future ocean, such as carbonate chemistry, will affect different organisms differently. Although there is increasing knowledge on the responses of marine bacteria and phytoplankton to increasing CO_2 concentrations, little is known about their interactions. Photoautotrophs play a crucial role in the carbon cycle while fixing carbon by photosynthesis. Much of the organic matter formed is then recycled by microbes in the surface ocean, the so-called microbial loop [71,72]. In fact, in the present ocean 50 to 96% of net marine primary production is remineralised in the surface ocean within the microbial loop by bacteria [71,73]. Enhanced CO_2 concentrations is expected to affect planktonic communities, with consequences for the cycling of organic matter, namely by enhancing photosynthesis of several marine algae analysed (e.g., [73,74]) as well as the rate of organic carbon and TEP production [74,75], with important repercussions for organic matter production and export to the seafloor [76]. However, the biotic interactions are highly dependent on other environmental conditions, such as nutrient availability and organisms' abundance, which affect physiological conditions of the phytoplankton cells.

The present study shows that the impact of increasing CO_2 concentrations is affected by the interactions established between *E. huxleyi* and specific bacteria, which are key for the aggregate dynamics and cycling of organic matter [67]. The growth of both bacteria was enhanced when the bacteria were grown with the coccolithophore in the future ocean CO_2 scenario, potentially increasing recycling of organic matter produced and, therefore, minimising the negative feedback to the atmosphere resulting from photosynthesis.

Supplementary Materials: The following are available online at https://www.mdpi.com/article/10.3390/microorganisms10122461/s1, Figure S1. Phylogenetic tree inferred from 16S rDNA sequences, from *Idiomarina abyssalis* PhyBa_CO2_1 (in bold) and references strains (accession numbers in parenthesis); Figure S2. Phylogenetic tree inferred from 16S rDNA sequences, from *Brachybacterium* sp. PhyBa_CO2_2 (in bold) and references strains (accession numbers in parenthesis); Table S1: Carbonate chemistry at the beginning, after 4 days of incubation and through (average) the experiments; Table S2: pHt after 13/14 days of incubation; Table S3: Average nutrient drawdown for all conditions with *E. huxleyi* during the incubation period. Nutrients concentrations were measured (*) at the beginning and on day 4. From day 4 to the end of the experiment data refers to estimates assuming constant drawdown per cell and *E. huxleyi* buildup. Total nutrients concentrations used, corresponds to estimates considering constant drawdown per cell and *E. huxleyi* number of cells at the end of the experiment; Table S4: Results of fermentation of carbohydrates by strains *I. abyssalis* and *Brachybacterium* sp. Change in colour was expressed as (++) for clear positive, (+) for positive and (−) for negative reaction; Table S5: Percentage of particulate matter attributed to each species in the co-cultures based on average cellular quotas from the single cultures (white) or estimated for each condition (grey); Table S6: Genome properties and quality metrics of the strains sequenced in this study; Table S7: Overview of phylogenetic comparisons between selected pairs of strains; Table S8: Number of genes associated with general COG functional categories; Table S9: Functional genome profiling of *Brachybacterium* sp. PhyBa_CO2_2 genome according to COG and Pfam annotations; Table S10: Functional genome profiling of *Idiomarina abyssalis* PhyBa_CO2_1 genome according to COG and Pfam annotations.

Author Contributions: Conceptualization, J.B.e.R.; methodology, J.B.e.R., S.C.R., K.G.S., A.C., F.J.R.D.C.C., V.O., N.C.M.G.; experiment/formal analysis, J.B.e.R., S.C.R., F.J.R.D.C.C., V.O.; investigation, J.B.e.R., C.B., U.P.; writing—original draft preparation, J.B.e.R.; writing—review and editing, J.B.e.R., S.C.R., K.G.S., A.C., F.J.R.D.C.C., V.O., N.C.M.G., C.B., U.P., E.B.d.A.; project administration, J.B.e.R., E.B.d.A.; funding acquisition, J.B.e.R., K.G.S., E.B.d.A. All authors have read and agreed to the published version of the manuscript.

Funding: We gratefully acknowledge the support provided by DRCT through the project PhybaCO2 (ACORES-01-0145-FEDER-000038) and (M1.1.a/008/Funcionamento/2018/RTF/010), as well as the FCT—Fundação para a Ciência e a Tecnologia, I.P. projects (UID/CVT/0153/2016 and /2019 and UIDB/00153/2020). ARM Program supported by DOE, in the framework of the ENA project through the agreement between LANL and the University of the Azores. Thanks are also due to FCT/MCTES for the financial support to CESAM (UIDP/50017/2020+UIDB/50017/2020+LA/P/0094/2020), through national funds. The research contract of FJRCC was funded by national funds through FCT under the Scientific Employment Stimulus—Individual Call—reference CEECIND/00070/2017. VO was funded by National funds (OE), through FCT, in the scope of the framework contract foreseen in the numbers 4, 5 and 6 of the article 23, of the Decree-Law 57/2016, of 29 August, changed by Law 57/2017, of July 19. SEM use was possible due to the programmatic support from FCT, IP to CIBIO-Açores, InBIO, BIOPOLIS, University of the Azores.

Data Availability Statement: The genome sequence of *Brachybacterium* sp. PhyBa_CO2_2 and *Idiomarina abyssalis* PhyBa_CO2_1 were deposited in GenBank database under accession number JAKJJW000000000 and JAOXLS000000000, respectively. 16S rRNA gene sequences of *Brachybacterium* sp. PhyBa_CO2_2 and *Idiomarina abyssalis* PhyBa_CO2_1 were deposited to GenBank under the accession numbers OM338106 and OP648247, respectively.

Acknowledgments: We would like to thank CIBIO, namely Luís Silva, Roberto Resendes and Paulo Melo for taking photographs with the SEM.

Conflicts of Interest: The authors declare no conflict of interest. The funders had no role in the design of the study; in the collection, analyses or interpretation of data; in the writing of the manuscript; or in the decision to publish the results.

References

1. Azam, F. Microbial Control of Oceanic Carbon Flux: The Plot Thickens. *Science* **1998**, *280*, 694–696. [CrossRef]
2. Falkowski, P.G.; Fenchel, T.; Delong, E.F. The Microbial Engines That Drive Earth's Biogeochemical Cycles. *Science* **2008**, *320*, 1034–1039. [CrossRef]
3. Tada, Y.; Taniguchi, A.; Nagao, I.; Miki, T.; Uematsu, M.; Tsuda, A.; Hamasaki, K. Differing Growth Responses of Major Phylogenetic Groups of Marine Bacteria to Natural Phytoplankton Blooms in the Western North Pacific Ocean. *Appl. Environ. Microbiol.* **2011**, *77*, 4055–4065. [CrossRef] [PubMed]
4. Díaz, S.; Cabido, M. Vive la différence: Plant functional diversity matters to ecosystem processes. *Trends Ecol. Evol.* **2001**, *16*, 646–655. [CrossRef]
5. IPCC. *Climate Change 2014: Synthesis Report. Contribution of Working Groups I, II and III to the Fifth Assessment Report of the Intergovernmental Panel on Climate Change*; Core Writing Team, Pachauri, R.K., Meyer, L.A., Eds.; IPCC: Geneva, Switzerland, 2014; p. 151.
6. Meyer, J.; Riebesell, U. Reviews and Syntheses: Responses of coccolithophores to ocean acidification: A meta-analysis. *Biogeosciences* **2015**, *12*, 1671–1682. [CrossRef]

7. Bach, L.; Riebesell, U.; Gutowska, M.A.; Federwisch, L.; Schulz, K. A unifying concept of coccolithophore sensitivity to changing carbonate chemistry embedded in an ecological framework. *Prog. Oceanogr.* **2015**, *135*, 125–138. [CrossRef]
8. Coelho, F.J.R.C.; Santos, A.L.; Coimbra, J.; Almeida, A.; Cunha, Â.; Cleary, D.F.R.; Calado, R.; Gomes, N.C.M. Interactive effects of global climate change and pollution on marine microbes: The way ahead. *Ecol. Evol.* **2013**, *3*, 1808–1818. [CrossRef]
9. Piontek, J.; Lunau, M.; Händel, N.; Borchard, C.; Wurst, M.; Engel, A. Acidification increases microbial polysaccharide degradation in the ocean. *Biogeosciences* **2010**, *7*, 1615–1624. [CrossRef]
10. Schulz, K.G.; Riebesell, U.; Bellerby, R.G.J.; Biswas, H.; Meyerhöfer, M.; Müller, M.N.; Egge, J.K.; Nejstgaard, J.C.; Neill, C.; Wohlers, J.; et al. Build-up and decline of organic matter during PeECE III. *Biogeosciences* **2008**, *5*, 707–718. [CrossRef]
11. Feng, Y.; Hare, C.E.; Leblanc, K.; Rose, J.M.; Zhang, Y.; DiTullio, G.R.; Lee, P.A.; Wilhelm, S.; Rowe, J.M.; Sun, J.; et al. The Effects of Increased pCO_2 and Temperature on the North Atlantic Spring Bloom: I. The Phytoplankton Community and Biogeochemical Response. *Mar. Ecol. Prog. Ser.* **2009**, *388*, 13–25. [CrossRef]
12. Behringer, G.; Ochsenkühn, M.A.; Fei, C.; Fanning, J.; Koester, J.A.; Amin, S.A. Bacterial Communities of Diatoms Display Strong Conservation Across Strains and Time. *Front. Microbiol.* **2018**, *9*, 659. [CrossRef] [PubMed]
13. Goecke, F.R.; Thiel, V.; Wiese, J.; Labes, A.; Imhoff, J.F. Algae as an important environment for bacteria—Phylogenetic relationships among new bacterial species isolated from algae. *Phycologia* **2013**, *52*, 14–24. [CrossRef]
14. Green, E.R.; Mecsas, J. Bacterial Secretion Systems: An Overview. *Microbiol. Spectr.* **2016**, *4*, 1. [CrossRef] [PubMed]
15. Krohn-Molt, I.; Wemheuer, B.; Alawi, M.; Poehlein, A.; Güllert, S.; Schmeisser, C.; Pommerening-Röser, A.; Grundhoff, A.; Daniel, R.; Hanelt, D.; et al. Metagenome survey of a multispecies and alga-associated biofilm revealed key elements of bacterial-algal interactions in photobioreactors. *Appl. Environ. Microbiol.* **2013**, *79*, 6196–6206. [CrossRef]
16. Seyedsayamdost, M.R.; Case, R.; Kolter, R.; Clardy, J. The Jekyll-and-Hyde chemistry of *Phaeobacter gallaeciensis*. *Nat. Chem.* **2011**, *3*, 331–335. [CrossRef]
17. Yilmaz, P.; Yarza, P.; Rapp, J.Z.; Glöckner, F.O. Expanding the World of Marine Bacterial and Archaeal Clades. *Front. Microbiol.* **2016**, *6*, 1524. [CrossRef]
18. Pelve, E.A.; Fontanez, K.M.; Delong, E.F. Bacterial Succession on Sinking Particles in the Ocean's Interior. *Front. Microbiol.* **2017**, *8*, 2269. [CrossRef]
19. Borchard, C.; Engel, A. Organic matter exudation by Emiliania huxleyi under simulated future ocean conditions. *Biogeosciences* **2012**, *9*, 3405–3423. [CrossRef]
20. Guillard RR, L.; Ryther, J.H. Studies of marine planktonic diatoms. I. Cyclotella nana Hustedt and Detonula confervacea Cleve. *Can. J. Microbiol.* **1962**, *8*, 229–239.
21. Dickson, A.G.; Afghan, J.D.; Anderson, G.C. Reference materials for oceanic CO_2 analysis: A method for the certification of total alkalinity. *Mar. Chem.* **2003**, *80*, 185–197. [CrossRef]
22. Clayton, T.D.; Byrne, R.H. Spectrophotometric seawater pH measurements: Total hydrogen ion concentration scale calibration of m-cresol purple and at-sea results. *Deep Sea Res.* **1993**, *40*, 2115–2129. [CrossRef]
23. DeGrandpre, M.D.; Spaulding, R.S.; Newton, J.O.; Jaqueth, E.J.; Hamblock, S.E.; Umansky, A.A.; Harris, K.E. Considerations for the measurement of spectrophotometric pH for ocean acidification and other studies. *Limnol. Oceanogr. Methods* **2014**, *12*, 830–839. [CrossRef]
24. Schulz, K.G.; Barcelos e Ramos, J.; Zeebe, R.E.; Riebesell, U. CO_2 perturbation experiments: Similarities and differences between dissolved inorganic carbon and total alkalinity manipulations. *Biogeosciences* **2009**, *6*, 2145–2153. [CrossRef]
25. Lewis, E.; Wallace, D.W.R. Program Developed for CO2 System Calculations. Available online: https://cdiac.ess-dive.lbl.gov/ftp/co2sys/CO2SYS_calc_DOS_v1.05/cdiac105.pdf (accessed on 30 August 2022).
26. Mehrbach, C.; Culberson, C.H.; Hawley, J.E.; Pytkowicx, R.M. Measurement of the apparent dissociation constants of carbonic acid in seawater at atmospheric pressure 1. *Limnol. Oceanogr.* **1973**, *18*, 897–907. [CrossRef]
27. Dickson, A.G. *Millero. Deep-Sea Research*; Pergamon Press: Oxford, UK, 1987; Volume 34, pp. 1733–1743.
28. Hansen, H.P.; Koroleff, F. *Determination of Nutrients*; Wiley-VCH: Hoboken, NJ, USA, 1999.
29. Sharp, J.H. Improved analysis for particulate organic carbon and nitrogen from seawater. *Limnol. Oceanogr.* **1974**, *19*, 984–989. [CrossRef]
30. Bach, L.T.; Bauke, C.; Meier KJ, S.; Riebesell, U.; Schulz, K.G. Influence of changing carbonate chemistry on morphology and weight of coccoliths formed by *Emilliana huxleyi*. *Biogeosciences* **2012**, *9*, 3449–3463. [CrossRef]
31. Hope, H.-G. Use of Fluorogenic model substracts for extracellular enzyme activity (EEA) measurements of bacteria. In *Handbook of Methods in Aquatic Microbial Ecology*; Chapter 48; Lewis Publishers: Boca Raton, FL, USA, 1993; p. 800. [CrossRef]
32. Malfatti, F.; Turk, V.; Tinta, T.; Mozetič, P.; Manganelli, M.; Samo, T.; Ugalde, J.; Kovač, N.; Stefanelli, M.; Antonioli, M.; et al. Microbial mechanisms coupling carbon and phosphorus cycles in phosphorus-limited northern Adriatic Sea. *Sci. Total Environ.* **2014**, *470–471*, 1173–1183. [CrossRef] [PubMed]
33. Bolger, A.M.; Lohse, M.; Usadel, B. Trimmomatic: A flexible trimmer for Illumina sequence data. *Bioinformatics* **2014**, *30*, 2114–2120. [CrossRef]
34. Prjibelski, A.; Antipov, D.; Meleshko, D.; Lapidus, A.; Korobeynikov, A. Using SPAdes de novo assembler. *Curr. Protoc. Bioinform.* **2020**, *70*, e102. [CrossRef]
35. Mikheenko, A.; Prjibelski, A.; Saveliev, V.; Antipov, D.; Gurevich, A. Versatile genome assembly evaluation with QUAST-LG. *Bioinformatics* **2018**, *34*, i142–i150. [CrossRef]

36. Parks, D.H.; Imelfort, M.; Skennerton, C.T.; Hugenholtz, P.; Tyson, G.W. CheckM: Assessing the quality of microbial genomes recovered from isolates, single cells, and metagenomes. *Genome Res.* **2015**, *25*, 1043–1055. [CrossRef] [PubMed]
37. Aziz, R.K.; Bartels, D.; Best, A.A.; DeJongh, M.; Disz, T.; Edwards, R.A.; Formsma, K.; Gerdes, S.; Glass, E.M.; Kubal, M.; et al. The RAST Server: Rapid annotations using subsystems technology. *BMC Genom.* **2008**, *9*, 75. [CrossRef] [PubMed]
38. Dereeper, A.; Guignon, V.; Blanc, G.; Audic, S.; Buffet, S.; Chevenet, F.; Dufayard, J.-F.; Guindon, S.; Lefort, V.; Lescot, M.; et al. Phylogeny. fr: Robust phylogenetic analysis for the non-specialist. *Nucleic Acids Res.* **2008**, *36*, W465–W469. [CrossRef] [PubMed]
39. Jain, C.; Rodriguez-RL, M.; Phillippy, A.M.; Konstantinidis, K.T.; Aluru, S. High throughput ANI analysis of 90K prokaryotic genomes reveals clear species boundaries. *Nat. Commun.* **2018**, *9*, 1–8. [CrossRef]
40. Richter, M.; Rosselló-Móra, R.; Oliver Glöckner, F.; Peplies, J. JSpeciesWS: A web server for prokaryotic species circumscription based on pairwise genome comparison. *Bioinformatics* **2016**, *32*, 929–931. [CrossRef] [PubMed]
41. Richter, M.; Rosselló-Móra, R. Shifting the genomic gold standard for the prokaryotic species definition. *Proc. Natl. Acad. Sci. USA* **2009**, *106*, 19126–19131. [CrossRef] [PubMed]
42. Wu, S.; Zhu, Z.; Fu, L.; Niu, B.; Li, W. WebMGA: A customizable web server for fast metagenomic sequence analysis. *BMC Genom.* **2011**, *12*, 444. [CrossRef]
43. Eddy, S.R. Accelerated profile HMM searches. *PLoS Comput. Biol.* **2011**, *7*, e1002195. [CrossRef]
44. Lombard, V.; Golaconda Ramulu, H.; Drula, E.; Coutinho, P.M.; Henrissat, B. The carbohydrate-active enzymes database (CAZy). *Nucleic Acids Res.* **2014**, *42*, D490–D495. [CrossRef]
45. Zhang, H.; Yohe, T.; Huang, L.; Entwistle, S.; Wu, P.; Yang, Z.; Busk, P.K.; Xu, Y.; Yin, Y. dbCAN2: A meta server for automated carbohydrate-active enzyme annotation. *Nucleic Acids Res.* **2018**, *46*, W95–W101. [CrossRef]
46. McCarren, J.; Becker, J.W.; Repeta, D.J.; Shi, Y.; Young, C.R.; Malmstrom, R.; Chisholm, S.; DeLong, E.F. Microbial community transcriptomes reveal microbes and metabolic pathways associated with dissolved organic matter turnover in the sea. *Proc. Natl. Acad. Sci. USA* **2010**, *107*, 16420–16427. [CrossRef] [PubMed]
47. Zorz, J.; Willis, C.; Comeau, A.M.; Langille, M.G.I.; Johnson, C.L.; Li, W.K.W.; LaRoche, J. Drivers of Regional Bacterial Community Structure and Diversity in the Northwest Atlantic Ocean. *Front. Microbiol.* **2019**, *10*, 281. [CrossRef] [PubMed]
48. Milici, M.; Vital, M.; Tomasch, J.; Badewien, T.; Giebel, H.-A.; Plumeier, I.; Wang, H.; Pieper, D.H.; Wagner-Döbler, I.; Simon, M. Diversity and community composition of particle-associated and free-living bacteria in mesopelagic and bathypelagic Southern Ocean water masses: Evidence of dispersal limitation in the Bransfield Strait. *Limnol. Oceanogr.* **2017**, *62*, 1080–1095. [CrossRef]
49. Barcelos e Ramos, J.; Müller, M.N.; Riebesell, U. Short-term response of the coccolithophore *Emiliania huxleyi* to an abrupt change in seawater carbon dioxide concentrations. *Biogeosciences* **2010**, *7*, 177–186. [CrossRef]
50. Teira, E.; Fernández, A.; Álvarez-Salgado, X.A.; García-Martín, E.E.; Serret, P.; Sobrino, C. Response of two marine bacterial isolates to high CO_2 concentration. *Mar. Ecol. Prog. Ser.* **2012**, *453*, 27–36. [CrossRef]
51. Endres, S.; Galgani, L.; Riebesell, U.; Schulz, K.-G.; Engel, A. Stimulated Bacterial Growth under Elevated pCO_2: Results from an Off-Shore Mesocosm Study. *PLoS ONE* **2014**, *9*, e99228. [CrossRef]
52. James, A.K.; Passow, U.; Brzezinski, M.A.; Parsons, R.J.; Trapani, J.N.; Carison, C.A. Elevated pCO_2 enhances bacterioplankton removal of organic carbon. *PLoS ONE* **2017**, *12*, e0173145. [CrossRef]
53. Simon, M.; Grossart, H.; Schweitzer, B.; Ploug, H. Microbial ecology of organic aggregates in aquatic ecosystems. *Aquat. Microb. Ecol.* **2002**, *28*, 175–211. [CrossRef]
54. Smith, D.C.; Steward, G.F.; Long, R.A.; Azam, F. Bacterial mediation of carbon fluxes during a diatom bloom in a mesocosm. *Deep Sea Res. Part II Top. Stud. Oceanogr.* **1995**, *42*, 75–97. [CrossRef]
55. Grossart, H.-P.; Ploug, H. Microbial degradation of organic carbon and nitrogen on diatom aggregates. *Limnol. Oceanogr.* **2001**, *46*, 267–277. [CrossRef]
56. Tran, N.-A.T.; Tamburic, B.; Evenhuis, C.R.; Seymour, J.R. Bacteria-mediated aggregation of the marine phytoplankton Thalassiosira weissflogii and Nannochloropsis oceanica. *J. Appl. Phycol.* **2020**, *32*, 3735–3748. [CrossRef]
57. Piontek, J.; Borchard, C.; Sperling, M.; Schulz, K.G.; Riebesell, U.; Engel, A. Response of bacterioplankton activity in an Arctic fjord system to elevated CO_2: Results from a mesocosm perturbation study. *Biogeosciences* **2013**, *10*, 297–314. [CrossRef]
58. Grossart, H.-P.; Engel, A.; Arnosti, C.; De La Rocha, C.L.; Murray, A.E.; Passow, U. Microbial dynamics in autotrophic and heterotrophic seawater mesocosms. III. Organic matter fluxes. *Aquat. Microb. Ecol.* **2007**, *49*, 143–156. [CrossRef]
59. Bergauer, K.; Fernandez-Guerra, A.; Garcia, J.A.L.; Sprenger, R.R.; Stepanauskas, R.; Pachiadaki, M.G.; Jensen, O.N.; Herndl, G.J. Organic matter processing by microbial communities throughout the Atlantic water column as revealed by metaproteomics. *Proc. Natl. Acad. Sci. USA* **2018**, *115*, E400–E408. [CrossRef]
60. Hou, S.; Saw, J.H.; Lee, K.S.; Freitas, T.A.; Belisle, C.; Kawarabayasi, Y.; Donachie, S.P.; Pikina, A.; Galperin, M.Y.; Koonin, E.V.; et al. Genome sequence of the deep-sea gamma-proteobacterium Idiomarina loihiensis reveals amino acid fermentation as a source of carbon and energy. *Proc. Natl. Acad. Sci. USA* **2004**, *101*, 18036–18041. [CrossRef]
61. Mari, X.; Passow UMigno, C.; Burd, A.B.; Legendre, L. Transparent exopolymer particles: Effects on carbon cycling in the ocean. *Prog. Oceanogr.* **2017**, *151*, 13–37. [CrossRef]
62. Grossart, H.-P. Interactions between marine bacteria and axenic diatoms (Cylindrotheca fusiformis, Nitzschia laevis, and Thalassiosira weissflogii) incubated under various conditions in the lab. *Aquat. Microb. Ecol.* **1999**, *19*, 1–11. [CrossRef]

63. Labare, M.P.; Bays, J.T.; Butkus, M.A.; Snyder-Leiby, T.; Smith, A.; Goldstein, A.; Schwartz, J.D.; Wilson, K.C.; Ginter, M.R.; Bare, E.A.; et al. The effects of elevated carbon dioxide levels on a Vibrio sp. isolated from the deep-sea. *Environ. Sci. Pollut. Res.* **2010**, *17*, 1009–1015. [CrossRef]
64. Brussaard, C.P.D.; Wilhelm, S.W.; Thingstad, F.; Weinbauer, M.G.; Bratbak, G.; Heldal, M.; A Kimmance, S.; Middelboe, M.; Nagasaki, K.; Paul, J.H.; et al. Global-scale processes with a nanoscale drive: The role of marine viruses. *ISME J.* **2008**, *2*, 575–578. [CrossRef]
65. Brussaard, C.P.D.; Noordeloos, A.A.M.; Witte, H.; Collenteur, M.C.J.; Schulz, K.; Ludwig, A.; Riebesell, U. Arctic microbial community dynamics influenced by elevated CO_2 levels. *Biogeosciences* **2013**, *10*, 719–731. [CrossRef]
66. Allgaier, M.; Riebesell, U.; Vogt, M.; Thyrhaug, R.; Grossart, H.-P. Coupling of heterotrophic bacteria to phytoplankton bloom development at different pCO_2 levels: A mesocosm study. *Biogeosciences* **2008**, *5*, 1007–1022. [CrossRef]
67. Grossart, H.-P.; Allgaier, M.; Passow, U.; Riebesell, U. Testing the effect of CO_2 concentration on the dynamics of marine heterotrophic bacterioplankton. *Limnol. Oceanogr.* **2006**, *51*, 1–11. [CrossRef]
68. Roy, A.-S.; Gibbons, S.M.; Schunck, H.; Owens, S.; Caporaso, J.G.; Sperling, M.; Nissimov, J.I.; Romac, S.; Bittner, L.; Mühling, M.; et al. Ocean acidification shows negligible impacts on high-latitude bacterial community structure in coastal pelagic mesocosms. *Biogeosciences* **2013**, *10*, 555–566. [CrossRef]
69. Yamada, N.; Suzumura, M. Effects of seawater acidification on hydrolytic enzyme activities. *J. Oceanogr.* **2010**, *66*, 233–241. [CrossRef]
70. Tanaka, T.; Thingstad, T.F.; Løvdal, T.; Grossart, H.-P.; Larsen, A.; Allgaier, M.; Meyerhöfer, M.; Schulz, K.G.; Wohlers, J.; Zöllner, E.; et al. Availability of phosphate for phytoplankton and bacteria and of glucose for bacteria at different pCO_2 levels in a mesocosm study. *Biogeosciences* **2008**, *5*, 669–678. [CrossRef]
71. Azam, F.; Fenchel, T.; Field, J.G.; Gray, J.S.; Meyer-Reil, L.A.; Thingstad, F. The ecological role of water-column microbes in the sea. *Mar. Ecol. Prog. Ser.* **1983**, *10*, 257–263. [CrossRef]
72. Liu, J.W.; Weinbauer, M.G.; Maier, C.; Dai, M.H.; Gattuso, J.-P. Effect of ocean acidification on microbial diversity and on microbe-driven biogeochemistry and ecosystem functioning. *Aquat. Microb. Ecol.* **2010**, *61*, 291–305. [CrossRef]
73. Del Giorgio, P.A.; Duarte, C.M. Respiration in the open ocean. *Nature* **2002**, *420*, 379–384. [CrossRef]
74. Beardall, J.; Raven, J.A. The potential effects of global climate change on microalgal photosynthesis, growth and ecology. *Phycologia* **2004**, *43*, 26–40. [CrossRef]
75. Anja, E.; Passow, U. Carbon and nitrogen content of transparent exopolymer particles (TEP) in relation to their Alcian Blue adsorption. *Mar. Ecol. Prog. Ser.* **2001**, *219*, 1–10.
76. Passow, U.; Shipe, R.; Murray, A.; Pak, D.; Brzezinski, M.; Alldredge, A. The origin of transparent exopolymer particles (TEP) and their role in the sedimentation of particulate matter. *Cont. Shelf Res.* **2001**, *21*, 327–346. [CrossRef]

microorganisms

MDPI

Article

Phytoplankton Community Structure in Highly-Mineralized Small Gypsum Karst Lake (Russia)

Alexander Okhapkin [1], Ekaterina Sharagina [1], Pavel Kulizin [1], Natalja Startseva [1] and Ekaterina Vodeneeva [1,2,*]

[1] Department of Botany and Zoology, Institute of Biology and Biomedicine, Lobachevsky State University, 603950 Nizhny Novgorod, Russia; okhapkin@bio.unn.ru (A.O.); ajugareptans@mail.ru (E.S.); KulizinPavel@yandex.ru (P.K.); startseva@bio.unn.ru (N.S.)

[2] Nizhny Novgorod Branch of the Federal State Budgetary Scientific Institution "Russian Research Institute of Fisheries and Oceanography", 603116 Nizhny Novgorod, Russia

* Correspondence: vodeneeva@mail.ru; Tel.: +7-9-10-3968-975

Abstract: Gypsum karst lakes are unique water ecosystems characterized by specific habitat conditions for living organisms, including phytoplankton species, as primary producers and mediating biogeochemical cycles in the water bodies. Studies of diversity and structure of phytoplankton communities can be used to identify the specific and typical lake features and plan basin-wide monitoring. The aim of this research was to analyze the structural variables of algocenoses in the small gypsum karstic Lake Klyuchik (Middle Volga basin), atypical for the subzone of mixed coniferous and deciduous forest zone high values of water mineralization (brackish water) and low temperatures. The lake has two water areas, connected by a shallow strait (ecotone zone) and differing from each other in the chemical compositions and physical properties of the water. A total of 133 species of phytoplankton with prevalence percentages of Bacillariophyta (46%), Chlorophyta (24%), and Ochrophyta (11%) were found; α-diversity varied from 4 to 30 specific and intraspecific taxa per sample. According to Spearman's correlation coefficients, the diversity indices (Shannon, Pielou, Simpson) were mainly determined by the number of dominant species. The uniquely high (up to 130 g/m^3) biomass of phytoplankton was noted in the ecotone, on the border between the water column and the bottom. The formation of mono- and oligo-dominant nannoplankton diatom communities with a predominance of the rare species *Cyclotella distinguenda* Hustedt was demonstrated there. The roles of flagellate algae and cyanobacteria were found to be less significant.

Keywords: phytoplankton; highly-mineralized water; gypsum karst lake; community; structural variables; diversity indices

Citation: Okhapkin, A.; Sharagina, E.; Kulizin, P.; Startseva, N.; Vodeneeva, E. Phytoplankton Community Structure in Highly-Mineralized Small Gypsum Karst Lake (Russia). *Microorganisms* **2022**, *10*, 386. https://doi.org/10.3390/microorganisms10020386

Academic Editors: Carmela Caroppo and Patrizia Pagliara

Received: 17 December 2021
Accepted: 27 January 2022
Published: 7 February 2022

1. Introduction

Karst lakes are widespread types of natural lentic aquatic ecosystems in the landscape world [1,2]. These lakes are mainly small, but relatively deep and often stratified. The hydrochemical facies of these water bodies are typical complexes of predominating solutes, pointing toward definite climatic and, accordingly, geochemical (weathering), soil, and hydrogeological and hydrobiological conditions, under which lake waters acquire their concentrations and compositions [3,4]. Gypsum karst lakes are characterized as sulphate lakes, as they are rich in gypsum or calcium sulphate, and they have elevated conductivity values [5]. These lakes are brackish (dissolve salts up to 2.0 g/L) and often cold water. Some have powerful springs of underground water pressure, specific water balances, amplified water exchanges, high transparency, and azure (ultramarine) water colors [4,6]. On the global scale, surface outcrops of gypsiferous strata appear quite limited [7]. In this regard, such lakes are often considered as endemic [8] or unique [4–6,9].

Therefore, these lakes are interesting model systems for the investigation of the different groups of microorganisms. It was found that karst lakes can even have different plankton community compositions and structures compared to other karst lakes of similar geneses and morphometric parameters, even within one catchment area [2,5,6,10,11] Such lakes are interesting biotopes for the formation of special biodiversity; however, its importance has not yet been thoroughly evaluated [5,12].

Phytoplankton is an essential part of water ecosystems, which plays a significant role in food web dynamics, energy flow, and nutrient cycling [13]. Studying the patterns in the species composition and abundance in phytoplankton communities helps to understand, in detail, the complex biogeochemical phenomena in water ecosystems [14]. In addition to environmental factors and the lake's age, the latitude position has significant effects on the diversity and development of the phytoplankton in lakes [15], including karstic ones.

The phytoplankton of karst lakes located in temperate zones are characterized by co-occurrences of chrysophytes (Chrysophyceae), dinoflagellates (Dinophyceae), and diatoms (Bacillariophyta) as the most diverse and abundant group [16], and, in some lakes, by Cyanobacteria [17] or Chlorophyta [10]. In lakes of the "warm belt", the dominant role is taken over by Chlorophyta and Cyanobacteria [12,18]. In the spring, high turbulence favors the development and maintenance of diatoms; during the summer, stratification—dinoflagellates, and Cryptophyta, mainly in metalimnion [12,19]. Most of the planktonic algae of karst lakes are cosmopolitan forms. Endemic species were found in the Plitvice lakes [16,20], rare ones—in karst lakes of Greece [12], Romania [21], etc.

At present, there is little information about phytoplankton community structures assessed by using standard biocenotic metrics [22–25]. Structural indicators of the phytoplankton community (e.g., species richness, diversity, evenness, dominance, size structure) are rarely described in detail. It reduces the opportunity to determine the main connections that are established in biological communities under certain abiotic conditions [26]. These structural parameters may reflect the influences of a variety of stressors, including climate change and the consequences related with it [27]. They are useful in understanding energy transfer and may be beneficial for more holistic measurements of the health and resilience of lake ecosystems, in general, to multiple stressors. In turn, a detailed study of structural community indicators allows highlighting the features of typicality or the uniqueness of water ecosystems. Such studies are also relevant from the point of view of identifying the biodiversity of aquatic ecosystems, studying the biology and ecology of rare species, and in planning the protection of unique landscapes or habitats [28].

The Middle Volga region in Russia is the zone of the classical manifestation of karst, presented there by various forms [6], including the rarest gypsum karst. Lake Klyuchik represents an example of gypsum karst lakes of sulfate water types. The unique features of the lake are its feeding (by waters of the underground Surin River) [29] and the presence of an ecotone, i.e., a transitional zone between two water areas of the lake where changes in different habitat parameters have been recorded [30]. Due to its unique characteristics the lake has been assigned as a nature sanctuary of regional value [31].

Previous studies carried out in Lake Klyuchik focused on taxonomical compositions and the development of phytoplankton [32]. Moreover, we assessed the morphological and ecological parameters of mass species in planktonic algocenoses, *Cyclotella distinguenda* Hustedt [33]. Yet, the structural variables of phytoplankton communities, as well as their spatial and temporal distributions, have not been sufficiently investigated.

The aim of this work was to analyze the various structural parameters of the phytoplankton community, and their spatial and vertical distributions in connection with environmental conditions in the small gypsum karst Lake Klyuchik with unique parameters.

2. Materials and Methods

2.1. Study Area and Sampling

Karstic Lake Klyuchik (middle Volga basin) is located in an active karst area of Central Russia, Pavlovsky district, Nizhny Novgorod region (56°58′ N, 43°20′ E). Klyuchik is a small lake with a surface area of 11 ha, a maximum depth of 13.5 m, and a mean depth of 3.8 m [34]. The lake is stretched from west to east in an oval shape, with a weakly sinuous coastline and moderately steep coastal slopes. The relief of the lake bottom is uneven, with pits, represented by a sandy-silty substrate [35]. The lake has two water areas, connected by a shallow strait (ecotone zone). Western and eastern parts of the lake differ from each other in the chemical compositions and physical properties of water (transparency, water temperature, oxygen concentration) [30].

The source of the lake feeding, the underground spring (the river Surin), is located in its western part and is unloaded in the voklina at a depth of 15 m [29]. The western part of the lake is cold water (temperature varies from 6–10 °C during year) and has low oxygen content (40% saturation) in the surface horizon. The eastern part of Lake Klyuchik is warmer (up to 22–25 °C in summer) and has a high oxygen concentration (more than 115%) [30]. Maximum changes in these parameters are observed in the transitional ecotone zone.

The western part of the lake is characterized by a uniform vertical distribution of temperature (unstratified), oxygen, and pH throughout the year [30]; it does not freeze in the winter. In the eastern part of the lake, the vertical distributions of environmental parameters are more pronounced, especially during the summer season. The upper warm layer of the epilimnion and the stretched metalimnion were found here. The hypolimnion layer with stable temperature, oxygen, and other parameters were not observed in the lake [30].

The lake is a brackish-water (salinity, ~2 g/dm^3, electrical conductivity, 1515–1640 μS/cm) [30]. The water color index gradually increases from 40 degree (Platinum-Cobalt scale) in the western part (bluish) of the lake to 62 and 80 degree in the central and eastern (greenish) parts, respectively (Figure 1). The content of phosphates in the water was at the level of permissible values and varied slightly over the lake (0.01–0.05 mg/L). Nitrogen was contained in the water in two forms: nitrate ions 5.2 (mg/L) and ammonium nitrogen (less than 0.01 mg/L) [29]. The sulfate content was high (90–160 mg/L) and could exceed the maximum permissible values up to 7 times.

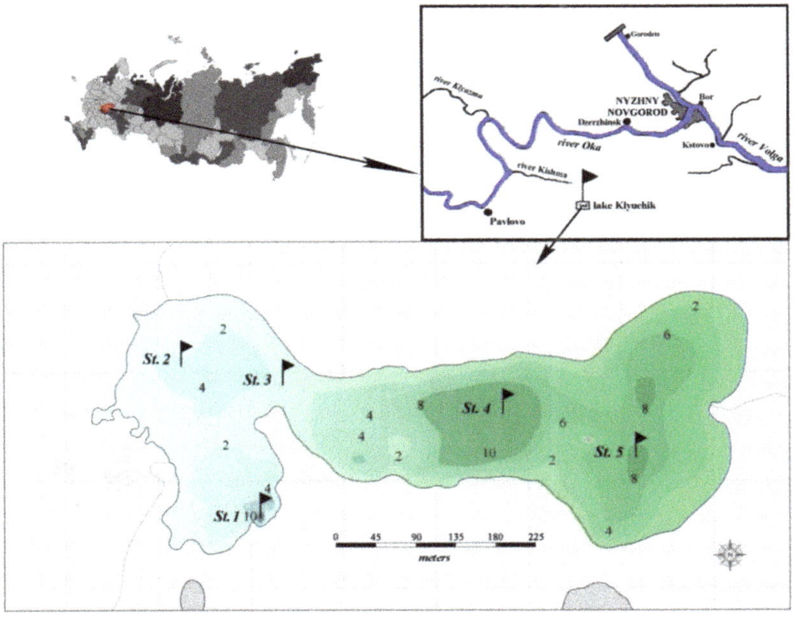

Figure 1. Map scheme of Lake Klyuchik. St. 1–St. 5—sampling stations; 2, 4, 6, 8, 10 are depths (meters) (according to www.lakemaps.org/ru, accessed on 20 November 2021, with some modifications).

Lake Klyuchik has an interesting regional importance because of the numerous recreational activities that take place there.

Water samples of phytoplankton (a total of 76 samples) were collected with a Ruttner bathometer and preserved with an iodine–formalin solution in June–August 2017 at five stations (Figure 1), which were chosen by taking into account the limnological and hydrochemical characteristics of the lake. Stations 1 and 2 were set in the deepest western part of the lake, where water has a bluish color. Stations 4 and 5 were located in the central and eastern parts, respectively, which have a greenish water color; station 3 – in the transitional zone. Integrated samples were taken at all stations. Additionally. in July and August, we collected the samples at stations 1, 3, and 5 in increments of 1 m from the surface to the bottom. The water temperature, and pH were measured in situ using a portable Testo sensor, model 206_pH1 (Company Testo SE & Co. KGaA, Lenzkirch, Germany). Transparency (m) was estimated with a Secchi disk (Papanin Institute for Biology of Inland Waters Russian Academy of Sciences, Borok, Russia).

2.2. Data Analysis

In the laboratory the samples were concentrated to 5 mL by combining the settling method and direct filtration [36], and examined under a light microscope (MEIJI Techno, Saitama, Japan) at 1000 magnification. Phytoplankton was analyzed using a 0.01 mL Najott chamber. We estimated the phytoplankton biovolume (mm^3/L) using geometric shapes closest to the cell shape, taking as a result the mean values of the measurements of 20 to 30 individuals [37,38]. The biomass of each species was calculated by multiplying the number of cells and its biovolume (g/m^3), total biomass—by summarizing each species biomass [39].

Identification of diatoms was possible due to preparation of permanent slides using Naphrax resin (Brunel Microscopes Ltd., Chippenham Wiltshire, United Kingdom) [40]. Centric diatoms were analyzed with the help of a JSM-25S scanning electron microscope (JEOL Ltd., Tokyo, Japan) [33]. Phytoplankton species were identified based on morphology. Nowadays, the optical method of phytoplankton analysis continues to be the principal approach in the ecological monitoring of the water quality, despite some limitations [28,41]. The list guides used for species identification were performed in previous studies [42]. The current names of taxa were also checked using the AlgaeBase website (https://www.algaebase.org/, accessed on 29 July 2021) [43].

The authors analyzed such parameters of the phytoplankton community (=coenocytic parameters) as follows: abundance (N), 10^6 cells/L, biomass (B), g/m^3, species richness (α diversity, Sp—number of species in one sample). Phytoplankton alpha diversity indices were evaluated using the Shannon–Weaver diversity index for abundance (HN, bit/Ex), and biomass (HB, bit/g) [44,45].

$$H = -\sum_{i=1}^{S} P_i \cdot log_2\, P_i \tag{1}$$

Pielou evenness index (EN and EB)

$$E = \frac{H}{H_{max}}, \tag{2}$$

where P_i is the relative abundance or biomass of the i-th species, S is the total number of species in the sample, H_{max} is the maximum Shannon–Weaver index for a given number of a species.

Simpson dominance index (DN and DB)

$$D = \frac{n_i(n-1)}{N(N-1)} \tag{3}$$

where n_i is the abundance or biomass of the *i*-th species; *N* is the total abundance or biomass of phytoplankton in the sample

Water saprobity was evaluated by Pantle–Buck indices, which were calculated according to abundance (SN) and biomass (SB) [46].

$$S = \frac{\Sigma S \cdot h}{\Sigma h} \quad (4)$$

where *S* is the indicator significance of the saprobic species-indicator, *h* is the abundance or biomass indicator.

Size structures of phytoplankton communities were estimated two ways—as arithmetic (by geometric shapes) volume (V_{am}) of the algae cell in a sample and a coenocytic volume estimated as B/N (Vc, μm^3). Moreover, we considered the share of a small cell fraction (<20 μm, nannoplankton, according to: [47]) in the total phytoplankton abundance (%N) and biomass (%B). In addition, the dynamics of flagellar forms of phytoplankton were analyzed. They were evaluated by the share of the monad forms in total abundance (%Nflagel) and biomass (%Bflagel) of algae.

The authors defined the dynamics of the number of dominant and co-dominant species, which were extracted by abundance (SDN and SMN, respectively) and biomass (SDB, SMB). The dominant species included species with an abundance or a biomass more than 10% of the total value and the co-dominant species included species with an abundance (biomass) of 5–10% [48].

2.3. Statistical Analysis

As data do not have normal distribution, the non-parametrical [49] Mann–Whitney criterion (U-criterion) was used to compare the variables; and the Spearman correlation (R_s) to estimate the relationship among the parameters of phytoplankton communities. Statistical processing was conducted using the Statistica 8.0 software package (Statsoft TIBCO, Palo Alto, CA, USA). The authors discussed reliable connections of parameters at the significance level of $p \leq 0.05$.

3. Results

3.1. Environmental Conditions

Table 1 shows the values of the environmental variables recorded at the sampling stations. The depths of sampling stations ranged from 2 m in the transition zone (St. 3) to 10.7 m in the voklina (St. 1), in the area of the main groundwater supply, and from 7.0 (St. 4) to 9.4 (St. 5) meters at deep-water stations in the eastern part of the lake. The transparency of lake waters in the zone of maximum depths decreased twice from station 1 to station 5. In the western part, the temperature varied from 7.5 to 8.0 °C in June to 10.5 °C in August.

Table 1. Limnological, physical, and chemical variables in different stations of Lake Klyuchik, mean values with standard deviations, summer 2017.

Parameter	St. 1	St. 2	St. 3	St. 4	St. 5
Depth (m)	10.7	3.0	2.0	9.4	7.0
Transparency (m)	8.0–8.5	to the bottom	to the bottom	3.4–5.4	3.0–4.5
Temperature (°C)	7.5–9.5 8.8 ± 0.9	8.0–10.5 9.2 ± 0.7	8.1–13.6 10.6 ± 1.6	11.0– 15.8 13.8 ± 1.3	13.2–17.3 15.6±1.2
pH	7.0–7.1 7.1 ± 0.04	7.1–7.2 7.1 ± 0.03	7.1–7.6 7.3 ± 0.16	7.3–7.4 7.4 ± 0.02	7.4–7.6 7.5 ± 0.06

The eastern part of the lake was warmer; the temperature changed there from 11.3 to 13.2 °C in June to 15.8–17.3 °C in August. During the study period, the pH value ranged between 7.0 and 7.6, with an implicit tendency of increasing from June to August, and a clearly manifested increase from the western part of the lake to the eastern one. In the same direction, the color of the lake's waters changed from sky-blue to greenish–blue.

3.2. Phytoplankton Community Composition

The phytoplankton community studied was composed of 133 species, including Bacillariophyta 46%, Chlorophyta 24%; Ochrophyta 11%; Cyanobacteria 8%; and Charophyta, Cryptophyta, Euglenozoa, and Myzozoa, less than 3% each. The number of phytoplankton species per sample varied from 4 to 30 taxa (specific and intraspecific) at different stations. We recorded 13–20 taxa in the integrated samples and 4–15 for particular depths in the western part of the lake. In the ecotone zone (St. 3) the number of taxa increased up to 26–30 and 10–29, respectively.

3.3. The Spatial Distribution of Phytoplankton Abundance, Biomass, Diversity Indices, Size Structure

Table 2 shows the average (M ± m) values of the structural (coenocytic) parameters of the phytoplankton community in different stations (spatial distribution). The abundance and biomass of phytoplankton fluctuated significantly during summer period.

Table 2. Structural variables of phytoplankton in Lake Klyuchik, mean values with standard deviation, summer 2017.

Structural Variables	St. 1	St. 2	St. 3	St. 4	St. 5	U Criterion St. 1 × St. 5
N	0.31 ± 0.07	3.57 ± 2.10	25.80 ± 11.40	1.90 ± 0.41	1.53 ± 0.65	
B	0.94 ± 0.12	11.20 ± 6.00	55.50 ± 21.90	8.32 ± 0.48	6.82 ± 2.96	$p \leq 0.05$
SN	1.59 ± 0.04	1.54 ± 0.02	1.94 ± 0.20	1.52 ± 0.01	1.64 ± 0.07	
SB	1.55 ± 0.04	1.75 ± 0.13	1.87 ± 0.05	1.62 ± 0.07	1.49 ± 0.08	
HN	2.25 ± 0.46	1.56 ± 0.21	1.01 ± 0.56	2.83 ± 0.02	2.22 ± 0.07	
HB	2.19 ± 0.42	1.61 ± 0.23	1.31 ± 0.70	3.31 ± 0.17	2.23 ± 0.18	
DN	0.37 ± 0.10	0.47 ± 0.04	0.71 ± 0.16	0.22 ± 0.01	0.34 ± 0.03	
DB	0.36 ± 0.11	0.50 ± 0.08	0.64 ± 0.20	0.14 ± 0.02	0.36 ± 0.06	
EN	0.58 ± 0.06	0.38 ± 0.04	0.21 ± 0.10	0.66 ± 0.01	0.51 ± 0.03	
EB	0.57 ± 0.09	0.40 ± 0.06	0.27 ± 0.13	0.77 ± 0.02	0.53 ± 0.07	
%N	74.00 ± 3.50	94.60 ± 1.60	91.60 ± 3.70	74.5 ± 2.60	73.40 ± 7.20	$p \leq 0.05$
%B	59.80 ± 8.50	79.70 ± 9.40	82.80 ± 9.00	38.80 ± 5.10	44.60 ± 18.50	$p \leq 0.05$
%Nflagel	0.60 ± 0.30	1.10 ± 0.20	0.20 ± 0.10	8.40 ± 5.90	11.30 ± 4.70	$p \leq 0.05$
%Bflagel	0.40 ± 0.30	6.10 ± 3.70	1.00 ± 0.90	0.90 ± 0.70	14.30 ± 12.30	$p \leq 0.05$
SDN	1.70 ± 0.30	2.00 ± 0.00	1.30 ± 0.30	3.00 ± 0.60	3.00 ± 0.60	
SDB	2.70 ± 0.30	2.30 ± 0.30	1.00 ± 0.00	2.70 ± 0.30	1.70 ± 0.30	
SMN	3.70 ± 0.30	2.30 ± 0.30	2.00 ± 0.60	3.30 ± 0.30	3.70 ± 0.30	
SMB	3.70 ± 0.30	3.70 ± 0.90	1.70 ± 0.30	6.70 ± 0.30	3.30 ± 0.70	
V_{am}	4.60 ± 1.10	12.50 ± 4.10	7.80 ± 1.90	16.20 ± 4.90	19.90 ± 4.70	
Vc	3.30 ± 0.50	3.30 ± 0.20	2.30 ± 0.50	4.80 ± 0.90	5.60 ± 1.90	
Sp	18.30 ± 5.30	17.30 ± 1.80	26.00 ± 4.60	20.00 ± 1.70	21.00 ± 3.20	

The lowest average abundance (0.31 million cells/L) and biomass (0.94 g/m^3) of phytoplankton was recorded at station 1 (voklina). According to the diversity indices in this part of the lake the oligo-dominant algocenoses with medium values of species richness, diversity, evenness, and dominance developed. The phytoplankton community was mainly formed by a small-cell species with an insignificant proportion of flagellar forms (%B flagel—0.4; %N flagel—0.6) among them.

The northwestern shallower water area of the lake (St. 2) was characterized by a ten-fold increase in the degree of phytoplankton development, with an unreliably pronounced tendency to increase the temperature and pH of the water. The development of 2–4 dominant algae species was accompanied by a decrease in the diversity and evenness of communities. At the same time, the proportion of small-cell fraction and flagellar forms increased (especially in terms of biomass %Bflagel—6.1). At this station, the arithmetic mean volume (V_{am}—12.5) of algal cells increased (almost three-fold), while the coenocytic volume (Vc) did not change. This may indicate the appearance in the phytoplankton communities of a few large forms of algae, which did not yet occupy a dominant position.

The development of phytoplankton in the eastern water area of the lake (Sts. 4, 5) did not significantly differ from the station 2, but it was higher, especially biomass (the statistical comparison from the U-criterion showed significant differences ($p < 0.05$)) than in the zone of the main groundwater supply (St. 1). In these sites of the lake, the distribution of the phytoplankton species, in terms of abundance and biomass, was more even. Accordingly, the Shannon indices (especially at St. 4) and the Pielou evenness indices were higher here, and the dominance was less pronounced. In the eastern water area of the lake, the role of large-celled algae species became more significant, whereas the proportions of low-sized components in phytoplankton communities decreased. In these stations, some representatives had sizes of more than 64 microns (net plankton, according to [48]). In addition, the composition of dominant forms became richer, and there was a clear tendency to increase the role of flagellar algae (especially on station 5).

In the ecotone zone (St. 3), the development of phytoplankton was the most productive. Due to the diatoms "blooming", monodominant (DN = 0.71, DB = 0.64) communities formed in this part of the lake (Table 2), in which the only one species of centric diatoms—*Cyclotella distinguenda*—dominated. Most of this diatom species had a cell diameter of less than 20 microns, so the share of small cell algae in the abundance (91.6%) and in the biomass (82.8%) in this area of the lake was the maximum.

3.4. The Vertical Distribution of Phytoplankton Abundance, Biomass, Dominant Species, Diversity Indices, Size Structure

Heterogeneity in the distribution of structural parameters of the phytoplankton community was also noted along the depth (Figures 2–5).

In the deepest part of the western area of the lake (station 1), two layers with noticeable phytoplankton richness were distinguished (0–1 and 8–9 m in July, 0–2 and 7–9 m in August) (Figures 1 and 2). In the first one (surface layer), the highest values of species richness (32 species), Shannon indices (HN = 3.52, HB = 3.36), and evenness index (EN = 0.70, EB = 0.67) were established in July. In this month, the phytoplankton abundance and biomass were typical for mesotrophic lakes (0.45 million cells/L, 1.37 g/m^3). Algocenoses were formed with a predominance of *Cyclotella distinguenda* and *Melosira varians* Ag., accompanied by large-celled benthic diatom forms (species of the genera *Pinnularia*, *Caloneis*, *Amphora*, *Navicula radiosa* Kütz., etc.). At the depths of 5–7 m, the development of phytoplankton significantly decreased, increasing again in the bottom layer. Near the bottom, there were the phytoplankton communities with the predominance of large-cell forms (especially in the formation of biomass) and the participation of *Melosira varians*, *Cyclotella distinguenda* and *Meridion circulare* (Grev.) Ag. A few flagellar algae (genera *Chrysococcus*, *Dinobryon*, *Euglena*, *Vacuolaria*) were found at depths from 0.5 to 4.0 m.

In August, at station 1, the vertical distribution of structural parameters of phytoplankton was the same (Figure 2). In the surface water layer, the development of algae was insignificant (the abundance reached 0.1–0.2 million cells/L, the biomass—0.1–0.26 g/m^3). Near the bottom (7–9 m), the abundant increased 1.5 times, and the biomass—6 times. In the intermediate layer (3.0–6.0 m), the decrease in species richness (up to 4–6 species in one sample) and diversity indices were found. In comparison with surface layers, the participation of the small-celled fraction of plankton in the formation of its abundance became more noticeable (by 1.5–2.2 times), and its contribution to the phytoplankton biomass was maximal (up to 79%). From a depth of 6 m to the bottom, the proportion of large-cell phytoplankton species increased because of the benthic species of the diatoms presence in the sample (species from genera *Nitzschia*, *Navicula*, *Cymbella*, *Gomphonema*, *Pinnularia*) (Figures 3 and 4). Thus, in the zone of the greatest depths of the western part of the lake, two maxima of phytoplankton development were formed at the surface, and near the bottom in July and at the bottom in August.

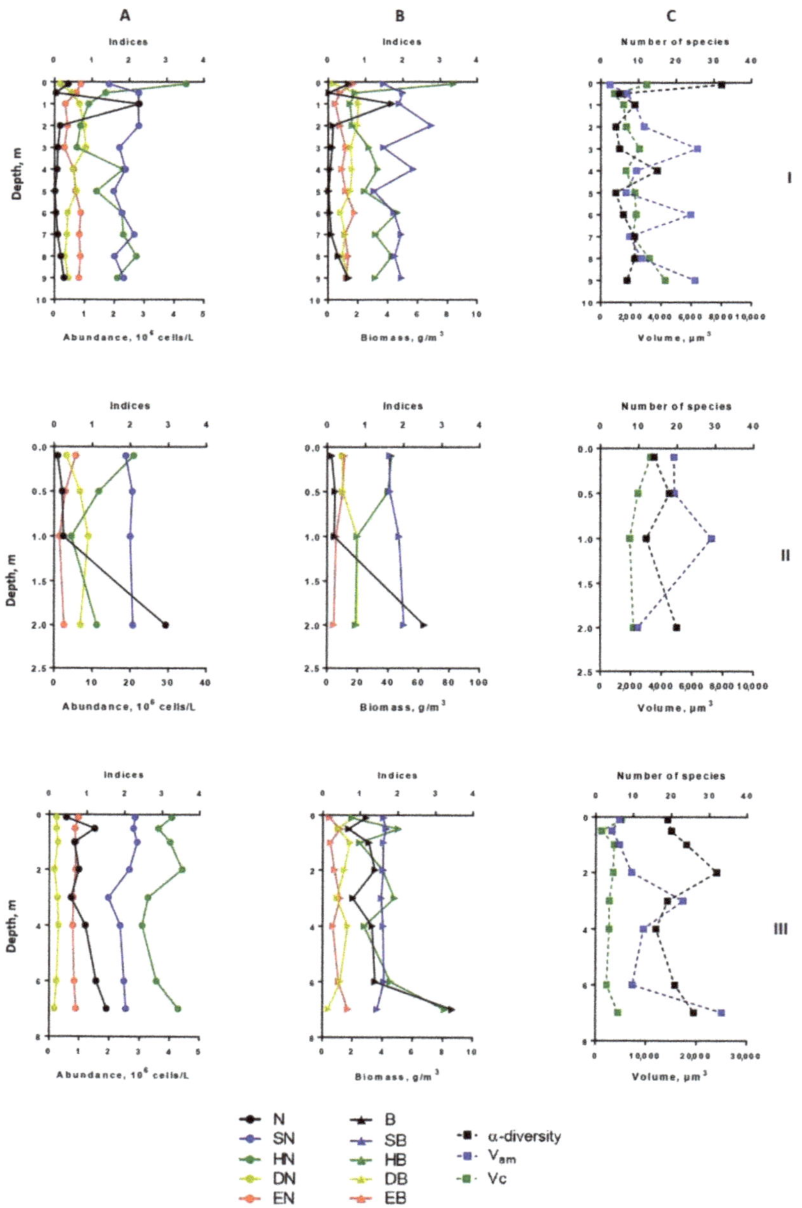

Figure 2. Vertical distribution of diversity indices and structural variables (α-diversity, abundance, biomass, saprobity, size structure) of phytoplankton in Lake Klyuchik in 2017 (July). A—Indicators for the abundance; B—indicators for biomass; C—the number of species and size structures, I—Station 1, II—Station 3, III—Station 5. N, 10^6 cells/L—abundance; B, g/m^3—biomass; SN–Pantle–Buck index, calculated to abundance; SB—Pantle–Buck index, calculated to biomass; HN bit/Ex—Shannon–Weaver diversity index for abundance; HB, bit/g—Shannon–Weaver diversity index for biomass; DN—Simpson dominance index for abundance; DB—Simpson dominance index for biomass; EN—Pielou evenness index for abundance; EB—Pielou evenness index for biomass; α-diversity—number of species per sample; V_{am}—arithmetic volume; Vc, µm^3—coenocytic volume.

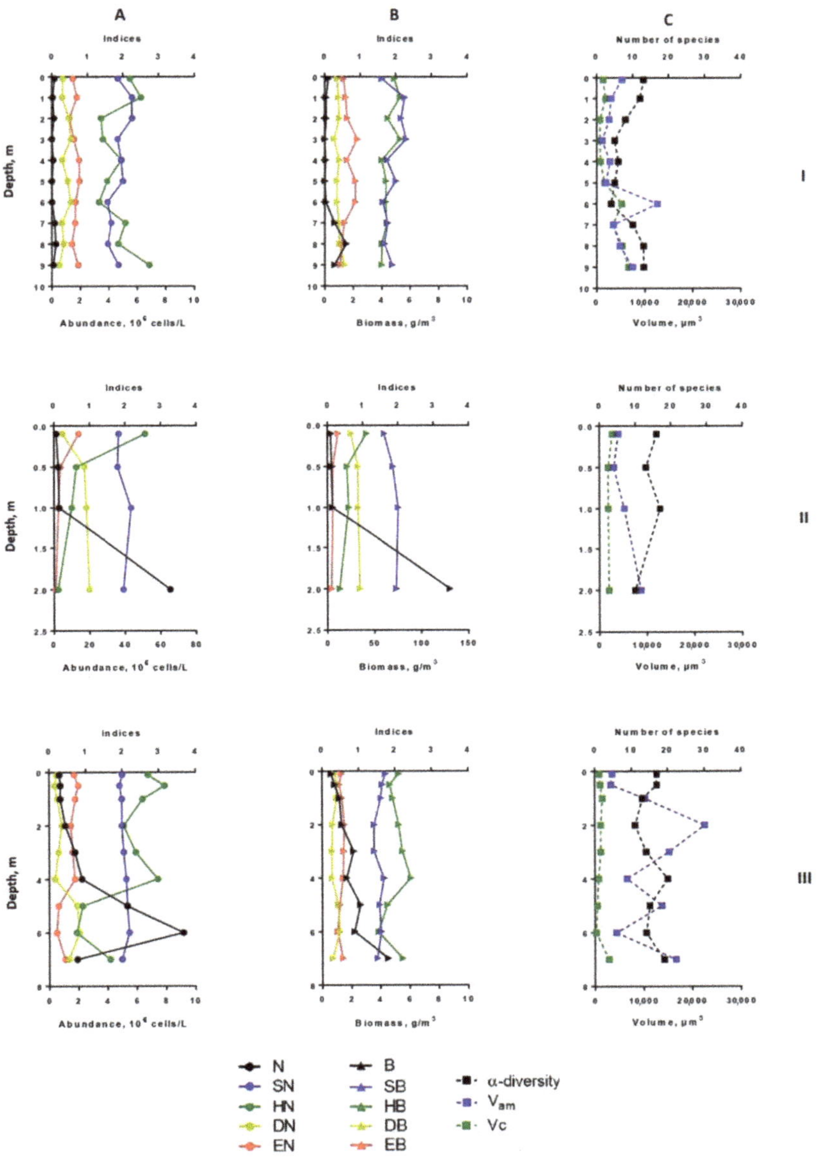

Figure 3. Vertical distribution of diversity indices and structural variables (α-diversity, abundance, biomass, saprobity, size structure) of phytoplankton in Lake Klyuchik in 2017 (August). A—Indicators for the abundance; B—indicators for biomass; C—the number of species and size structures, I—Station 1, II—Station 3, III—Station 5. N, 10^6 cells/L—abundance; B, g/m^3—biomass; SN—Pantle–Buck index, calculated to abundance; SB—Pantle–Buck index, calculated to biomass; HN bit/Ex—Shannon–Weaver diversity index for abundance; HB, bit/g—Shannon–Weaver diversity index for biomass; DN—Simpson dominance index for abundance; DB—Simpson dominance index for biomass; EN—Pielou evenness index for abundance; EB—Pielou evenness index for biomass; α-diversity—number of species per sample; V_{am}—arithmetic volume; Vc, μm^3—coenocytic volume.

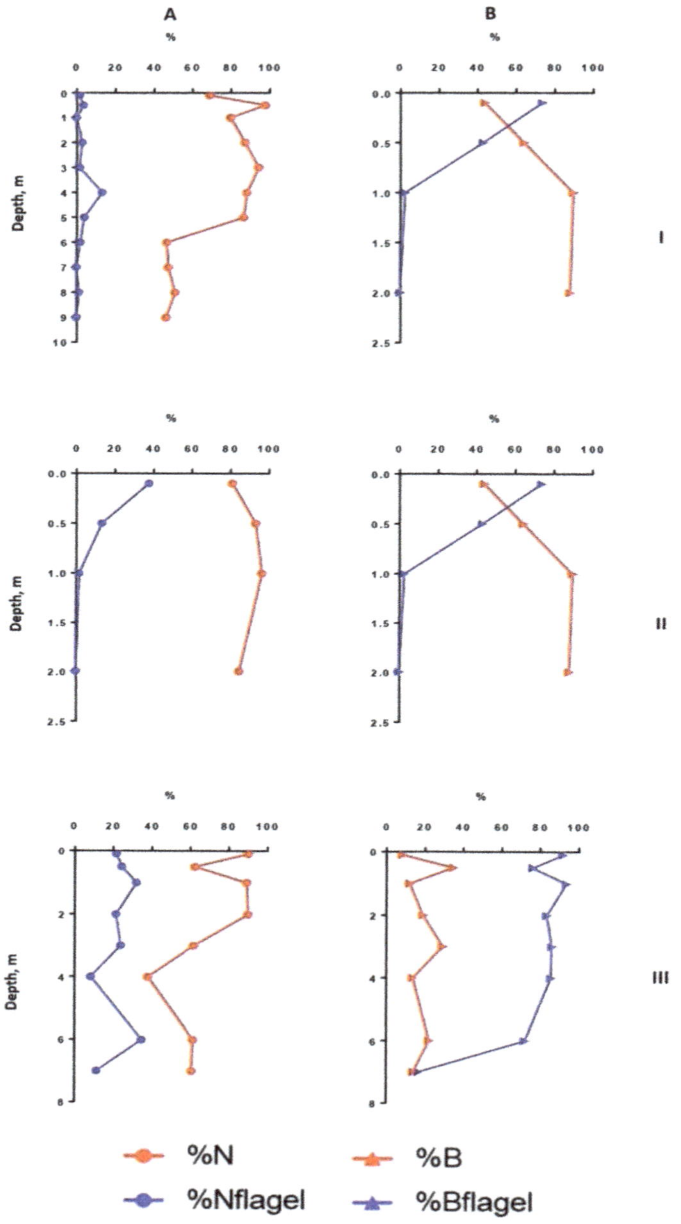

Figure 4. Vertical distribution of relative structural variables of phytoplankton in Lake Klyuchik in 2017 (July). A—Share of small-celled fraction and flagellar species proportion counted for the abundance, B—share of small-celled fraction and flagellar species proportion counted for the biomass; I—Station 1, II—Station 3, III—Station 5. %N—a share of a small cell fraction (<20 µm) in the total phytoplankton abundance; %B—a share of a small cell fraction (<20 µm) in the total phytoplankton biomass; %Nflagel—a share of the monad forms in total abundance of algae; %Bflagel—a share of the monad forms in total biomass of algae.

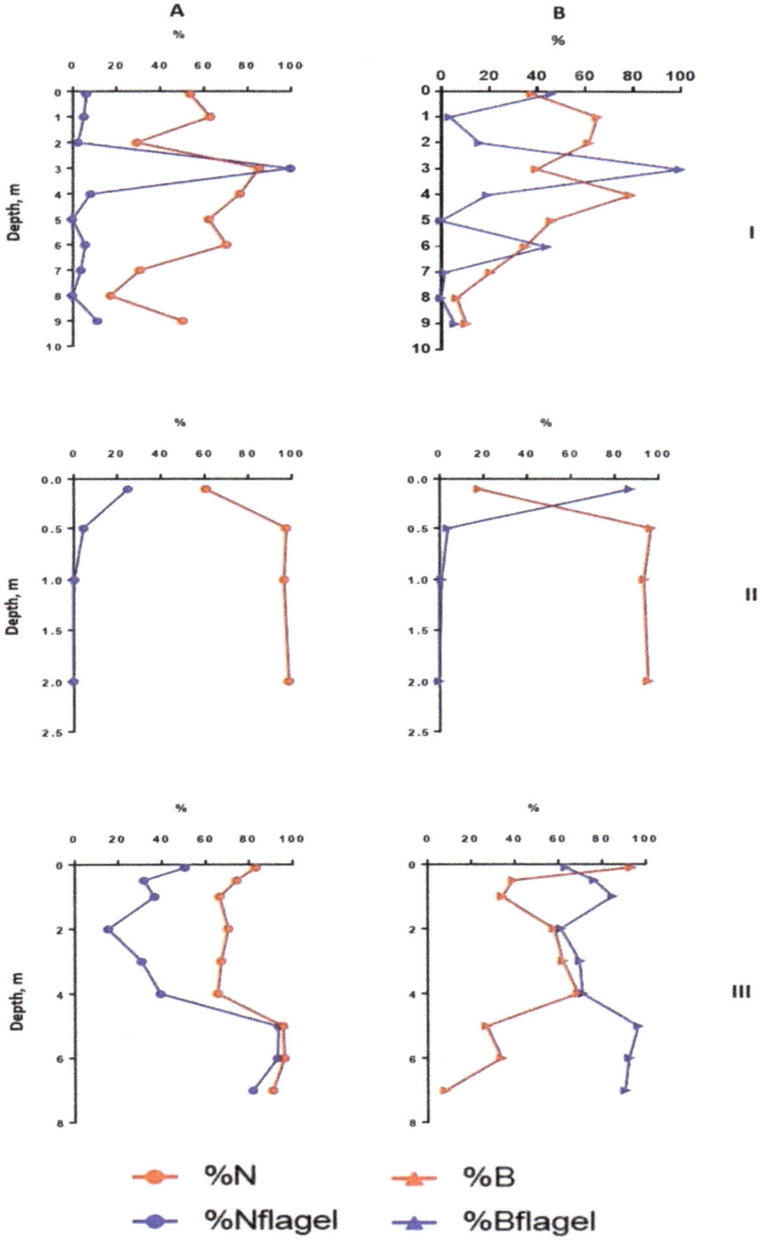

Figure 5. Vertical distribution of relative structural variables of phytoplankton in Lake Klyuchik in 2017 (August). A—Share of small-celled fraction and flagellar species proportion counted for the abundance, B—share of small-celled fraction and flagellar species proportion counted for the biomass; I—Station 1, II—Station 3, III—Station 5. %N – a share of a small cell fraction (<20 μm) in the total phytoplankton abundance; %B—a share of a small cell fraction (<20 μm) in the total phytoplankton biomass; %Nflagel—a share of the monad forms in total abundance of algae; %Bflagel—a share of the monad forms in total biomass of algae.

At station 5 (in the eastern part of the lake), the maximum phytoplankton development was clearly manifested near the bottom (Figure 2) with predominant large benthic species of diatoms (genera *Pinnularia*, *Amphora*, *Nitzschia*, *Cymbella*). In July, their relative biomass and abundance were more than 60% and 10%, respectively. In contrast to benthic communities, communities on the surface of this station were formed by small-cell species of phytoplankton, such as *Cyclotella distinguenda*, *Hadmannia comta* (Ehr.) Kociolek and Khursevich, *Monoraphidium minutum* (Nägeli) Komárk.-Legn., and *Dictyosphaerium subsolitarium* V. Goor with a significant relative abundance. The phytoplankton of this part of the lake was enriched with flagellate algae, which were found in each layer (22.5 $\pm$ 3.2% and 75.9 $\pm$ 8.8% of the total abundance and biomass, respectively, for the entire water column). Their relative biomass was the highest at the surface and gradually decreased at the bottom. Among the monad fraction, both a few large-celled species (*Peridinium cinctum* Pénard. and *Ceratium hirundinella* (O. Müll.) Dujard.) and numerous small-celled ones (*Komma caudata* (Geitler) D.R.A. Hill., *Kephyrion* gen. spp.) were distinguished. Dominant species *Peridinium cinctum* developed in a hole water column (from the surface to the bottom), forming for 61–89% of the phytoplankton biomass in the surface and intermediate layers; near the bottom—only 14%. Algocenoses that formed in the bottom layer (depth 7 m) were characterized by high Shannon indices (HN = 3.43; HB = 3.25) and Pielou evenness indices (EN = 0.72; EB = 0.68) of species and weak dominance in abundance (DN = 0.15) and biomass (DB = 0.16) of phytoplankton (Figures 2 and 4).

In August, the vertical distribution of phytoplankton structural parameters at this station was similar to that of July. In the surface layer, the biomass and abundance of phytoplankton corresponded to the oligotrophic state, in a depth of 6–7 m—to the mesotrophic level (9.2 million cells/L – 4.53 g/m^3) (Figure 3). Increasing the depth, there was a tendency in enlarging of α-diversity, although the number of dominant species in abundance in the bottom layers turned were lower than in the rest of the water column. The phytoplankton community was mainly formed by small-celled species. Their proportion in the total abundance was maximal between layers of 5 to 7 m, while the contribution in the biomass sharply decreased near the bottom (Figure 5). At the deepest layer upon contact with the bottom surface, the benthic species of algae (mainly large-cell pennate diatoms—*Navicula radiosa*, *Amphora ovalis* (Kütz.), *Pinnularia* sp.)—enriched the plankton communities and contributed the increase of the arithmetic volume of cells (V_{am} = 16640) (Figure 3). The relative abundance of flagellate algae increased more than twice in comparison with July, but their proportion in the total biomass remained the same. Dominant in abundance were *Komma caudata*, species of the genera *Cryptomonas*, *Chrysococcus*, *Dinobryon*, and other unidentified small-cell chrysomonads. These species vegetated (up to 8.92 million cells/L) at a depth of 4–5 m to the bottom forming more than 90% of the total abundance on the 6-m layer. Most notable in the creation of biomass were large-celled *Ceratium hirundinella* (up to 1.78 g/m^3) and *Peridinium cinctum* (up to 2.0 g/m^3), they prevailed in the bottom layer of water, and *Vacuolaria virescens* Cienk. (up to 0.7 g/m^3) at a depth of 4.0 m.

In the shallowest, station 3, in the ecotone zone during the summer, the highest integrated abundance and biomass (97.8 g/m^3) of phytoplankton were observed in August. The vertical stratifications of phytoplankton abundance and biomass were the most pronounced there. Phytoplankton was sharply concentrated in the bottom layer of water (the abundance of this group of living organisms was 32 and 49 times higher than in the surface in July and August, respectively, and the biomass was 21 and 36 times higher; the maximum value of biomass was 130 g/m^3) (Figures 2 and 4). In this part of the lake, oligo- and even monodominant phytoplankton communities, poor in species, were formed with an absolute prevalence of *Cyclotella distinguenda* near the bottom. Alfa-diversity indices (especially in terms of the abundance of species) had higher values in the surface layers and reached a minimum (Shannon indices: HN = 0.11; HB = 0.34 and Pielou evenness indices: EN = 0.04; EB = 0.11) at the bottom during the period of diatoms "blooming". In the vertical distribution, there was a reduction of the proportion of flagellar algae from the surface (25.2–37.9% of abundance, 73.9–87.1% of the biomass) to the bottom (0.11 and 0 accordingly), and opposite, gradually increasing the share of small-celled plankton from the surface to the bottom (by 2–5 times in terms of biomass), from the beginning of the summer period to its end (Figures 3 and 5).

3.5. Statistical Analysis

Phytoplankton abundance and biomass in Lake Klyuchik were positive correlated with each other (Rs = 0.89). The abundance tended to increase in more saprobic and trophic waters (the correlation coefficients between N and SB was 0.63). At the same time, no significant relations with the structural variables of phytoplankton communities were revealed, except for the abundance and its evenness index (EN), which had a negative relation (Rs = −0.54). The abundance of phytoplankton positively correlated (Rs = 0.52) with the proportion of the small-cell species (%N) in its formation and negatively correlated with the proportion of the dominant species, in terms of biomass (Rs = −0.59 and −0.69, the relationship of this parameter with the abundance and biomass, respectively).

The correlation analysis showed that the Shannon–Weaver indices (HN, HB) positively correlated with the Pielou evenness indices, and they had a significant negative correlation with the Simpson dominant indices (Table 3).

Table 3. Spearman correlation matrix of the diversity indices of phytoplankton in Lake Klyuchik ($p \leq 0.05$).

	HB	HN	DB	DN	EB	EN
HB	1.00	0.88	−0.99	−0.89	0.86	0.88
HN	0.88	1.00	−0.87	−0.96	0.74	0.95
DB	−0.99	−0.87	1.00	0.87	−0.88	−0.85
DN	−0.89	−0.96	0.87	1.00	−0.73	−0.91
EB	0.86	0.74	−0.88	−0.73	1.00	0.78
EN	0.85	0.95	−0.85	−0.91	0.78	1.00

The Shannon indices had a negative relation with the proportion of the small-celled species in the abundance and biomass, saprobity, and a weakly positive correlation, with a relative abundance of flagellar forms and a diversity of dominant and mass species (Table 4). A similar tendency was recorded with Pielou, which had a positive correlation with a majority of structural variables. The Simpson dominant indices had the opposite trend (Table 4).

Table 4. The correlation coefficients found by correlation analysis of α-diversity indices and some structural variables of phytoplankton communities in Lake Klyuchik ($p \leq 0.05$).

	HB	HN	DB	DN	EB	EN
N	$p \geq 0.05$	$p \geq 0.05$	$p \geq 0.05$	$p \geq 0.05$	$p \geq 0.05$	−0.54
SB	$p \geq 0.05$	−0.52	$p \geq 0.05$	$p \geq 0.05$	$p \geq 0.05$	−0.55
%N	−0.62	−0.76	0.59	0.76	−0.66	−0.82
%B	−0.71	−0.85	0.73	0.74	−0.67	−0.81
%Nflagel	0.57	0.64	−0.54	−0.77	$p \geq 0.05$	0.61
SDN	0.70	0.64	−0.65	−0.78	0.56	0.57
SMN	0.56	0.77	$p \geq 0.05$	−0.75	$p \geq 0.05$	0.74
SDB	*p*≥*0.05*	$p \geq 0.05$	$p \geq 0.05$	$p \geq 0.05$	$p \geq 0.05$	0.53
SMB	0.74	0.67	−0.77	−0.74	0.68	0.76

No significant relationship was found between species richness and other structural variables of phytoplankton communities. The number of dominant species by biomass was negatively related to the abundance (Rs = −0.69) and biomass (Rs = −0.59) of phytoplankton and positively to the coenocytic volume (Rs = −0.57). The number of dominant and subdominant species in abundance had negative correlations with the share of the small-cell species in the abundance (Rs = −0.51 and −0.85 respectively) and biomass (Rs = −0.51

and −0.53) of phytoplankton, i.e., with its increase, the composition of coenose-forming components with a size of less than 20 microns became poorer. Positive correlations were also noted with the share of flagellates, in abundance (Rs = 0.53–0.86).

The number of dominant species was not determined by the species richness ($p \geq 0.05$). Usually, it was not notable (2–4 species in a sample, rarely more – up to 7, station 4), as well the total number of species (average for the lake 15 ± 1, maximum 32–35).

There was a positive correlation (Rs = 0.65) between the arithmetic volume of algal cells (V_{am}) and coenocytic volume (Vc) in phytoplankton communities of Lake Klyuchik; the connection between these parameters and the abundance and biomass was negative, but unreliable ($p \geq 0.05$). The low positive correlation of the small-cell species proportion in abundance with the biomass of phytoplankton (Rs = 0.52) and the significant contribution in those parameters (over 40–70%) (Table 4) reflected the predominance of the nannoplankton component in phytoplankton communities.

Flagellate algae were poorly represented in the phytoplankton communities of the lake. Their share on average for all stations was 4.3–4.5% of the abundance or biomass; at some stations (St. 5) it increased to 11–14% (Table 2). The presence of monadic algae to a certain depth was also unclear. At St. 1, they concentrated at a depth of 3 m (August) and were the only group in that layer, or their noticeable development (up to 87% of the total biomass) was noted in the surface layer (St. 3—ecotone zone); sometimes they created a maximum (4.53 g/m^3—up to 97% of the total biomass) in the bottom layer (St. 5) (Figure 5). The proportion of flagellate algae in the abundance showed a positive correlation with the arithmetic volume of the algal cell (Rs = 0.71), since the large-sized representatives (*Ceratium*, *Peridinium*, *Gonyostomum*, *Vacuolaria*) or colonial (*Dinobryon*) species of algae were mainly observed there.

A review of relations of phytoplankton structural parameters with environmental parameters demonstrated a significant role of the depth. The depth of the lake had a positive correlation with the number of dominant species (Rs = 0.62–0.66), diversity Shannon–Weaver indices (Rs = 0.76 for HN and Rs = 0.61 for HB), and Pielou evenness (Rs = 0.82 for EN and Rs = 0.61 for EB), and negative with Simpson dominant indices (Rs = −0.68 for DN and Rs = −0.61 for DB). A significant negative correlation was also noted between the depth and water saprobity index by biomass s (Rs = −0.61).

The value of water transparency had a negative correlation with the abundance (Rs = −0.97) and biomass (Rs = −0.90) of phytoplankton, the water saprobity index by biomass (Rs = −0.77), and the proportion of small-celled algae that played a role in the formation biomass (Rs = −0.80). Transparency correlated positively with diversity index based on abundance (Rs = 0.62), evenness (Rs = 0.59), and the number of dominant species (Rs = 0.73). The temperature factor had positive correlations with the volume of algae cells (Rs = 0.53) and the proportion of the flagellar species in the forming of abundance (Rs = 0.51) only. The value of pH showed a positive relationship only with the proportion of flagellate algae, in terms of abundance (Rs = 0.57).

4. Discussion

Lake Klyuchik is a small gypsum karst lake with calcium sulphate water. Like most gypsum reservoirs [4–6], it has brackish water (dissolved salts up to 2.0 g/L), high transparency, and an azure (ultramarine) water color. The lake also has some specific characteristics. Firstly, it has an ecotone zone that connects two parts of the lake differing from each other in the chemical composition and physical properties of the water. Secondly, in the western water area of the lake, there is an underground spring, the flow rate of which varies from 1.79 m^3/s to 4.21 m^3/s [34]. According to the flow-based classification [50], it belongs to the second-magnitude group of springs; from this point of view, lake Klyuchik can be considered a unique hydrological object. The presence of an underground spring ensures a constant low temperature in the western part of the lake (it does not freeze during the winter) throughout the year, and a lack of temperature and oxygen stratification [30].

With this combination of factors, Lake Klyuchik turned out to be similar to Lake Goluboe (Samara Region, Russia) [17] and Lake Ochiul Beiului (Romania) [2].

The water transparency in Lake Klyuchik was in the range of 3 to 8 m and was peculiar to oligotrophic–mesotrophic types of water bodies [51]. The values of transparency were less in the ecotone zone due to the slight depth at this part of the lake (up to 2 m). The coefficient of relative transparency, estimated as the ratio of the average transparency to the average depth, was 1.1. It is typical to the class of optically deep water bodies, in which the water transparency is 1–2 times greater than the average depth [51], favoring the development of phytoplankton at considerable depths. Lake Klyuchik can be classified as neutral or oligo-alkaline, according to the pH value [51].

We regard this lake as a model of an aquatic system, where the effect of the environment "severity" is clearly expressed. According to Bigon et al. [52], the environment "severity" means the predominance among environmental conditions of one or few limiting factors (acidification, pollution, thermification, etc.), which are responsible for influencing the structural variables of the biotic communities and suppressing the effects of other factors. In Lake Klyuchik, two main factors (temperature and mineralization, which have great influence on the structure of algocenoses) are combined in an unusual way, not typical for the lakes of the temperate forest zone. Although the amount of solar energy supplied per unit of the lake surface area is typical for the forest zone, the thermal regime of the lake is more similar for the northern lakes of the coniferous–deciduous forests. Lake Klyuchik, consistent with the thermal regime, is similar to water bodies of the tundra and forest–tundra, for example, the subarctic lakes of Fennoscandia [53]. On the other hand, the values of mineralization are more typical for southern lakes of the steppe zone [6,10,17].

The continual inflow of underground cold, blue-colored waters, increases the albedo of the lake in its western part. It contributes to a significant reflection of light energy from the surface, weak water heating (average summer water temperature was 8.8 ± 0.5 °C, seasonal temperature variations were 2.5–3.0 °C), and the formation of stable low-temperature conditions in the summer season. The eastern part of the lake is more productive because its waters are warmer in the summer (average temperature was 14.7 ± 0.9 °C, seasonal temperature variations were 4.1–4.5 °C). The transparency of the waters is about two times lower than in the western part, the waters are greenish–blue in color because of the development of autotrophic plankton. With the same amount of incoming sun energy, its absorption and, accordingly, water heating, are more efficient in the western part of the lake. It contributes to the compensation of the low temperature background and stimulates the development of thermophilic algae species.

The floristic list of algae in Lake Klyuchik was formed by nine taxonomical groups where diatoms significantly prevailed in terms of the number of species (more than 40%). The proportions of green algae and cyanoprokaryotes were lower, while in the majority of water bodies in the temperate zone, they usually predominated [10,54,55]. The same proportion of large taxa (divisions) of algae were recorded in Lake Goluboe, Russia [17], and in Lake Ochiul Beiului, Romania [2], with similar habitat conditions. The species composition of phytoplankton of Lake Klyuchik was represented mainly by benthic, littoral, and truly planktonic forms with dominance of the cosmopolitan species (87.1%) and a low proportion of boreal ones (12.9%) [32]. In this regard, this water bodies are similar to most of the lakes [15].

Species richness is one of the most important parameters of the phytoplankton community [52,56,57]. The specific combination of environmental factors in Lake Klyuchik was weakly beneficial for the formation of species-rich phytoplankton communities there. The alpha diversity of phytoplankton (18–20 species per sample in summer season) was significantly lower than in other waters bodies (e.g., in the lakes of the Pustynsky Reserve—up to 40–50 species per sample) of the same karst zone of the Volga basin [54]. Similarly, low values of α-diversity were noted for highly-humified and acidic water bodies [58–60]; in this term, it can be considered a common answer, in regard to biota communities and stress conditions.

Analysis of the spatial and vertical distribution of the main parameters of the phytoplankton structure of Lake Klyuchik and their correlations revealed some patterns. The distributions of the taxonomic composition, quantitative development, and diversity indices of phytoplankton were characterized by spatial heterogeneity. The abundance and the biomass varied significantly during the summer period and generally characterized Lake Klyuchik as a eutrophic lake [51] with the average biomass being more than 5.0 g/m^3, except for the area of voklina (St. 1). At this station, the phytoplankton abundance and biomass were not high, peculiar to oligotrophic or the weakly mesotrophic state. In terms of temporary changes, the maximum biomass values at stations 2 and 5 were recorded in July, and at stations 3 and 4—in mid-August. At the deepest, station 1, biomass values were similar during the summer. Successions of the dominant species are presented in previous studies in more detail [32].

The dominant groups of phytoplankton were similar to those noted for other karst lakes of temperate zones, including gypsum ones [2,10,16,21]. The phytoplankton of Lake Klyuchik is characterized by co-occurrences of chrysophytes (Chrysophyceae—up to 20% of the total abundance), dinoflagellates (Dinophyceae—up to 40% of the total biomass), and diatoms (Bacillariophyta) in the warm water part of the lake, and practically complete (50–100%) dominance of diatoms in the cold water part, where they often develop near the bottom. In high transparency conditions (up to 8.5 m), and a lack of light limitation, photosynthesis was possible throughout the entire water column, including the area near the benthal zone. This area became available for the normal functioning of planktonic or benthic algae, including representatives of the "shadow" flora, mainly from the diatoms. Avoidance by diatoms of well-lit surface layers of highly transparent water bodies (both marine and fresh) is a well-known fact in the ecology of phytoplankton and diatoms [47] (the phenomenon of a deepened maximum of photosynthesis).

The proportion of dominance groups and individual development values of dominant species were rather specific. In this regard, the ecotone zone of the lake is of particular interest, where the most noticeable phytoplankton development was noted on the border between the water column and bottom sediments. The phytoplankton biomass was uniquely high there (up to 130 g/m^3) and was typical for hypertrophic water bodies [51]. Such development of phytoplankton sharply distinguished Lake Klyuchik from other water bodies, the Middle Volga basin, including the highly eutrophic Cheboksary reservoir [61], the Oka rivers [36], and others.

Despite the higher alpha diversity (26 taxa) in the ecotone zone, compared to the other stations, monodominant phytoplankton communities were formed here, in which only one species of centric diatoms—*Cyclotella distinguenda*—had prevailed. A small part of the population of this diatom had a cell diameter of more than 20 microns [33]. It affected the proportion of large and small cell components in the algocenoses. The share of small cell algae in the biomass in this area of the lake was maximum.

Cyclotella-species "blooming" are often phenomena in the karstic lake, in both the temperate zone [16,20] and in the lakes of the warm belt [12,18,19], especially during spring and autumn turbulence.

Cyclotella distinguenda is a rare species for the algoflora of the Volga basin, as well as for river systems in Hungary [62]. This species of centric diatoms, according to its ecological preferences, prefers mesotrophic conditions, and is sensitive to stratification [63]. In accordance with the literature data [16], similarly high values of biomass and abundance of this species were not observed in other lakes. We suppose that the unique combination and dynamics of the environmental factors of the lake (high mineralization, favorable light conditions, low temperature background, and lack of thermal stratification) were optimal for the mass development of this species and allowed it to regulate the structure and productivity of algocenoses in weak competition conditions.

Such combinations of factors were likely not beneficial for the other groups of phytoplankton. Cyanobacteria are frequent components of phytoplankton communities in highly mineralized water bodies of the steppe and semi-desert zones [64–66], and in the subtropical zone [18,19], especially in the midsummer stratification period. In Lake Klyuchik, despite the high values of mineralization, the development and "blooming" of cyanobacteria were limited by low temperatures and by an absence in the western part of the reservoir, or weakly expressed stratification (in the eastern part). The cyanobacteria found in plankton were represented by small coccoid and filamentous forms (*Aphanocapsa* spp., *Pseudanabaena* spp.), but their abundance was not significant.

A similar trend can be noted for coccoid green algae, the diversity, growth rate, and productivity of which are stimulated by elevated temperatures [66,67].

The role of phytoflagellates (species from different taxonomic groups), frequent inhabitants of lentic water bodies [12,16,68], turned out to be less significant than other groups. Flagellar algae are weakly dependent on light conditions due to their greater tendencies to mixotrophic feeding and heterotrophic carbon assimilation [60,69,70], but they are sensitive to turbulence [19]. Since there was no pronounced stratification in the lake, accumulations of mobile monad algae at different stations of the lake could be formed in any part of the water column and their preference for a certain depth was not clearly manifested. However, the proportion of flagellate algae was statistically higher in the warm eastern part of the lake.

Among structural parameters, species richness did not show significant relations with other structural variables, whereas the number of dominant and subdominant species turned out to be more indicative and showed correlations. It means that the alpha diversity of phytoplankton (Shannon diversity index, Pielou evenness index and Simpson dominance index) were mainly determined by the number of the dominant species (not by the species richness of phytoplankton) and their ecological preferences (the ability to vegetate at low summer temperatures in mesohaline conditions).

The size structure of the phytoplankton community, as an important indicator of water ecosystem eutrophication [47], reflected the predominance of nannoplankton components in phytoplankton communities of Lake Klyuchik. The proportion of small-celled organisms increased in communities with more pronounced dominance and decreased with an increase in species diversity and evenness.

The saprobity of water bodies within the limnosaprobic levels of their pollution with organic matter changes, in parallel with the levels of trophicity, can be considered as structural indicators of communities [46]. The variations in the average values of the saprobity indices in Lake Klyuchik was within the framework of oligo-β-mesosaprobic water pollution, with higher values in the ecotone zone of the lake. As the saprobity enhanced, we observed a tendency of simplification of phytoplankton communities (a decrease in diversity and evenness) and an increase in the role of the small-cell fraction.

5. Conclusions

The investigation of the gypsum karstic lake with the unique abiotic environmental factors made it possible to identify the most important features of the structure and dynamics of phytoplankton communities developing in such conditions. These patterns expand our understanding about the diversity of phytoplankton communities, in general, and provide clarifications about their organizations under certain environmental scenarios. The ideas will be useful for studying the biodiversity of phytoplankton communities for lakes with unusual—as well as typical—parameters, to assist in the planning (i.e., in the protection of unique landscapes or habitats) and in the assessments of the ecological statuses of these lakes.

Author Contributions: Conceptualization: E.V. and A.O. Laboratory investigation: P.K., N.S., E.S. and E.V. Contributed to the sample collection: E.V., P.K. and N.S. Manuscript preparation: A.O. and E.V. Methodology: P.K. and N.S. Contributed to the visualization: E.S. All authors have read and agreed to the published version of the manuscript.

Funding: This research was funded by the Russian Foundation for Basic Research, project number 20-04-01005A.

Institutional Review Board Statement: Not applicable.

Informed Consent Statement: Not applicable.

Data Availability Statement: No new data were created or analyzed in this study. Data sharing is not applicable to this article.

Conflicts of Interest: The authors declare no conflict of interest. The funders had no role in the design of the study; in the collection, analyses, or interpretation of data; in the writing of the manuscript, or in the decision to publish the results.

References

1. Ryanzhin, S.V. Size distribution of world lakes and rivers derived from WORLDLAKE database. In Proceedings of the 4th International Lake Ladoga Symposium, Saint-Petersburg, Russia, 2–6 September 2002; pp. 435–441.
2. Ciorca, A.M.; Momeu, L.; Battes, K.P. Same karstic substratum, different aquatic communities? Case study: Three water bodies from western Romania. *Studia Univ. Babeş Bolyai Biol.* **2017**, *62*, 67–85. [CrossRef]
3. Maksimovich, G.A. *Basics of the Karst Studies*; Perm Book Publishing House: Perm, Russia, 1963; Volume 1, p. 444. (In Russian)
4. Krevs, A.; Kucinskiene, A. Vertical distribution of bacteria and intensity of microbiological processes in two stratified gypsum Karst Lakes in Lithuania. *Knowl. Manag. Aquat. Ecosyst.* **2011**, *4*, 1–12. [CrossRef]
5. Chalkia, E.; Zacharias, I.; Thomatou, A.A.; Kehayias, G. Zooplankton dynamics in a gypsum karst lake and interrelation with the abiotic environment. *Biologia* **2012**, *67*, 151–163. [CrossRef]
6. Alimova, A.F.; Mingazova, N.M. (Eds.) *Unique Ecosystems of Brackish-Water Karst Lakes in the Middle Volga Region*; Kazan University Publishing House: Kazan, Russia, 2001; p. 256. (In Russian)
7. Klimchouk, A.; Fotri, P.; Cooper, A. Gypsum karst of the world: A brief overview. *Int. J. Speleol.* **1996**, *25*, 159–181. [CrossRef]
8. Šulčius, S.; Alzbutas, G.; Juknevičiūtė, V.; Šimoliūnas, E.; Venckus, P.; Šimoliūnienė, M.; Paškauskas, R. Exploring viral diversity in a Gypsum Karst lake ecosystem using targeted single-cell genomics. *Genes* **2021**, *12*, 886. [CrossRef]
9. Stachura, M.; Wieczorek, D.; Zieliński, A. An attempt at a typology of karst lakes in the Połaniec Basin (Małopolska Upland). *Bull. Geography. Phys. Geogr. Ser.* **2018**, *15*, 63–74. [CrossRef]
10. Palagushkina, O.V. Ecology of Phytoplankton of Karst Lakes in the Middle Volga Region. Extended Abstract of Cand. Sci. (Biol.). PhD Thesis, Kazan State University, Kazan, Russia, 14 December 2004; p. 25. (In Russian)
11. Gusev, E.S. Phytoplankton primary production in several karst lakes in central Russia. *Inland Water Biol.* **2008**, *1*, 356–361. [CrossRef]
12. Danielidis, D.B.; Spartinou, M.; Economou-Amilli, A. Limnological survey of Lake Amvrakia, western Greece. *Hydrobiologia* **1996**, *318*, 207–218. [CrossRef]
13. Meng, F.; Li, Z.; Li, L.; Lu, F.; Liu, Y.; Lu, H.; Fan, Y. Phytoplankton alpha diversity indices response the trophic state variation in hydrologically connected aquatic habitats in the Harbin Section of the Songhua River. *Sci. Rep.* **2020**, *10*, 1–13. [CrossRef]
14. Sevindik, T.O.; Altundal, E.; Kucuk, F. The seasonal and spatial distribution of the phytoplankton of a Mesotrophic Lake related to certain physical and chemical parameters. *Ekoloji* **2015**, *24*, 14–23. [CrossRef]
15. Pollingher, U. Effects of latitude on phytoplankton composition and abundance in large lakes. In *Large Lakes. Brock*; Tilzer, M.M., Serruya, C., Eds.; Springer Series in Contemporary Bioscience; Springer: Berlin/Heidelberg, Germany, 1990; pp. 368–402. [CrossRef]
16. Udovič, M.G.; Cvetkoska, A.; Žutinić, P.; Bosak, S.; Stanković, I.; Špoljarić, I.; Mršić, G.; Kralj Borojević, K.; Ćukurin, A.; Plenković-Morja, A. Defining centric diatoms of most relevant phytoplankton functional groups in deep karst lakes. *Hydrobiologia* **2017**, *788*, 169–191. [CrossRef]
17. Tarasova, N.G. The composition of the algal flora of the plankton of Lake Goluboe (Samara region). *Samara Luka Probl. Reg. Glob. Ecol.* **2010**, *19*, 157–161. (In Russian)
18. Valadez, F.; Rosiles-González, G.; Almazán-Becerril, A.; Merino-Ibarra, M. Planktonic cyanobacteria of the tropical karstic lake Lagartos from the Yucatan Peninsula, Mexico. *Rev. Biol. Trop.* **2013**, *61*, 971–979. [CrossRef]
19. Miracle, M.R.; Vicente, E.; Pedrós-Alió, C. Biological studies of Spanish meromictic and stratified karstic lakes. *Limnetica* **1992**, *8*, 59–77. [CrossRef]
20. Petar, Ž.; Marija, G.U.; Koraljka, K.B.; Anđelka, P.M.; Judit, P. Morpho-functional classifications of phytoplankton assemblages of two deep karstic lakes. *Hydrobiologia* **2014**, *740*, 147–166. [CrossRef]
21. Momeu, L.; Ciorca, A.; László, O.T.; Segedi, C.; Battes, K.P.; Cîmpean, M. The karstic lake Iezerul Ighiel (Transylvania, Romania): Its first limnological study. *Studia Univ. Babes Bolyai Biol. LX* **2015**, *2*, 39–60.
22. Korneva, L.G. *Diversity and Structure of Phytoplankton of Some Low-Mineralized Forest Lakes in the Vologda Oblast, in Gidrobiologicheskie Voprosy (Hydrobiological Issues)*; Yakutsk University: Yakutsk, Russia, 2000; p. 94, Part 2. (In Russian)
23. Okhapkin, A.G. Dynamics of the species structure of potamophytoplankton in watercourses of different types. *Inland Water Biol.* **2000**, *1*, 53–61. (In Russian)
24. Okhapkin, A.G.; Startseva, N.A. Dynamics of species structure of phytoplankton in small waterbodies of urbanized territories: Species diversity and size structure of communities. *Inland Water Biol.* **2005**, *2*, 29–33. (In Russian)

25. Korneva, L.G. *Phytoplankton of Volga River Basin Reservoirs*; Kostroma Printing House: Kostroma, Russia, 2015; p. 284. (In Russian)
26. Alimov, A.F. *Elements of Aquatic Ecosystem Function Theory*; Nauka: Saint Petersburg, Russia, 2000; p. 147. (In Russian)
27. Carvalho, L.; Poikane, S.; Solheim, A.L.; Phillips, G.; Borics, G.; Catalan, J.; de Hoyos, C.; Drakare, S.; Dudley, B.J.; Järvinen, M.; et al. Strength and uncertainty of phytoplankton metrics for assessing eutrophication impacts in lakes. *Hydrobiologia* **2013**, *704*, 127–140. [CrossRef]
28. Borics, G.; Abonyi, A.; Salmaso, N.; Ptacnik, R. Freshwater phytoplankton diversity: Models, drivers and implications for ecosystem properties. *Hydrobiologia* **2021**, *848*, 53–75. [CrossRef] [PubMed]
29. Kozlov, A.V.; Tarasov, I.A.; Dedyk, V.E. Ecological and hydrochemical characteristics of the water area of the Klyuchik Lake in Pavlovsky district of the Nizhny Novgorod region. *Mod. Probl. Sci. Educ.* **2017**, *1*, 126–134. (In Russian)
30. Bayanov, N.G. Klyuchik Lake is a unique natural object of the Nizhny Novgorod Volga region. *Work. State Nat. Biosph. «Kerzhnesky»* **2019**, *9*, 65–72. (In Russian)
31. Bakka, S.V.; Kiseleva, N.Y. *Specially Protected Natural Areas of the Nizhny Novgorod Region*; Ministry of Ecology Natural Resources: Nizhny Novgorod, Russia, 2009; p. 560.
32. Vodeneeva, E.L.; Okhapkin, A.G.; Genkal, S.I.; Kulizin, P.V.; Sharagina, E.M.; Skamejkina, K.O. Composition, structure and distribution of phytoplankton of a highly mineralized karst lake. *Inland Water Biol.* **2020**, *13*, 573–582. [CrossRef]
33. Genkal, S.I.; Okhapkin, A.G.; Vodeneeva, E.L. On the morphology and taxonomy of *Cyclotella distinguenda* (Bacillariophyta). *Nov. Sist. Nizsh. Rast.* **2019**, *5*, 247–253. [CrossRef]
34. Petrov, M.S.; Astashin, A.E. Dynamics of the consumption of water in the biggest in the Nizhny Novgorod region spring Surin. Orfanov readings—2017. In Proceedings of the Materials All-Russian Scientific and Practical Conference Dedicated to the 80th Anniversary of Higher Geographical Education in the Nizhny Novgorod Region and the 70th Anniversary of the Nizhny Novgorod Branch of the All-Russian Public Organization Russian Geographical Society, Minin University, Nizhny Novgorod, Russia, 24–25 November 2017; pp. 29–32. (In Russian)
35. Bakanina, F.M.; Vorotnikov, V.P.; Lukina, E.V.; Fridman, B.I. *Lakes of the Nizhny Novgorod Region*; VOOP Publishing House: Nizhny Novgorod, Russia, 2001; p. 165. (In Russian)
36. Okhapkin, A.G.; Genkal, S.I.; Scharagina, E.M.; Vodeneeva, E.L. Structure and dynamics of phytoplankton in the Oka river mouth at the beginning of the 21th century. *Inland Water Biol.* **2014**, *7*, 357–365. [CrossRef]
37. Sun, J.; Liu, D. Geometric models for calculating cell biovolume and surface area for phytoplankton. *J. Plankton Res.* **2003**, *25*, 1331–1346. [CrossRef]
38. Hillebrand, H.; Dürselen, C.D.; Kirschtel, D.; Pollingher, D.; Zohary, T. Biovolume calculation for pelagic and benthic microalgae. *J. Phycol.* **1999**, *35*, 403–424. [CrossRef]
39. Abakumov, V.A. (Ed.) *Guide on Methods for Hydrobiological Analysis of Surface Waters and Bottom Sediments*; Gidrometeoizdat: Leningrad, Russia, 1983; p. 239. (In Russian)
40. USSR Science Academy. *Methods for Studying Biogeocenoses of Inland Water Bodies*; Morduhai-Boltovskoi, F.D., Ed.; Nauka: Moscow, Russia, 1975; p. 240. (In Russian)
41. Malashenkov, D.V.; Dashkova, V.; Zhakupova, K.; Vorobjev, I.A.; Barteneva, N.S. Comparative analysis of freshwater phytoplankton communities in two lakes of Burabay National Park using morphological and molecular approaches. *Sci. Rep.* **2021**, *11*, 16130. [CrossRef]
42. Vodeneeva, E.L.; Kulizin, P.V. *Algae of the Mordovian Nature Reserve. Annotated List of Species*; Flora and Fauna of Nature Reserves: Moscow, Russia, 2019; Volume 134, p. 62. (In Russian)
43. Guiry, M.D.; Guiry, G.M. *AlgaeBase. World-Wide Electronic Publication*; National University of Ireland: Galway, Ireland, 2021; Available online: http://www.algaebase.org (accessed on 29 July 2021).
44. Shannon, C.E.; Weaver, W. *The Mathematical Theory of Communication*; The University of Illinois Press: Urbana, IL, USA, 1949; p. 117.
45. Pielou, E.C. The measurement of diversity in different types of biological collections. *J. Theor. Biol.* **1966**, *13*, 131–144. [CrossRef]
46. Sládeček, V. System of water quality from biological point of view. *Arch. Hydrobiol. Erg. Limnol.* **1973**, *7*, 218.
47. Reynolds, C.S. *Ecology of Phytoplankton*; Cambrige University Press: New York, NY, USA, 2006; p. 535.
48. Okhapkin, A.G. Species composition of phytoplankton as an indicator of living conditions in streams of different types. *Bot. Zh.* **1998**, *83*, 1–13. (In Russian)
49. Mann, H.B.; Whitney, D.R. On a test of whether one of two random variables is stochastically larger than the other. *Ann. Math. Stat.* **1947**, *18*, 50–60. [CrossRef]
50. Stevens, L.E.; Schenk, E.R.; Springer, A.E. Springs ecosystem classification. *Ecol. Appl.* **2021**, *31*, e2218. [CrossRef]
51. Kitaev, S.P. *Basics of Limnology for Hydrobiologists and Ichthyologists*; Karelian Science Center Press RAN: Petrozavodsk, Russia, 2007; p. 394. (In Russian)
52. Bigon, M.; Harper, D.; Tausend, K. *Ecology. Individuals, populations and communities*; Mir Publishers: Moscow, Russia, 1989; Volume 2, p. 477. (In Russian)
53. Weckström, J.; Korhola, A.; Blom, T. Diatoms as quantitative indicators of ph and water temperature in subarctic Fennoscandian lakes. *Hydrobiologia* **1997**, *347*, 171–184. [CrossRef]

54. Okhapkin, A.G. Phytoplankton of the system of Pustynskie Lakes of the Gorky Region. In *Biological Bases for Increasing the Productivity and Protection of Plant Communities in the Volga Region*; Gorky State University Press: Gorky, Russia, 1981; pp. 106–109. (In Russian)
55. Gusev, E.S. *Vertical Distribution of Phytoplankton in Small Karst Lakes in Central Russia. Algae: Taxonomy, Ecology, Use in Monitoring*; Ural Branch of the Russian Academy of Sciences: Ekaterinburg, Russia, 2011; pp. 160–167. (In Russian)
56. Odum, U. *Ecology*; Mir Publishing: Moscow, Russia, 1986; Volume 1, p. 328. (In Russian)
57. Magarran, E. *Ecological Diversity and Its Measurement*; Mir Publishing: Moscow, Russia, 1992; p. 181. (In Russian)
58. Korneva, L.G. Impact of acidification on structural organization of phytoplankton community in the forest lakes of the north-western Russia. *Water Sci. Technol.* **1996**, *33*, 291–296. [CrossRef]
59. Lessmann, D.; Fyson, A.; Nixdorf, B. Experimental eutrophication of a shallow acidic mining lake and effects on the phytoplankton. *Hydrobiologia* **2003**, *506*, 753–758. [CrossRef]
60. Okhapkin, A.G.; Vodeneeva, E.L.; Yulova, G.A. Phytoplankton of reservoirs of the "Kerzhensky" nature reserve (Nizhny Novgorod region). *Bot. Zh.* **2004**, *89*, 1264–1275.
61. Okhapkin, A.G.; Sharagina, E.M.; Bondarev, O.O. Phytoplankton of the Cheboksary reservoir at the present state of its existence. *Povolzhskiy J. Ecol.* **2013**, *2*, 190–199. (In Russian)
62. Kiss, K.T.; Klee, R.; Ector, L.; Ács, É. Centric diatoms of large rivers and tributaries in Hungary: Morphology and biogeographic distribution. *Acta Bot. Croat.* **2012**, *71*, 311–363. [CrossRef]
63. Padisák, J.; Crossetti, L.; Naselli-Flores, L. Use and misuse in the application of the phytoplankton functional classification: A critical review with updates. *Hydrobiologia* **2009**, *621*, 1–19. [CrossRef]
64. Dokulil, M.T.; Teubner, K. Cyanobacterial dominance in lakes. *Hydrobiologia* **2000**, *438*, 1–12. [CrossRef]
65. Downing, J.A.; Watson, S.B.; McCauley, E. Predicting cyanobacteria dominance in lakes. *Can. J. Fish. Aquat. Sci.* **2001**, *58*, 1905–1908. [CrossRef]
66. Bouterfas, R.; Belkoura, M.; Dauta, A. Light and temperature effects on the growth rate of the three freshwater algae isolated from a eutrophic lake. *Hydrobiologia* **2002**, *489*, 207–217. [CrossRef]
67. Padisák, J. *Phytoplankton. The Lakes Handbook. Limnology and Limnetic Ecology*; O'Sullivan, P.E., Reynolds, C.S., Eds.; Blackwell Publishing: Hoboken, NJ, USA, 2004; Volume 1, pp. 251–308.
68. Trifonova, I.S. *Ecology and Succession of Lake Phytoplankton*; Science: Leningrad, Russia, 1990; p. 183. (In Russian)
69. Rottberger, J.; Gruber, A.; Boenigk, J.; Kroth, P.G. Influence of nutrients and light on autotrophic, mixotrophic and heterotrophic freshwater chrysophytes. *Aquat. Microb. Ecol.* **2013**, *71*, 179–191. [CrossRef]
70. Saad, J.F.; Unrein, F.; Tribelli, P.M.; López, N.; Izaguirre, I. Influence of lake trophic conditions on the dominant mixotrophic algal assemblages. *J. Plankton Res.* **2016**, *38*, 818–829. [CrossRef]

microorganisms

MDPI

Article

Filtration of the Microalga *Amphidinium carterae* by the Polychaetes *Sabella spallanzanii* and *Branchiomma luctuosum*: A New Tool for the Control of Harmful Algal Blooms?

Loredana Stabili [1,2,*], Margherita Licciano [1], Adriana Giangrande [1] and Carmela Caroppo [2,*]

1 Department of Biological and Environmental Sciences and Technology, University of Salento, 73100 Lecce, Italy; margherita.licciano@unisalento.it (M.L.); adriana.giangrande@unisalento.it (A.G.)
2 Water Research Institute (IRSA), Taranto Section CNR, 74121 Taranto, Italy
* Correspondence: loredana.stabili@irsa.cnr.it (L.S.); carmela.caroppo@irsa.cnr.it (C.C.); Tel.: +39-0832-298972 (L.S.); +39-099-4542211 (C.C.)

Abstract: Harmful algal blooms (HABs) are extreme biological events representing a major issue in marine, brackish, and freshwater systems worldwide. Their proliferation is certainly a problem from both ecological and socioeconomic contexts, as harmful algae can affect human health and activities, the marine ecosystem functioning, and the economy of coastal areas. Once HABs establish, valuable and environmentally friendly control actions are needed to reduce their negative impacts. In this study, the influence exerted by the filter-feeding activity of the two sabellid polychaetes *Branchiomma luctuosum* (Grube) and *Sabella spallanzanii* (Gmelin) on a harmful dinoflagellate was investigated. Clearance rates (C) and retention efficiencies were estimated by employing the microalga *Amphidinium carterae* Hulburt. The C_{max} was 1.15 ± 0.204 L h^{-1} g^{-1} DW for *B. luctuosum* and 0.936 ± 0.151 L h^{-1} g^{-1} DW for *S. spallanzanii*. The retention efficiency was 72% for *B. luctuosum* and 68% for *S. spallanzanii*. Maximum retention was recorded after 30 min for both species. The obtained results contribute to the knowledge of the two polychaetes' filtration activity and to characterize the filtration process on harmful microalgae in light of the protection of water resources and human health. Both species, indeed, were extremely efficient in removing *A. carterae* from seawater, thus suggesting their employment as a new tool in mitigation technologies for the control of harmful algae in marine environments, as well as in the aquaculture facilities where HABs are one of the most critical threats.

Keywords: harmful algal blooms; bioremediation; filtration activity; polychaetes; dinoflagellates

Citation: Stabili, L.; Licciano, M.; Giangrande, A.; Caroppo, C. Filtration of the Microalga *Amphidinium carterae* by the Polychaetes *Sabella spallanzanii* and *Branchiomma luctuosum*: A New Tool for the Control of Harmful Algal Blooms? *Microorganisms* **2022**, *10*, 156. https://doi.org/10.3390/microorganisms10010156

Academic Editor: Assaf Sukenik

Received: 10 December 2021
Accepted: 10 January 2022
Published: 12 January 2022

1. Introduction

Marine phytoplankton forms the base of the food webs, but some species are considered "harmful" as they produce toxins that negatively affect human health, or blooms with deleterious effects on aquatic ecosystems, fisheries, aquaculture, and tourism [1,2]. The term "Harmful Algal Blooms" (HABs) is used to describe all of these events (e.g., [3]). HABs are natural phenomena that, in recent years, have increased in occurrence and spread worldwide [4], and also in relation to an enhanced scientific knowledge and an intensification of monitoring programs, especially in aquaculture sites [5].

Determining the causative factors for HABs is complex. Many studies have been carried out to understand this ecological phenomenon, but the mechanisms responsible for the bloom formation are still not fully understood and their prevention is a highly difficult task [6]. In fact, not only eutrophication, but also other causes, can be responsible for the expansion and intensification of HABs, such as human-mediated introduction of alien harmful species, climatic variability, expanded utilization of coastal waters for aquaculture, and habitat modification [7,8]. Preventive measures, such as nutrient input control, are necessary to preclude the occurrence of high biomass HABs. However, HABs may continue to increase due to warming and rising CO_2, according to the 2019 IPCC [9]. While the exact scenarios are unknown and more studies should be conducted, the fact that HABs are, in part, natural events requires us to address methods to mitigate their impacts. Effective and environmentally friendly control actions have become a priority to suppress them [4,5]. Biological methods represent promising options to control HABs, as they usually do not cause secondary pollution and can be species-specific [10]. They usually use enzymolysis or parasitism effects on HAB species and nutrient competition, or allelopathic effects of macroalgae or seagrass, or grazing by marine protozoa, zooplankton, and filter-feeding shellfish [4]. Marine invertebrate filter feeders have been shown to extract large quantities of phytoplankton from the water, thus having a pronounced grazing impact on the phytoplankton communities and biomass in many shallow marine areas. The diversity of organisms involved in the filtration process thus ensures water filtration, the use of plankton, and water self-purification [11–13]. In coastal marine ecosystems, these filter feeders constitute a large component both in terms of biomass and abundance and, due to their intense filtering capacity, these organisms could potentially process large volumes of water daily. Several researchers have studied the filtration activities of different zoological groups, such as ascidians, polychaetes, bivalves, and sponges [14–24], with different types and concentrations of particles [11,25].

Polychaetes species belonging to the Sabellidae family are characterized by the presence of a branchial crown, which protrudes out of the tube they inhabit to collect and sort material of different sizes. Small particles are ingested, while large ones are rejected into the water [26,27]. Several studies have been focused on the filtering organs and particle capture mechanisms in sabellids [14–16,20–24,28–35]. Most of the field and laboratory studies on filter feeder polychaetes have been carried out employing phytoplankton as a trophic source [29,33,34]. In particular, Clapin [29] has shown that the sabellid *Sabella spallanzanii* (Gmelin), when settled on artificial structures at about 5 m depth, is able to influence the phytoplankton level through the filtration process, filtering 23.4×10^3 L m^{-2} day^{-1} (biomass = 258 g DW m^{-2}), and, finally, to ensure, for almost five times a day, a turnover of the overlying water column.

In the present paper, we investigate the filtration process of two sabellid species, *S. spallanzanii* and *Branchiomma luctuosum*, on the potential toxic dinoflagellate species *Amphidinium carterae*. We characterized the filtration process of these two sabellid species and evaluated their clearance rate and retention efficiency. The filtration capability of these two species was previously studied by using bacterioplankton as a trophic source [14–16].

Amphidinium carterae is a high-biomass producer [36–38], as well as a producer of more than 20 secondary metabolites that have demonstrated haemolytic, cytotoxic, ichthyotoxic, and antifungal activities [39,40].

Until now, mitigation measures tested for the blooms of this microalga were chemical controls using modified sands, clay, and soil [41–43]. The obtained results are discussed in light of the potential employment of the investigated polychaete species as bioremediators of harmful microalgae, effective in protecting water resources and human health in coastal waters, as well as in the field of the aquaculture.

2. Material and Methods

2.1. Polychaete Sampling

Adult specimens of *S. spallanzanii* and *B. luctuosum* were sampled in the Gulf of Taranto (Mediterranean Sea, Ionian Sea, Italy), where rearing trials were realized in an integrated multitrophic aquaculture (IMTA) system equipped with fish cages within the framework of the Remedia Life Project (LIFE16 ENV/IT/000343). The experimental design of the project consisted of an IMTA system including six fish cages, where the waste restoration was realized by the presence of some marine organisms, including the selected filter feeder polychaetes, acting as bioremediators. Polychaetes were obtained from natural recruitment on plastic nets, used as collectors, immersed in the fish farm around the fish cages suspended at about 12 m depth (Figure 1). After their collection, the worms were transferred to the laboratory and prepared for the filtration experiments. Prior to experiments, as already reported by Licciano et al. [14], epibionts were removed from the worm tubes, then 12 individuals of similar tube length were selected for each species, placed in an aerated aquarium equipped with 25 L of sterile-filtered sea water (SFSW) (0.22 µm pore size filters, Millipore), which was daily replaced, and kept in a temperature-controlled room (22 °C) for 48 h. After starvation, individuals of *S. spallanzanii* and *B. luctuosum* were separately utilized for the filtration experiments.

Figure 1. (**A**) Map of the sampling site in the Gulf of Taranto (Mediterranean Sea, Ionian Sea, Italy). (**B**) Specimen of the polychaete *Sabella spallanzanii*. (**C**) Specimens of the polychaete *Branchiomma luctuosum*. (**D**) Rearing trials of the polychaetes in the integrated multitrophic aquaculture (IMTA) system realized in the Gulf of Taranto. Polychaetes were arranged in polypropylene nets, which were hung vertically within a typical long-line system.

2.2. Microalgal Culture Conditions

The experiments were carried out using a strain of the dinoflagellate, *A. carterae*, previously isolated from the Gulf of Taranto [44,45]. Microalgal cultures were grown in f/2 medium [46]. Culture media was prepared with Gulf of Taranto seawater (salinity of 38‰) filtered through glass fiber filters (Whatman GF/C) and then autoclaved. All microorganisms were grown in 1000 mL Erlenmeyer flasks with 500 mL of f/2 medium and kept in optimal growth conditions in a thermostatic chamber at 20 ± 1 °C under an illumination cycle of 12D:12N (100 micromol photons m^{-2} s^{-1}) and stirred at 100 rpm by means of a mechanical stirrer.

For the filtration experiments, microalgae were collected after thirty days of growth (exponential phase). Before biological tests, microalgal cultures were counted using a 1-mL Sedgwick Rafter chamber, and then appropriate aliquots were added to the experimental beakers to obtain the required final concentrations of *A. carterae* corresponding to 6×10^3 cells mL^{-1}.

2.3. *Amphidinium carterae* Retention and Clearance Rate Calculation

In order to evaluate the clearance rate (C) and the retention efficiency (R) of *A. carterae* by *S. spallanzanii* and *B. luctuosum*, two separate experiments were carried out.

For each species, six starved worms were individually placed in six beakers filled with 10 L of SFSW (treatments), and six beakers were filled only with SFSW and employed as controls, for a total of 12 beakers. Just prior to the beginning of the experiments, an *A. carterae* suspension of a known concentration (2.2×10^5 cells mL^{-1}) was added to both the treatment and control beakers so that the final concentration was about 6×10^3 cells mL^{-1}

for both sabellid species. The retention of *A. carterae* cells by the two investigated filter feeders was assessed by evaluating the removal of algal cells from SFSW over 4 h at a temperature of 22 °C. Aliquots of SFSW (10 mL) were aseptically collected from the treatment and control beakers every 15 min within the first hour and every 30 min during the following 3 h, for a total of 11 sampling times. In order to determine the phytoplankton abundances, from each beaker at the end of each sampling time, the aliquots of SFSW were immediately fixed with an acid Lugol's iodine solution to a final concentration of 1% and stored at 4 °C until analysis. *Amphidinium* counts were performed by using an inverted microscope (Labovert FS Leitz) equipped with phase contrast and following the Utermöhl method [47]. Three replicates were prepared from each aliquot. At each time, a mean value from the three replicates was computed. For each species, the retention efficiency R was calculated as a percentage for the difference in algal abundances at each sampling time by the following equation:

$$R\ (\%) = 100 \times [(c_0 - c_t)/c_0] \quad (1)$$

where c_0 is the initial algal concentration in the experimental beakers and c_t is the algal concentration at the end of time t.

In order to determine the dry weights (g) of the employed filter feeders, at the end of the experiments, each worm was drawn out of the tube, dried at 60 °C for 24 h, and then weighed.

Clearance rates were assessed according to Coughlan [48] by evaluating the algal removal in the treatment beakers as a function of time. Taking into account the predictable shifts between feeding activity and rest during the filtration process of the worms, recognizable by the intrusion of the branchial crown within the tube, for each species, C was calculated within each sampling time as the mean ± standard deviation (SD) of the C values recorded in all the treatment beakers at each time point and expressed in liters per hour, per gram, of dry worm tissue ($L\ h^{-1}\ g^{-1}$ DW). Furthermore, according to Navarro and Widdows [17], the maximum clearance rate (C_{max}) was also calculated for each species as the mean ± SD of the highest C values recorded in each treatment beaker during the entire experimental period (4 h).

2.4. Statistical Analysis

Analysis of variance was used to assess significant variations in *A. carterae* abundance in the treatment and control beakers. The experimental design consisted of two factors: polychaetes (Po, two levels, i.e., absent and present, fixed) and time (Ti, 11 levels, fixed and crossed with Po), with $n = 6$ per combination of factors, for a total of 132 observation units.

Prior to analysis, the homogeneity of variance was tested using the Cochran test. The Student–Newman–Keuls test (SNK) was used for post hoc comparisons among means [49]. Analysis was done using the GMAV 5 computer program (Statistical software for Windows developed by Institute of Marine Ecology; Marine Ecology Labs, A11, University of Sydney, NSW 2006 AUSTRALIA, 1997).

3. Results

3.1. Filtration Activity on *Amphidinium carterae* by *Sabella spallanzanii*

The mean concentrations of *A. carterae* in the control and treatment beakers during the filtration experimental observations employing the polychaete *S. spallanzanii* are shown in Figure 2A.

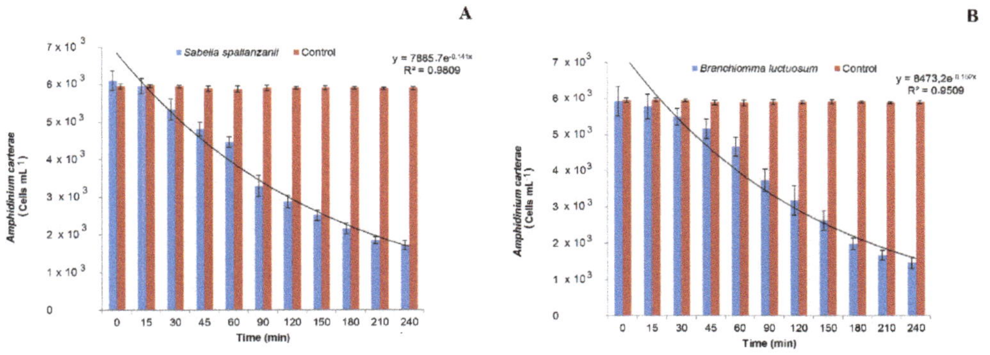

Figure 2. Changes in *Amphidinium carterae* abundances measured in the control and treatment beakers with (**A**) *Sabella spallanzanii* and (**B**) *Branchiomma luctuosum*.

The algal density remained almost unchanged in the control beakers and a mean value of $5.92 \pm 0.493 \times 10^3$ cells mL^{-1} was measured for the entire duration of the experiment. By contrast, an exponential reduction in phytoplankton concentration was detected in the treatment beakers ($y = 7885.7\ e^{-0.141x}$; $R^2 = 0.9809$). Analysis of variance revealed a significant Po × Ti interaction (Table 1) and post hoc comparisons among means showed that *A. carterae* concentration in the treatment beakers was always significantly lower than in the control beakers ($p < 0.01$), except for T0 and T15. In particular, the algal density decreased with time in the treatment beakers, starting from T30, when the mean value of $5.33 \pm 0.279 \times 10^3$ cells mL^{-1} was recorded ($p < 0.01$). The lowest algal concentration of $1.72 \pm 0.130 \times 10^3$ cells mL^{-1} was measured at time T 240. This last value, however, was not significantly different from the previous one recorded at T 210.

Table 1. Summaries of ANOVA testing for differences in average *Amphidinium carterae* abundances measured at the different sampling times in the control and treatment beakers with *Sabella spallanzanii*.

Source of Variation	DF	MS	F	*p*
Po	1	157,564,720.3712	6543.83	***
Ti	10	8,376,261.4636	347.88	***
Po × Ti	10	8,036,449.4879	333.76	***
Residual	110	24,078.35		
TOT	131			
Cochran Test		C = 0.1854 ($p < 0.05$)		
Transformation		None		
SNK Test		AT < AC		
Po(Ti)		(AT T0 = AC T0)		
		(AT T1 = AC T1)		
		AT T0 > AT T1 > AT T2 > AT T3 > AT T4 > AT T5 > AT T6 > AT T7 > AT T8 > AT T9 > AT T10		
Ti(Po)		(AT T10 = AT T11)		
		AC T0 = AC T1= AC T2 = AC T3 = AC T4 = AC T5 = AC T6 = AC T7 = AC T8 = AC T9 = AC T10 = AC11		

Reported are: Po: polychaetes; Ti: time; AT: algal concentration in the treatment beakers; AC: algal concentration in the control beakers; AT T0: algal concentration in the treatment beakers at T0; AT T1: algal concentration in the treatment beakers at T1; AT T*n*: algal concentration in the treatment beakers at T*n* (with $n = 1,2,3 \ldots \ldots 11$); AC T0: algal concentration in the control beakers at T0; AC T1: algal concentration in the control beakers at T1; AC T*n*: algal concentration in the control beakers at T*n* (with $n = 1,2,3 \ldots \ldots 11$); ***: $p < 0.001$.

The retention efficiency of *A. carterae* by *S. spallanzanii* gradually increased with time, with the highest value occurring 210 min after the beginning of the experiment, when the value of R = 68% was recorded (Figure 3A).

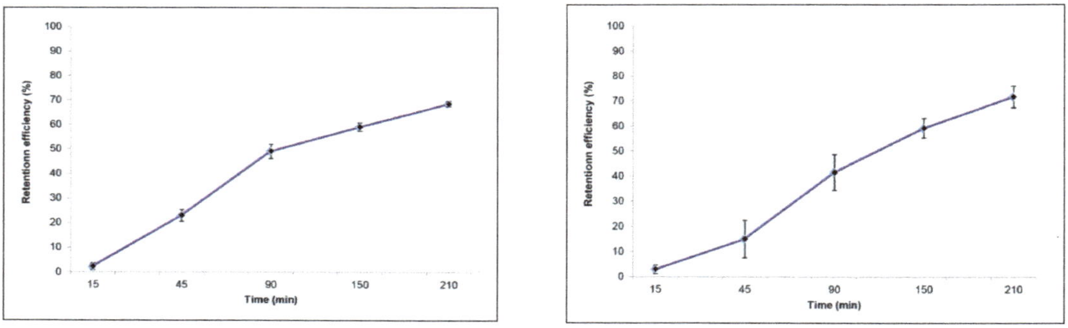

(A) (B)

Figure 3. Retention efficiency calculated within each sampling time for (**A**) *Sabella spallanzanii* and (**B**) *Branchiomma luctuosum*.

The clearance rate values measured for *S. spallanzanii* within each sampling time are shown in Figure 4A. At T 90, the highest C value was recorded for this species (C = 0.75 L h^{-1} g^{-1} DW), while the lowest value was measured at T 15 (C = 0.11 L h^{-1} g^{-1} DW). The C values significantly differed among sampling times ($p < 0.001$). The C_{max}, calculated as the mean value of all the single maximal values recorded for each individual, was 0.936 ± 0.151 L h^{-1} g^{-1} DW.

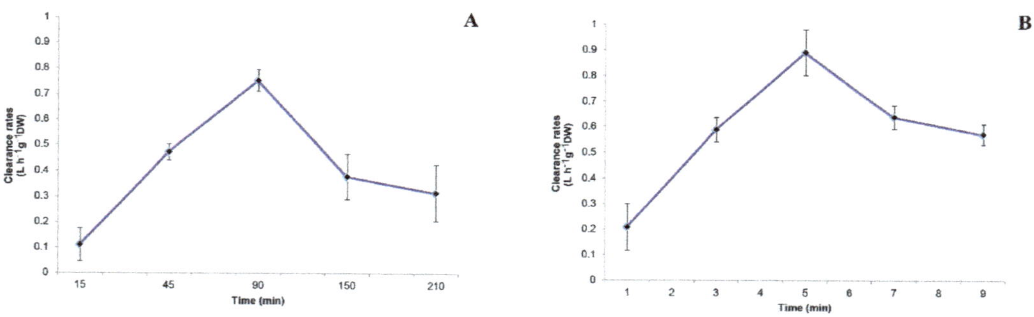

Figure 4. Clearance rates (C) calculated within each sampling time for (**A**) *Sabella spallanzanii* and (**B**) *Branchiomma luctuosum*.

3.2. *Filtration activity on Amphidinium carterae by Branchiomma luctuosum*

In Figure 2B, the abundance of *A. carterae* recorded at the different experimental times in the control and treatment beakers employing the polychaete *B. luctuosum* is shown.

After the beginning of the experiment, the phytoplankton concentrations in the treatment beakers lowered exponentially as function of time ($y = 8473.2\ e^{-0.152x}$; $R^2 = 0.9509$). By contrast, the algal concentration in the control beakers remained almost unvaried during the experimental observations, with the mean value of 5.92 ± 0.493 × 10^3 cells mL^{-1}.

The SNK test performed on the Po × Ti interaction after ANOVA (Table 2) revealed that *A. carterae* concentration in the treatment beakers was always significantly lower than in the control ones ($p < 0.01$), except for the initial times of the experimental observations T0 and T15.

Table 2. Summaries of ANOVA testing for differences in average *Amphidinium carterae* abundances measured at the different sampling times in the control and treatment beakers with *Branchiomma luctuosum*.

Source of Variation	DF	MS	F	p
Po	1	149,460,080.9	2990.75	***
Ti	10	8,872,472.823	177.54	***
Po × Ti	10	8,561,803.017	171.32	***
Residual	110	49,974.1591		
TOT	131			
Cochran Test	C = 0.1902 ($p < 0.05$)			
Transformation	None			
SNK Test	AT < AC			
Po(Ti)	(AT T0 = AC T0)			
	(AT T1 = AC T1)			
	AT T0 > AT T1 > AT T2 > AT T3 > AT T4 > AT T5 > AT T6 > AT T7 > AT T8 > AT T9 > AT T10			
Ti(Po)	(AT T10 = AT T11)			
	AC T0 = AC T1= AC T2 = AC T3 = AC T4 = AC T5 = AC T6 = AC T7 = AC T8 = AC T9 = AC T10 = AC11			

Reported are: Po: polychaetes; Ti: time; AT: algal concentration in the treatment beakers; AC: algal concentration in the control beakers; AT T0: algal concentration in the treatment beakers at T0; AT T1: algal concentration in the treatment beakers at T1; AT T*n*: algal concentration in the treatment beakers at T*n* (with *n* = 1,2,3 . . . . . . 11); AC T0: algal concentration in the control beakers at T0; AC T1: algal concentration in the control beakers at T1; AC T*n*: algal concentration in the control beakers at T*n* (with *n* = 1,2,3 . . . . . . 11); ***: $p < 0.001$.

In particular, a significant reduction of microalgae in the treatment beakers was observed at T 30 ($p < 0.01$) when the value of $5.49 \pm 0.23 \times 10^3$ cells mL^{-1} was recorded. At the following sampling times, the algal density further decreased, with time in the treatment beakers reaching the lowest value at T 240 ($1.46 \pm 0.155 \times 10^3$ cells mL^{-1}), although no significant difference was detected comparing this value with that recorded at the previous time T 210.

As shown in Figure 3B, the retention efficiency measured for *B. luctuosum* within each sampling time increased with time. At the end of the experimental observations (T 210), the highest value of R = 72% was reached.

The clearance rates calculated for *B. luctuosum* within each sampling time are reported in Figure 4B. The C values ranged from the maximum of 0.86 L h^{-1} g^{-1} DW measured at T 90 min to the minimum of 0.13 L h^{-1} g^{-1} DW at T 15, with significant differences among values over time ($p < 0.001$). The C_{max} calculated as the mean of all the single highest C values measured in each treatment beaker with *B. luctuosum* was 1.15 ± 0.204 L h^{-1} g^{-1} DW.

4. Discussion

The problem of harmful algae is crucial due to their effects, including risks for human health, as microalgae are the link in the food chain responsible for the transfer of plankton toxicity to humans [50]. The blooms of harmful microalgae also have a negative influence both on marine resources, directly poisoning wild and farmed fish and invertebrates, and on tourism, because of the effects related to water discoloration and the production of repellent odors and foams. This translates into damage to the marine ecosystem, which inevitably affects the economy of coastal activities and fishing communities directly linked to coastal resources.

The proliferation of potentially harmful microalgae is, therefore, also a problem from an economic point of view. In particular, serious economic losses of several million pounds are caused by HABs in different countries, such as the UK, Australia, and the USA. Thus, the advance in management strategies and mitigation technologies aimed at their removal is of paramount importance in the protection of a significant fraction of the world's water resources, human health, and economic growth [51,52]. In this framework, phytoplankton filtration by macroinvertebrates could constitute a potential remedy.

This work represents a preliminary attempt to evaluate the ability of the filter feeder polychaetes *Sabella spallanzanii* and *Branchiomma luctuosum* to remove, in relation to their filtration process, a microalgal potentially toxic species, such as *Amphidinium carterae*, representing a danger to human health when present at high concentrations in seawater [36–38].

From our results, some interesting issues can be inferred.

- The potentially harmful microalgal species *A. carterae*, here used, when present in the surrounding environment, can be an integral part of the natural diet of the two investigated polychaetes. The C_{max} and highest retention efficiencies obtained are 0.936 ± 0.151 L h^{-1} g^{-1} DW and 68% for *S. spallanzanii*, and 1.15 ± 0.204 L h^{-1} g^{-1} DW and 72% for *B. luctuosum*, respectively.

Clearance rates of other polychaete species were always estimated by using phytoplankton in laboratory and field experiments [22–24,29,30,33,34,53–56]. In particular, the filtering activity of *S spallanzanii* has been investigated by Clapin [29] and Lemmens et al. [56], who reported that this species has a filtering capacity similar to that of the macro-filter-feeder community inhabiting the *Posidonia sinuosa* meadows of Southern Flats in Cockburn Sound (Australia). They estimated a mean filtering capacity of about 12 m^3 d^{-1} m^{-2} habitat and a feeding efficiency of 13 L mg^{-1} O_2 consumed.

Clearance rates on phytoplankton have also been recorded by employing other filter feeders. For example, due to their importance from an economic point of view, a group of filter feeders that has been widely studied in this regard is represented by mussels. From a paper by Cugier et al. [57] on the filtration capacity of phytoplankton of some bivalves, it appears that the clearance rates of these organisms vary from a minimum of 0.1 L h^{-1} g^{-1} DW to a maximum of 4 L h^{-1} g^{-1} DW. Thus, because of these considerations, it is clear that filter feeders, including polychaetes, play a crucial role and have a significant impact on the concentration of phytoplankton in their surrounding environment.

It is also important to underline that filter feeder polychaetes play a very crucial role in regulating the abundance of some components of the ecosystem, which not only include microalgae, as demonstrated by the present study, but also bacteria. In particular, some studies undertaken in order to investigate the effect of the filtration activity of *S. spallanzanii* and *B. luctuosum* on bacterioplankton density showed that the clearance rates and retention efficiencies for these two sabellid species were 12.4 L h^{-1} g^{-1} DW and 70% for the former, and 43.2 L h^{-1} g^{-1} DW and 98% for the latter [14–16]. The different values of clearance rates obtained using bacteria or phytoplankton as trophic sources is not surprising since several studies have shown that clearance rate values vary in relation to the size of the particles used. The recorded values of clearance rate obtained in the present study are particularly remarkable as a phytoplanktonic component with dimensions around 12–18 μm in length and 8–10 μm in width was supplied to the polychaetes. *Amphidinium carterae*, indeed, was considerably larger than bacteria that also constitute a significant constituent of the polychaetes' diet. Furthermore, several works have shown that the higher the concentration of suspended particles (algal cells, etc.), the lower the filtering activity and the food pressure on the plankton [11]. This means that, at relatively low particle concentrations, the relationship between the filtration rate and the quantity of particles is linear, or almost linear. Above a certain concentration threshold, however, this relationship becomes non-linear, [25]. Based on these considerations, presumably employing different concentration of *A. carterae* could give rise to a variability in the filtering activity on phytoplankton, leading to a relative decrease in the abundance of harmful algae in aquatic ecosystems.

- The here-obtained results on the filtration capability of the polychaetes *S. spallanzanii* and *B. luctuosum* on the harmful dinoflagellate *A. carterae* suggest that these invertebrates represent a new tool to reduce the impact of harmful algae on marine life and could constitute a potential advantage in comparison to the up-to-now used bioremediation methodologies in the bioremediation process. Current strategies to reduce the potential economic and human health effects from HABs to fisheries and aquaculture include chemical (e.g., flocculation [43]), physical, and biological measures [4]. Among these strategies, flocculation has been classified as the most cost-effective and convenient way to rapidly remove algae [58–60]. Flocculation by the sole use of natural clay, however, requires a high dosage in order to attain a fairly high (> 90%) removal efficiency [61,62]. New methods were tested to further increase the removal efficiency of natural clay minerals for the mitigation, for example, of *A. carterae* [41,43], but in general, the use of clay remains often too rudimental, confined, or problematic for large-scale implementation and could potentially have detrimental effects on filter-feeding invertebrates [63,64]. For these reasons, in many countries, severe environmental controls preclude the use of these techniques [3]. As regards physical methods, often used to minimize HABs' impacts on aquaculture plants, they consist, for example, of enhanced flushing and aeration of cages and/or feeding cessation that would help to manage the problem, but only in a small volume of water [10]. Finally, biological methods are environmentally friendly and can be specific for algal cells, without impact on non-HAB species. These include the use of bacteria, fungi, phages, macroalgae or seagrass, protozoa, zooplankton, and filter-feeding shellfish [4].

The employment of these two species in the control of harmful algae could also be recommended in the case of aquaculture production, where HABs are one of the most critical threats. Evidence on the occurrence and impacts of HABs on marine fisheries and mariculture is being gathered by ongoing regional programs (e.g., [65]), national programs (e.g., UK FSA, https://www.food.gov.uk/business-guidance/biotoxin-and-phytoplankton-monitoring, accessed on 9 December 2021), and global programs (Global HAB Status Report) [66,67]. In the Gulf of Taranto, we realized an integrated multitrophic aquaculture (IMTA) system involving these polychaetes and macroalgae as bioremediators, reared/cultivated in association with fish cages, obtaining a high production of bioremediator biomass. At the same time, a restoration of the aquaculture-rearing environment was achieved in terms of microbial contamination (i.e., total coliforms and *Escherichia coli*) and nutrient concentrations (i.e., phosphorous and nitrogen salts were also realized) [68], as well as an amelioration of the benthic communities under the cages. It is noteworthy to underline that *B. luctuosum* is not originally in the Mediterranean Sea and can be considered an invasive species. However, in the long term, this species has reached a "naturalized" condition, so becoming a part of the fouling community everywhere along the Mediterranean coast without any negative impact [69]. The here-obtained results represent an added value indicating the potentiality of the reared polychaetes to remediate the surrounding farming environment also in terms of harmful algae, thus contributing to the realization of an eco-friendly approach. This is useful to guarantee food safety and environmental quality, taking into account that managing global food security is one of the greatest challenges of the twenty-first century. The realization of an IMTA system including polychaetes is also noteworthy considering that, despite the many reported cases of laboratory success in controlling HABs, only a few field applications are described [70]. In fact, there are great difficulties in cultivating and transporting the control agents when these technologies are applied in the water, and they cannot be applied in the case of a rapid response to HAB events. Therefore, as already suggested by Yu et al. [4], the main challenge of the biological control for HABs mitigation is to transfer the laboratory experiments into the environment that seems achievable in the case of the two considered polychaete species.

5. Conclusions

In conclusion, although our results are preliminary, they indicate a conspicuous filtration capability of the toxic microalga *A. carterae* by the two considered filter feeder polychaetes *S. spallanzani* and *B. luctuosum*. However, further studies will be carried out to confirm these results in the field. Furthermore, in the future, other experiments will be conducted using other typically planktonic algal species of different sizes to evaluate the effectiveness of the proposed technology in different conditions.

Since harmful algae have become an important global issue, the idea of employing the two considered polychaetes as bioremediatiors for purification of the water column is a sustainable, eco-friendly novelty also considering that, over the past several decades, researchers have made great efforts to develop an integrated management approach for HABs control.

The employment of *S. spallanzanii* and *B. luctuosum* in the control of HABs could be recommended not only also in the case of the bioremediation of marine ecosystems, but also in aquaculture facilities where HABs are one of the most critical threats.

Author Contributions: Conceptualization, L.S., M.L., A.G. and C.C.; methodology, L.S., M.L. and C.C.; validation, L.S., M.L. and C.C.; investigation, L.S., M.L. and C.C.; writing—review and editing, L.S., M.L., A.G. and C.C.; supervision, L.S., M.L., A.G. and C.C.; funding acquisition, L.S. and A.G. All authors have read and agreed to the published version of the manuscript.

Funding: This research was funded by the European Community under Grant Agreement No. LIFE16 ENV/IT/000343 project "REMEDIA Life".

Institutional Review Board Statement: Not applicable.

Informed Consent Statement: Not applicable.

Data Availability Statement: Not applicable.

Acknowledgments: Thanks are due to Giuseppe Portacci for his help in the field.

Conflicts of Interest: The authors declare no conflict of interest.

References

1. Karlson, B.; Andersen, P.; Arneborg, L.; Cembella, A.; Eikrem, W.; John, U.; West, J.J.; Klemm, K.; Kobos, J.; Lehtinen, S.; et al. Harmful algal blooms and their effects in coastal seas of Northern Europe. *Harmful Algae* **2021**, *102*, 101989. [CrossRef]
2. Zingone, A.; Escalera, L.; Aligizaki, K.; Fernandez-Tejedor, M.; Ismael, A.; Montresor, M.; Mozetič, P.; Taş, S.; Totti, C. Toxic marine microalgae and noxious blooms in the Mediterranean Sea: A contribution to the Global HAB Status Report. *Harmful Algae* **2020**, *102*, 101843. [CrossRef] [PubMed]
3. Wells, M.L.; Karlson, B.; Wulff, A.; Kudela, R.; Trick, C.; Asnaghi, V.; Berdalet, E.; Cochlan, W.; Davidson, K.; De Rijcke, M.; et al. Future HAB science: Directions and challenges in a changing climate. *Harmful Algae* **2019**, *91*, 101632. [CrossRef] [PubMed]
4. Yu, Z.; Song, X.; Cao, X.; Liu, Y. Mitigation and Control of Harmful Algal Blooms. In *Global Ecology and Oceanography of Harmful Algal Blooms*; Glibert, P.M., Berdalet, E., Burford, M.A., Pitcher, G.C., Zhou, M., Eds.; Springer International Publishing: Cham, Switzerland, 2018; pp. 403–424.
5. Hallegraeff, G.M.; Anderson, D.M.; Belin, C.; Dechraoui Bottein, M.-Y.; Bresnan, E.; Chinain, M.; Enevoldsen, H.; Iwataki, M.; Karlson, B.; McKenzie, C.H.; et al. Perceived global increase in algal blooms is attributable to intensified monitoring and emerging bloom impacts. *Commun Earth Env.* **2021**, *2*, 1–10. [CrossRef]
6. Anderson, D.M. HABs in a changing world: A perspective on harmful algal blooms, their impacts, and research and management in a dynamic era of climactic and environmental change. In Proceedings of the Harmful Algae 2012, 15th International Conference on Harmful Algae, Changwon Gyeongnam, Korea, 29 October–2 November 2012.
7. Koike, K. A red tide off the Myanmar coast: Morphological and genetic identification of the dinoflagellate composition. *Harmful Algae* **2013**, *27*, 149–158.
8. Kudela, R.M.; Berdalet, E.; Bernard, S.; Burford, M.; Fernand, L.; Lu, S.; Roy, S.; Tester, P.; Usup, G.; Magnien, R.; et al. *Harmful Algal Blooms—A Scientific Summary for Policy Makers*; IOC/INF-1320; UNESCO: Paris, France, 2015; p. 20.
9. Pörtner, H.O.; Roberts, D.C.; Masson-Delmotte, V.; Zhai, P.; Tignor, M.; Poloczanska, E.; Mintenbeck, K.; Alegría, A.; Nicolai, M.; Okem, A.; et al. *IPCC Special Report on the Ocean and Cryosphere in A Changing Climate*; IPCC Intergovernmental Panel on Climate Change: Geneva, Switzerland, 2019; Volume 1.
10. Kim, H.G. Mitigation and controls of HABs. In *Ecology of Harmful Algae*; Granéli, E., Turner, J., Eds.; Springer: Berlin/Heidelberg, Germany, 2006; pp. 327–338.
11. Ostroumov, S. Some aspects of water filtering activity of filter feeders. *Hydrobiologia* **2005**, *542*, 275–286. [CrossRef]
12. Ostroumov, S.A. Water Quality and Conditioning in Natural Ecosystems: Biomachinery Theory of Self-Purification of Water. *Russ. J. Gen. Chem.* **2017**, *87*, 3199–3204. [CrossRef]
13. Ostroumov, S.; Widdows, J. Inhibition of mussel suspension feeding by surfactants of three classes. *Hydrobiologia* **2006**, *556*, 381–386. [CrossRef]
14. Licciano, M.; Stabili, L.; Giangrande, A. Clearance rates of *Sabella spallanzanii* and *Branchiomma luctuosum* (Annelida: Polychaeta) on a pure culture of *Vibrio alginolyticus*. *Water Res.* **2005**, *39*, 4375–4384. [CrossRef]

15. Licciano, M.; Stabili, L.; Giangrande, A.; Cavallo, R.A. Bacterial accumulation by *Branchiomma luctuosum* (Annelida: Polychaeta): A tool for biomonitoring marine systems and restoring polluted waters. *Mar. Environ. Res.* **2007**, *63*, 291–302. [CrossRef] [PubMed]
16. Licciano, M.; Terlizzi, A.; Giangrande, A.; Cavallo, R.A.; Stabili, L. Filter-feeder macroinvertebrates as key players in bacterioplankton biodiversity control: A case of study with *Sabella spallanzanii* (Polychaeta: Sabellidae). *Mar. Environ. Res.* **2007**, *64*, 504–513. [CrossRef] [PubMed]
17. Navarro, J.M.; Widdows, J. Feeding physiology of *Cerastoderma edule* in response to a wide range of seston concentrations. *Mar. Ecol. Prog. Ser.* **1997**, *152*, 175–186. [CrossRef]
18. Riisgård, H.U.; Larsen, P.S. Filter-feeding in marine macro-invertebrates: Pump characteristics, modelling and energy cost. *Biol. Rev. Camb. Philos. Soc.* **1995**, *70*, 67–106. [CrossRef] [PubMed]
19. Riisgård, H.U.; Larsen, P.S. Particle capture mechanisms in suspension-feeding invertebrates. *Mar. Ecol. Prog. Ser.* **2010**, *418*, 255–293. [CrossRef]
20. Stabili, L.; Licciano, M.; Giangrande, A.; Fanelli, G.; Cavallo, R.A. *Sabella spallanzanii* filter-feeding on bacterial community: Ecological implications and applications. *Mar. Environ. Res.* **2006**, *61*, 74–92. [CrossRef] [PubMed]
21. Stabili, L.; Licciano, M.; Lezzi, M.; Giangrande, A. Microbiological accumulation by the Mediterranean invasive alien species *Branchiomma bairdi* (Annelida, Sabellidae): Potential tool for bioremediation. *Mar. Pollut. Bull.* **2014**, *86*, 325–331. [CrossRef]
22. Riisgård, H.U. Properties and energy cost of the muscular piston pump in the suspension feeding polychaete *Chaetopterus variopedatus*. *Mar. Ecol. Prog. Ser.* **1989**, *56*, 157–168. [CrossRef]
23. Riisgård, H.U. Suspension feeding in the polychaete *Nereis diversicolor*. *Mar. Ecol. Prog. Ser.* **1991**, *70*, 29–37. [CrossRef]
24. Riisgård, H.U.; Vedel, A.; Boye, H.; Larsen, P.S. Filternet structure and pumping activity in the polychaete *Nereis diversicolor*: Effects of temperature and pump modelling. *Mar. Ecol. Prog. Ser.* **1992**, *83*, 79–89. [CrossRef]
25. Shulman, G.E.; Finenko, G.A. *Bioenergetics of Hydrobionts*; Naukova Dumka Press: Kiev, Ukraine, 1990.
26. Nicol, E.A.T. The Feeding Mechanism, Formation of the Tube, and Physiology of Digestion in *Sabella pavonina*. *Trans. Roy. Soc. Edinb.* **1930**, *56*, 537–598. [CrossRef]
27. Mayer, S. Particle capture in the crown of the ciliary suspension feeding polychaete *Sabella penicillus*: Videotape recordings and interpretations. *Mar. Biol.* **1994**, *119*, 571–582. [CrossRef]
28. Bonar, D.B. Feeding and tube construction in *Chone mollis* Bush (Polychaeta, Sabellidae). *J. Exp. Mar. Biol. Ecol.* **1972**, *9*, 1–18. [CrossRef]
29. Clapin, G. The filtration rate, oxygen consumption and biomass of the introduced polychaete *Sabella spallanzanii* Gmelin within Cockburn Sound: Can They Control Phytoplankton Level and is it An Efficient Filter Feeder? Honours Thesis, Edith Cowan University, Joondalup, Western Australia, Australia, 1996; p. 90.
30. Dales, R.P. Preliminary observations on the role of the coelomic cells in food storage and transport in certain polychaetes. *J. Mar. Biol. Ass. UK* **1957**, *36*, 91–110. [CrossRef]
31. Fitzsimons, G. Feeding and tube-building in *Sabellastarte magnifica* (Shaw) (Sabellidae: Polychaeta). *Bull. Mar. Sci.* **1965**, *15*, 642–671.
32. Lewis, D.B. Feeding and tube-building in the Fabriciinae, (Annelida, Polychaeta). *Proc. Linnean Soc. London* **1968**, *179*, 37–49. [CrossRef]
33. Shumway, S.E.; Bogdanowicz, C.; Dean, D. Oxygen consumption and feeding rates of the sabellid polychaete, *Myxicola infundibulum* (Renier). *Comp. Biochem. Physiol.* **1988**, *90*, 425–428. [CrossRef]
34. Riisgård, H.U.; Ivarsson, N.M. The crown-filament pump of the suspension feeding polychaete *Sabella penicillus*: Filtration, effects of temperature, and energy cost. *Mar. Ecol. Prog. Ser.* **1990**, *62*, 249–257. [CrossRef]
35. Wells, G.P. The respiratory significance of the crown in the polychaete worms *Sabella* and *Myxicola*. *Proc. Roy. Soc. B* **1952**, *140*, 70–82.
36. Gárate-Lizárraga, I. Proliferation of *Amphidinium carterae* Gymnodiniales, Gymnodiniacea in Bahía de La Paz, Gulf of California. *CICIMAR Oceán* **2012**, *272*, 37–49. [CrossRef]
37. Gárate-Lizárraga, I.; González-Armas, R.; Verdugo-Díaz, G.; Okolodkov, Y.B.; Pérez-Cruz, B.; Díaz-Ortíz, J.A. Seasonality of the dinoflagellate *Amphidinium* cf. *carterae* (Dinophyceae: Amphidiniales) in Bahía de la Paz, Gulf of California. *Mar. Pollut. Bull.* **2019**, *146*, 532–541. [CrossRef]
38. Murray, S.A.; Kohli, G.S.; Farrell, H.; Spiers, Z.B.; Place, A.R.; Dorantes-Aranda, J.J.; Ruszczyk, J. A fish kill associated with a bloom of *Amphidinium carterae* in a coastal lagoon in Sydney, Australia. *Harmful Algae* **2015**, *49*, 19–28. [CrossRef] [PubMed]
39. Aquino-Cruz, A.; Okolodkov, Y.B. Impact of increasing water temperature on growth, photosynthetic efficiency, nutrient consumption, and potential toxicity of Amphidinium cf. carterae and Coolia monotis (Dinoflagellata). *Rev. Biol. Mar. Oceanogr.* **2016**, *51*, 565–580. [CrossRef]
40. Yan, M.; Leung, P.T.Y.; Gu, J.; Veronica, T.T.; Lam, V.T.T.; Murray, J.S.; Harwood, D.T.; Wai, T.C.; Paul, K.S.; Lam, P.K.S. Hemolysis associated toxicities of benthic dinoflagellates from Hong Kong Waters. *Mar. Pollut. Bull.* **2020**, *155*, 111114. [CrossRef]
41. Chen, J.; Pan, G. Harmful algal blooms mitigation using clay/soil/sand modified with xanthan and calcium hydroxide. *J. Appl. Phycol.* **2012**, *24*, 1183–1189. [CrossRef]
42. Liu, S.; Yu, Z.; Song, X.; Cao, X. Effects of modified clay on the physiological and photosynthetic activities of Amphidinium carterae Hulbert. *Harmful Algae* **2017**, *70*, 64–72. [CrossRef] [PubMed]

43. Pan, G.; Chen, J.; Donald, M.; Anderson, D.M. Modified local sands for the mitigation of harmful algal blooms. *Harmful Algae* **2011**, *10*, 381–387. [CrossRef]
44. Caroppo, C.; Bisci, A.P. First data on the benthic assemblages of harmful microalgal species in the Gulf of Taranto (Northern Ionian Sea). *Rapp. Commun. Int. Mer Médit.* **2010**, *39*, 341.
45. Pagliara, P.; Caroppo, C. Toxicity assessment of *Amphidinium carterae, Coolia* cfr. monotis and Ostreopsis cfr. ovata (Dinophyta) isolated from the northern Ionian Sea (Mediterranean Sea). *Toxicon* **2012**, *60*, 1203–1214. [CrossRef] [PubMed]
46. Guillard, R.R.L. Culture of phytoplankton for feeding marine invertebrates. In *Culture of Marine Invertebrates. Select Readings*; Berg, C.J., Ed.; Hutchinson Ross Publishing Co: Stroudsberg, PA, USA, 1983; pp. 108–132.
47. Edler, L.; Elbrächter, M. The Utermöhl method for quantitative phytoplankton analysis. In *Microscopic and Molecular Methods for Quantitative Phytoplankton Analysis*; Karlson, B., Cusack, C., Bresnan, E., Eds.; UNESCO IOC Manuals and Guides n. 55; UNESCO: Paris, France, 2010; pp. 13–20.
48. Coughlan, J. The estimation of filtering rate from the clearance of suspensions. *Mar. Biol.* **1969**, *2*, 356–358. [CrossRef]
49. Underwood, A.J. *Experiments in Ecology: Their Logical Design and Interpretation Using Analysis of Variance*; Cambridge University Press: Cambridge, UK, 1997; pp. 1–504.
50. Shumway, S.E.; Egmond, H.P.V.; Hurst, J.W.; Bean, L.L. Management of shellfish resources. In *Manual on Harmful Marine Microalgae*; Hallegraeff, G.M., Anderson, D.M., Cembella, A.D., Eds.; UNESCO: Paris, France, 1995; pp. 433–474.
51. Berdalet, E.; Fleming, L.E.; Gowen, R.; Davidson, K.; Hess, P.; Backer, L.C.; Moore, S.K.; Hoagland, P.; Enevoldsen, H. Marine harmful algal blooms, human health and wellbeing: Challenges and opportunities in the 21st century. *J. Mar. Biol. Assoc. UK* **2016**, *96*, 61–91. [CrossRef]
52. Dodds, W.K.; Bouska, W.W.; Eitzmann, J.L.; Pilger, T.J.; Pitts, K.L.; Riley, A.J.; Schloesser, J.T.; Thornbrugh, D.J. Eutrophication of U.S. Freshwaters: Analysis of Potential Economic Damages. *Environ. Sci. Technol.* **2009**, *43*, 12–19. [CrossRef]
53. Buhr, K.J. Suspension-feeding and assimilation efficiency in *Lanice conchilega* (Polychaeta). *Mar. Biol.* **1976**, *38*, 373–383. [CrossRef]
54. Dales, R.P. Observations on the respiration of the sabellid polychaete *Schizobranchia insignis*. *Biol. Bull.* **1961**, *121*, 82–91. [CrossRef]
55. Klockner, K. Zur Ökologie von *Pomatoceros triqueter* (Serpulidae, Polychaeta). *Helgol. Wiss. Meeresunters* **1978**, *31*, 219–223. [CrossRef]
56. Lemmens, J.W.T.J.; Clapin, G.; Lavery, P.; Cary, J. Filtering capacity of seagrass meadows and other habitats of Cockburn Sound, Western Australia. *Mar. Ecol. Prog. Ser.* **1996**, *143*, 187–200. [CrossRef]
57. Cugier, P.; Struski, C.; Blanchard, M.; Mazurié, J.; Pouvreau, S.; Olivier, F.; Trigui, J.R.; Thiébaut, E. Assessing the role of benthic filter feeders on phytoplankton production in a shellfish farming site: Mont Saint Michel Bay, France. *J. Mar. Syst.* **2010**, *82*, 21–34. [CrossRef]
58. Kim, H.G. An overview on the occurrences of harmful algal blooms (HABs) and mitigation strategies in Korean coastal waters. In *Coastal Environmental and Ecosystem Issues of the East China Sea*; Ishimatsu, A., Lie, H.J., Eds.; Terrapub and Nagasaki University Publication: Tokyo, Japan, 2010; pp. 121–131.
59. Pan, G.; Miao, X.; Bi, L.; Zhang, H.; Wang, L.; Wang, L.; Wang, Z.; Chen, J.; Ali, J.; Pan, M.; et al. Modified local soil (MLS) technology for harmful algal bloom control, sediment remediation, and ecological restoration. *Water* **2019**, *11*, 1123. [CrossRef]
60. Pierce, R.H.; Henry, M.S.; Higham, C.J.; Blum, P.; Sengco, M.R.; Anderson, D.M. Removal of harmful algal cells (*Karenia brevis*) and toxins from seawater culture by clay flocculation. *Harmful Algae* **2004**, *3*, 141–148. [CrossRef]
61. Pan, G.; Zhang, M.M.; Chen, H.; Zou, H.; Yan, H. Removal of cyanobacterial blooms in Taihu Lake using local soils. I. Equilibrium and kinetic screening on the flocculation of *Microcystis aeruginosa* using commercially available clays and minerals. *Environ. Pollut.* **2006**, *141*, 195e200. [CrossRef]
62. Sengco, M.R.; Li, A.; Tugend, K.; Kulis, D.; Anderson, D.M. Removal of red- and brown-tide cells using clay flocculation. I. Laboratory culture experiments with *Gymnodinium breve* and *Aureococcus anophagefferens*. *Mar. Ecol. Prog. Ser.* **2001**, *210*, 41–53. [CrossRef]
63. Anderson, D.M. Approaches to monitoring, control and management of harmful algal blooms (HABs). *Ocean Coast Manag.* **2009**, *52*, 342–347. [CrossRef] [PubMed]
64. Shumway, S.E.; Frank, D.M.; Ewart, L.M. Effect of yellow clay on clearance rate in seven species of benthic, filter-feeding invertebrates. *Aquac. Res.* **2003**, *34*, 1391–1402. [CrossRef]
65. Maguire, J.; Cusack, C.; Ruiz-Villarreal, M.; Silkeb, J.; McElligotta, D.; Davidson, K. Applied simulations and integrated modelling for the understanding of toxic and harmful algal blooms (ASIMUTH): Integrated HAB forecast systems for Europe's Atlantic Arc 2016. *Harmful Algae* **2016**, *53*, 160–166. [CrossRef]
66. Hallegraeff, G.M.; Bresnan, E.; Enevoldsen, H.; Schweibold, L.; Zingone, A. Call to contribute to Global HAB Status Report. *Harmful Algae News* **2017**, *58*, 1–3.
67. Zingone, A.; Enevoldsen, H.; Hallegraeff, G.M. Are HABs and their societal impacts expanding and intensifying? A call for answers from the HAB scientific community. In *Proceedings of the Marine and Fresh-Water Harmful Algae, 17th International Conference on Harmful Algae 2016*; Proenca, L.A.O., Hallegraeff, G.M., Eds.; International Society for the Study of Harmful Algae and Intergovernmental Oceanographic Commission of UNESCO: Paris, France, 2017; pp. 14–17.
68. Stabili, L.; Cecere, E.; Licciano, M.; Petrocelli, A.; Sicuro, B.; Giangrande, A. Integrated Multitrophic Aquaculture By-Products with Added Value: The Polychaete *Sabella spallanzanii* and the Seaweed *Chaetomorpha linum* as Potential Dietary Ingredients. *Mar. Drugs* **2019**, *17*, 677. [CrossRef] [PubMed]

69. Giangrande, A.; Pierri, C.; Del Pasqua, M.; Gravili, C.; Gambi, M.C.; Gravina, M.F. Mediterranean Sea in check: The other side of biological invasions. *Mar. Ecol.* **2020**, *41*, e12583. [CrossRef]
70. Pal, M.; Yesankar, P.J.; Dwivedi, A.; Asifa Qureshi, A. Biotic control of harmful algal blooms (HABs): A brief review. *J. Environ. Manag.* **2020**, *268*, 110687. [CrossRef] [PubMed]

microorganisms

Review

Versatile Applications of Cyanobacteria in Biotechnology

Ewa Żymańczyk-Duda *, Sunday Ocholi Samson *, Małgorzata Brzezińska-Rodak and Magdalena Klimek-Ochab

Faculty of Chemistry, Department of Biochemistry, Molecular Biology, and Biotechnology, Wroclaw University of Science and Technology, Wybrzeże Wyspiańskiego 29, 50-370 Wrocław, Poland
* Correspondence: ewa.zymanczyk-duda@pwr.edu.pl (E.Ż.-D.); sunday.samson@pwr.edu.pl (S.O.S.)

Abstract: Cyanobacteria are blue-green Gram-negative and photosynthetic bacteria which are seen as one of the most morphologically numerous groups of prokaryotes. Because of their ability to fix gaseous nitrogen and carbon dioxide to organic materials, they are known to play important roles in the universal nutrient cycle. Cyanobacteria has emerged as one of the promising resources to combat the issues of global warming, disease outbreaks, nutrition insecurity, energy crises as well as persistent daily human population increases. Cyanobacteria possess significant levels of macro and micronutrient substances which facilitate the versatile popularity to be utilized as human food and protein supplements in many countries such as Asia. Cyanobacteria has been employed as a complementary dietary constituent of feed for poultry and as vitamin and protein supplement in aquatic lives. They are effectively used to deal with numerous tasks in various fields of biotechnology, such as agricultural (including aquaculture), industrial (food and dairy products), environmental (pollution control), biofuel (bioenergy) and pharmaceutical biotechnology (such as antimicrobial, anti-inflammatory, immunosuppressant, anticoagulant and antitumor); recently, the growing interest of applying them as biocatalysts has been observed as well. Cyanobacteria are known to generate a numerous variety of bioactive compounds. However, the versatile potential applications of cyanobacteria in biotechnology could be their significant growth rate and survival in severe environmental conditions due to their distinct and unique metabolic pathways as well as active defensive mechanisms. In this review, we elaborated on the versatile cyanobacteria applications in different areas of biotechnology. We also emphasized the factors that could impede the implementation to cyanobacteria applications in biotechnology and the execution of strategies to enhance their effective applications.

Keywords: cyanobacteria; bioactive compounds; bioremediation; photobiocatalysis

Citation: Żymańczyk-Duda, E.; Samson, S.O.; Brzezińska-Rodak, M.; Klimek-Ochab, M. Versatile Applications of Cyanobacteria in Biotechnology. *Microorganisms* **2022**, *10*, 2318. https://doi.org/10.3390/microorganisms10122318

Academic Editor: Stefan Junne

Received: 24 October 2022
Accepted: 21 November 2022
Published: 23 November 2022

1. Introduction

Cyanobacteria are blue-green algae that generate biomasses via the conversion of carbon dioxide using solar energy [1]. Cyanobacteria have a wide variety of colors such as pink, red, yellow, green and brown [2], and they exist in diverse ecosystems such as in rock surfaces, oceans, soil and freshwater [3]. It is important to know that cyanobacteria are the first universal oxygen-producing photosynthetic microbes and for many billions of years, they have contributed to the universe's oxygen generation [4]. Cyanobacteria are microscopic organisms; however, they could be observed via unaided eyes in the form of blooms or colonies [5,6]. The potential of cyanobacteria to adapt to several environmental conditions, ability to generate oxygen and simple genomes (enabling easy manipulations), as sources of bioactive compounds, and their fast growth rate have prompted many researchers to perform thorough investigations on cyanobacteria as supplements in human food, animal feed, in industry and in medicine [7,8]. However, alterations in climate change, global warming, nutrients availability and biotic factors determine the rapid growth rate of cyanobacteria [9–11].

Cyanobacteria are abundant sources of bioactive compounds. Their secondary metabolites are rich sources of vitamins, enzymes and toxins which are crucial in many biotechnological industries [7], for instance, in the production of bioplastics from cyanobacteria's polyhydroxy-alkanoates (PHA) [12,13]. PHA accumulates intracellularly in many cyanobacteria species, which can be utilized in bioplastics production with properties such as polyethylene and polypropylene [12]. A wide variety cyanotoxins have been generated from cyanobacteria [14]. These bioactive compounds among others include saxitoxins, microcystins, and anatoxins [15]. This review emerged to broadly elaborate the important roles of cyanobacteria in biotechnology.

2. Cyanobacterial Metabolites

Cyanobacteria exhibit metabolic processes such carotenogenesis and photosynthesis, which generate highly valued primary and secondary metabolites. Primary metabolites are those metabolites that are strictly involved in processes such as reproduction, growth and cell division, which are referred to as developmental processes [15]. Primary metabolites include antioxidants, primary proteins and lipids [16]. These metabolites could be re-engineered to produce important biotechnological products such as dyes, biofertilizers, food supplements and bioplastics [17–20]. However, secondary metabolites are not used by cyanobacterial cells and are not directly engaged in their normal growth, reproduction or development [21]. Secondary metabolites are typically distinctive to specific organisms and are not regularly available in all environmental conditions. Hence, they are produced for defensive purposes [22]. Secondary metabolites from cyanobacteria include toxins, phenolic compounds, essential oils, alkaloids, steroids and antibiotics [23–29].

Previous epidemiological research revealed that cyanobacteria have a quite simple genetic material [30], enabling easy manipulations and modifications for the discovery of new important metabolites. These metabolites could be produced naturally from cyanobacteria via their responses to environmental stresses such as alteration of light intensity and deficiencies in micro and macro nutrients, which could alter their photosynthetic cellular metabolism [31].

Chen et al. [32] elucidated the manipulation of cyanobacteria growth conditions by comparing two phases. In the first phase, the cyanobacteria were allowed to grow in suitable and favorable conditions, whereas in the second phase, cyanobacteria were allowed to grow in unfavorable environmental conditions such as nitrogen deficiency, elevated intensity of light and deficiency of nutrients. The results revealed significant levels of secondary metabolites such as essential oils, starch granules, lipids and biopolymers from the second phase when compared to the first stage [32,33]. Rachana et al. [34] uncovered the exploitation of the cyanobacteria metabolic pathway (Methylerythritol 4-phosphate, MEP) to generate valuable secondary metabolites such as chlorophyll, fatty acids, and hormones. These products possess antimicrobial, antibiotic and anticarcinogenic properties that facilitate their active roles in pharmaceutical industries [34,35].

2.1. *Phenolic Acids*

Phenolic acids are aromatic rings consisting of one or more hydroxyl groups and one carboxyl group. Phenolic acids produced by cyanobacteria are essential for the defense against oxidative destruction that could result from hydroxyl radicals and the reactive oxygen species [36]. The buildup of phenolic acids in cyanobacteria ensures the adaptability and tolerance of cyanobacteria against several environmental stresses such as interaction with heavy metals and UV light [37,38], which could result in the deposition of free radicals in cells, damage to deoxyribose and chemical destruction of DNA [39].

As an example, Singh et al. [40] investigated how phenolic acids produced by cyanobacteria played an important role in the rummaging of free radicals and ensuring tolerance and adaptability when the cultures of various cyanobacteria species (*Oscillatoria acuta*, *Plectonema boryanum*, *Anabaena doliolum* and *Haplosiphon intricatus*) were subjected to a high sodium chloride concentrations (considered as unfavorable condition). The experimental results revealed significant accumulation of phenolic acids including vanillic acid, gallic acid, chlorogenic acid, caffeic acid and ferulic acid [40]. In another instance, Tanutcha et al. [41] revealed that the treatment of the cyanobacteria specie, *Halothece* sp. PCC7418 cells with temperature shock allowed for the robust production of phenolic compounds and phycobiliproteins. The quantification of phenolic compounds and phycobiliproteins in aqueous extracts showed that the amounts of the metabolites were regulated by the hot and cold temperature shocks [41].

Phenolic acids produced by cyanobacteria with great scavenging of free radicals and detoxification of reactive oxygen species (ROS) facilitated their relevance as therapeutic agents. For instance, anti-winkle and skin whitening facilitated by ferulic acids [42], anti-aging effect facilitated by gallic acids [43], effective heart failure recovery facilitated by syringic, gentisic and gallic acids [44], as well as antimicrobial effects of other phenolic compounds were generated by cyanobacteria [45–47].

2.2. Vitamins

Vitamins participate in many metabolic pathways, acting as coenzymes. They also engage in diverse processes such as antioxidants, controlling and regulating cell functioning and growth of tissues [48]. There are various categories of vitamins utilized by living organisms. These include lipid-soluble vitamins (vitamin A, D, E, and K) and water-soluble vitamins (vitamin B1, B2, B3, B5, B6, B7, B9, B12 and C). Cyanobacteria produce vitamin A, B, C, E and K in response to environmental stress, as elucidated by Helliwell et al. [49] and Asensi-Fabado et al. [50]. Deficiency of vitamins in humans is a significant menace globally and necessitates immediate attention. In the notion for bioactive metabolites generation, cyanobacteria denote one of the most capable potentials with many biotechnological applications and for allowing the advancement of an eco-sustainable creation of natural bioactive metabolites. It is important to know that not all vitamins are generated by all plants and some vitamin (B, D and K) are scarce in many plants [48,50]. Hence, cyanobacteria have been demonstrated to generate these vitamins that are scarce in several vascular plants [48]. Individuals acquire significant number of vitamins via the consumption of fruits, vegetables and dairy products such meat, fish and eggs. However, an individual with some vitamins' deficiency could be treated with artificial or synthetic vitamin, through oral or intramuscular delivery of vitamins: B_{12} (oral and intramuscular), D1 (oral ergocalciferol), C (oral) and K1 (intramuscular) [51–55].

Cyanobacterial products are fascinating and have gained increasing attention because of their high contents in vitamins, minerals, essential amino acids and protein [56]. As an example, *Spirulina* products are extensively promoted due to their high contents of vitamin B_{12}, high contents of protein (>60%) and additional micronutrients [56]. *Spirulina* is known to be an important vitamin B_{12} source for vegans and vegetarians, as chemical analyses showed that vitamin B_{12} dry weight levels range from 127 to 244 μg/100 g [57]. The range could be rendered into a day-to-day *Spirulina* intake of 1.6–3.2 g to reach the satisfactory consumption of 4 μg/day for cobalamin (vitamin B_{12}) according to the EFSA [58]. However, the suppliers of *Spirulina* suggested a regular intake of 3–9 g/day would be efficient to enhance vitamin B_{12} requirements [59]. So far, *Arthrospira* sp. (*Spirulina*) has been made accessible for human intake as a nutritional supplementto combat the deficiency of Vitamin B_{12} especially, thereby improving bone strength and stiffness as well as inhibiting the formation of ulcers [60,61]. In addition, numerous studies have showed the medicinal important of *Spirulina* as a rich source vitamin E and β-carotene. Previous investigation by Carcea et al. [62] revealed the generation of vitamin E from *Spirulina* with the content ranging between 2.8 and 12.5 mg/100 g in improved techniques and higher in traditionally dried *Spirulina* [62]. From the experimental samples, th etraditionally dried technique was recorded to be 30.9 mg/100 g content of vitamin E [62]. Research also showed a significant amount of β-carotene in *Spirulina* where the content varied between 33.5 and 231.6 mg/100 g) [62]. High contents of β-carotene were also examined in *Synechoccus* sp. and *Anabaena cylindrica* and in some fruits such as oranges, potatoes, broccoli and carrots [48,63].

Excluding the vascular plants, *Anabaena cylindrica* is also an exceptional source of vitamin C, an antioxidant that could offer a defensive mechanism against oxidants in cyanobacteria [64]. In response to UV radiation stress and preventing destruction of cell membrane, cyanobacteria (*Arthrospira* sp.) could generate a low amount of vitamin D [65,66]. Tarento et al. [67]. emphasized that a significant amount of vitamin K1 could be experimentally examined in the marine plant *Anabaena cylindrica* compared to the content of vitamin K1 in parsley and spinach [67].

2.3. Peptides

Peptides are small proteins which can be categorized based on their biosynthesis into ribosomal peptides (gene encoded) and non-ribosomal peptides (non-gene encoded). They could also be categorized according to their 3D-structure, covalent binding and modification, source, properties and function [68]. Peptides and polyketides biosynthesis in microbes have exceptional segmental pathways controlled by non-ribosomal peptides (NRPs) and polyketides (PKs). Polyketides (PKs) and non-ribosomal peptides (NRPs) are extensively useful as drugs today, and one possible source for new PKs and NRPs are cyanobacteria. However, there is limited information on the varieties of microorganisms and their PKs and NRPs biosynthetic genes in the marine deposit [69]. NRPS includes components, each of which joins proteinogenic amino acids with non-proteinogenic amino acids, carbohydrates, fatty acids and additional building blocks to form peptide chains [69]. Natural products of PKs and NRPs are secondary metabolites of microbes, used by microbes in adaptation and resistance to environmental stresses [69]. So far, over twenty three thousands natural products of non-ribosomal peptides and polyketides have been recognized and categorized, and they had been extensively applied in medicine as antitumor and antibiotic therapies [70,71].

Biological synthesis of PKs and NRPs are catalyzed by polyketide synthases (PKS) and non-ribosomal peptide synthases (NRPS) [72]. Remarkably, the gene clusters of NRPS and PKS commonly exist in microbes, such as cyanobacteria, compared with eurkarya or archaea [72]. Previous investigations revealed that these gene clusters exist in *Anabaena, Nostoc, Lyngbya, Planktothrix, Microcystism, Pleurocapsa* and *Nodularia* genera. However, *Nostoc* and *Pleurocapsa* species are known as the common NRP- and PK-producing cyanobacteria [73].

Ribosomal peptides (RPs) are proteinogenic amino acids that can be observed on ribosomes that are used as building blocks in ribosomal peptides' biosynthesis [68,74]. Cyanobacteria also generates lanthipeptides, one of the major ribosomal peptide families including microviridins and cyanobactin. The potential of cyanobacteria to generate lanthipeptides is due to the availability of class II lenthipeptidase and prochlorosins in their genome (such as the marine *Synechococcus* and *Prochlorococcus*) [75–77].

2.4. Terpenoids

Terpenoids are known to be the largest group of bioactive compounds. They are categorized into carotenoids, sesterterpenoids, steroids, hemiterpenoids, diterpenoids, sesquiterpenoids and monoterpenoids. Research shows that over fifty five thousands terpenoids have been discovered so far [78,79]. Triterpenoids were found in a significant amount both in natural cyanobacterium-dominated microbial mats and in laboratory cyanobacterial cultures, for instance, in 2-methylhopanoids (2-MeBHPs) [80,81]. 2-methylhopanoids (2-MeBHPs) belong to pentacyclic triterpenoids, which act as biomarkers for enhanced cyanobacteria in some ecological sites [81]. In addition, 2-MeBHP encourages thawing and freezing resistance, pH stress resistance and osmotic pressure resistance in cyanobacteria to ensure their existence in unfavorable environments such as hot springs, arctic and Antarctic soils, desert soil crusts and in high saline lakes [82,83].

Carotenoids, also known as tetraterpenoids, are important for energy dissipation and light-harvesting during the process of photosynthesis [83,84]. Examples of common carotenoids synthesized by cyanobacteria include zeaxanthin, echinenone and β-carotene. Kusama et al. [85] highlighted the importance of zeaxanthin and echinenone to shield PSII (photosystem II) against singlet oxygen [85]. In addition, for protection against direct UV-B radiation, *Pseudanabaena* sp. CCNU1 generates β-carotene, zeaxanthin, echinenone, myxoxanthophyll and canthaxanthin [84]. Carotenoids are necessary to provide protection against photooxidative destruction resulting from direct UV light during the photosynthesis process. Carotenoids are secondary metabolites (lipophilic) generated from the pathways of isoprenoids [84]. Filamentous *Calothrix* PCC 7507 and *Synechocystis* sp. PCC 6803 have been shown to generate sesquiterpenoid geosmin, a sesquiterpene with no isopropyl group [86,87].

Recent studies revealed the biological activities of cyanobacterial carotenoids as a source of medicine for the treatment of several infections. For example, carotenoids and their derivatives obtained from cyanobacteria proved high superoxide anion radical ($O_2^{\bullet -}$) anti-inflammatory and scavenging effects that facilitate psoriasis treatment [88]. A significant level of carotenoids was observed in *Tychonema* sp. LEGE 07175 and *Cyanobium* sp. LEGE 07175. The extracts revealed a strong antiaging result by hindering hyaluronidase synthesis, the enzyme that promotes hyaluronic acid depolymerization [89]. Antimicrobial sesterterpene from a *Scytonema* sp. (UTEX 1163) culture demonstrated growth inhibition against several pathogenic microbes such as *Mycobacterium tuberculosis*, *Bacillus anthracis*, *Candida albicans*, *Staphylococcus aureus* and *Escherichia coli* [89,90].

3. Biotechnological Applications of Cyanobacteria

Cyanobacteria are remarkably the most crucial group of microbes in the universe as they satisfy important environmental roles in the universe, as an important global supplier of oxygen, nitrogen and carbon [91]. Owing to their enormous array of industrial applications, they have been the subject of quite a lot of research. They are important sources of biofertilizers, biofuels, food additives and coloring dyes [92]. Cyanobacteria are used in water treatments, bioplastics production, cosmetics, forestry, feed for animals, production of hydrogen, bioethanol production and biogas production [93–99]. Here, we review the highly demanded applications of cyanobacteria in biotechnology.

3.1. Cyanobacteria as Food Supplements

Cyanobacterial carotenoids including zeaxanthin, beta-carotene, canthaxanthin and nostoxanthin are great sources used in food supplements, colorants, food additives and animal feed. The productions of these metabolites are on the rise. The supplements are sold in the form of tablets, granules and capsules. As an example, β-carotene, riboflavin, vitamin B_{12} and thiamine are greatly used supplements generated by cyanobacteria such as *Spirulina* [60,66]. In addition, cyanobacteria are known to be used as whole food or as dietary supplements such as minerals, amino acids, proteins, complex sugar, carbohydrate, phycocyanin, active enzymes, essential fatty acids and chlorophyll [100].

Arthrospira platensis (a filamentous, gram-negative cyanobacterium) is frequently used as a whole food supplement. It is grown globally and applied as an animal feed supplement in aquariums, poultry, aquariums and many agricultural industries worldwide. Dried *Spirulina* contains 8% fat, 5% water, 51–71% protein and 24% carbohydrate. It is a valuable supplier of several important nutrients and nutritional minerals, including iron and vitamin B_{12}. Vitamin B_{12} is important in the production of hemoglobin, maintains the nervous system and participates in DNA synthesis [101,102]. Previous research revealed that several nutritional supplements of cyanobacterial origin such as *Spirulina*, *chlorella* and *Aphanizomenon flos-aquae* are readily available in the consumer markets in the United States [103].

Several dietary supplements are frequently generated from the biomass of cyanobacterial species and eaten whole, unlike extracts utilized in pharmaceutical productions [103].

For example, ketocarotenoid (astaxanthin) is seen as a powerful antioxidant compared to vitamin A and vitamin C as well as several carotenoids which perform a crucial role in preventing destruction in human cells via photooxidation. *Haematococcus pluvialis* has been known to generate astaxanthin, which is a strong inhibitor of protease known for the treatment of several diseases including the human immunodeficiency virus disease (the virus that is responsible for the acquired immunodeficiency syndrome (AIDS) which is the final stage of HIV disease) [34,104,105].

3.2. *Cyanobacteria in Medical and Pharmaceutical Biotechnology*

Cyanobacteria is comprise of several secondary metabolites that are useful in the field of medical biotechnology. These microbes have achieved tremendous attention from researchers because of the generation of bioactive compounds that are incredibly useful in medical settings [106]. Although they generate effective toxins, they also generate various metabolites that are vital in terms of their anticancer, antibiotic, anti-inflammatory, immunosuppressant and antimicrobial effects [106–110]. Several global investigations have explored various bioactive compounds from cyanobacteria as anticancer potentials. For instance, Aurilide (isolated from *Dolabella auricularia*) revealed varying cytotoxicity from picomolar (pM) in nanomolar (nM) concentrations against numerous cancer cell lines. Aurilide aids mitochondrial-induced apoptosis by selectively binding to prohibitin 1 (PHB1) in the mitochondria and triggering the proteolytic dispensation of optic atrophy 1 (OPA1) [111,112]. Biselyngbyaside is another drug obtained from *Lyngbya* sp. Biselyngbyaside A shows cytotoxicity against HeLa S3 cells with an IC_{50} value of 0.1 μg/mL. However, Biselyngbyolide B, C, E and F have an antiproliferative impact in HeLa and HL-60 cells, whereas Biselyngbyolide C prompts endoplasmic reticulum (ER) stress and apoptosis in HeLa cells [113,114]. In addition, cryptophycin isolated from *Nostoc* sp. var. ATCC 53789 and GSV 224 is an excellent anticancer agent. Cryptophycin prevents microtubule formation and demonstrates anti-tumorigenic action against various solid tumors implanted in mice involving multidrug-resistant cancer cells. The IC_{50} value of cryptophycin was found to be lower than 50 pM for cell lines multidrug-resistant cancer [115]. In recent studies on cancer therapy, cryptophycin copulated with peptides and antibodies had been developed for targeted drug delivery [116]. The antiproliferative actions of Arg-Gly-Asp (RGD)–cryptophycin and *iso*Asp−Gly−Arg (*iso*DGR)-cryptophycin conjugates were experimented against human melanoma cell lines (M21 and M21-L). The investigation revealed that the conjugations exhibit anticancer efficacy at nanomolar concentrations with diverse expression integrin $\alpha_v\beta_3$ (a type of integrin that is a receptor for vitronectin) levels [116,117].

Apratoxin and its derivatives, developed from several types of marine cyanobacteria (such as *Moorea producens* strain (RS05), *Lyngbya bouillonii*, *L. sordida*, *L. majuscula*, genera *Phormidium*, and genera *Neolyngbya*), are known to combat diverse forms of cancer cell lines [106,118,119]. Curacin A islolated from Lyngbya *majuscula* is effective against cancer of the breast [120]. Consequently, in addition to the natural resources, cyanobacteria provide a favorable means, presenting a comprehensive variety of substances for new drug discovery and development [121]. Secondary metabolites of cyanobacteria can be applied as natural compositions in cosmetology [122]. For instance, they can be used as photoprotective Mycosporine-like Amino Acids (MAAs) in sunscreens to shield the skin from destructive UVR. In addition, cyanobacteria natural pigments such as phycobiliproteins and carotenoids may be employed as natural colorants as well as antioxidants to shield the skin from destruction resulting from exposure to UV irradiation [123]. Therefore, the field of cyanobacteria exploration is very crucial research.

3.3. *Cyanobacteria in Bioplastic and Biofuel Production*

Cyanobacteria features such remarkable approaches for fixation and absorption of atmospheric nitrogen and CO_2, and employing them it for growth in unfavorable climatic environments, such as unfertile soils and saline waters, make them extremely suitable for production of biodegradable plastics and biofuels [124]. Although numerous means are available for bioplastic and biofuel production, cyanobacteria have been studied as energy-rich suppliers due to the triacylglycerol (TAG) and diacylglycerol (DAG) production, which can be employed as biodiesel precursors [125]. Cyanobacteria such as *Synechocystis, Spirulina, Anabaena* and *Nostoc muscorum* can function as bio-factories for biofuel and bioplastic generation. For instance, they have the metabolism for generating cost-effective and sustainable biopolymer polyhydroxyalkanoates (PHAs), and polyhydroxybutyrate (PHB), among other copolymers [34]. Biopolymeric PHB presents material features such as polypropylene, a standard plastic obtained from petroleum (fossil fuels). However, in comparison to standard plastics, PHB is biodegradable, and its application as a complementary of standard plastics can assist in alleviating the critical ecological influences of fossil fuels and plastics in nonbiodegradable overconsumption [126,127].

Species of cyanobacteria such as *Synechococcus* and *Synechocystis* species can generate lactate and succinate which are important chemicals in the production of bioplastics. Succinate is an essential biotechnological chemical, a precursor of adipic acid, 1,4-butanediol, and other four-carbon chemicals [128]. In recent study by Durall et al. [128], the highest succinate titer was achieved in dark incubation (compared to the light and anoxic darkness conditions) of an engineered cyanobacteria strain (*Synechocystis* PCC 6803), coupled with a limited glyoxylate shunt (*aceA* and *aceB*) overexpressing isocitrate lyase with phosphoenolpyruvate carboxylase, with supplemented medium using 2-thenoyltrifluoroacetone [128]. Furthermore, a research team at Kobe University [129] illustrated the method by which *Synechocystis* sp. PCC 6803 (one of the most global researched cyanobacterial strains known to be the model microbe for photosynthate production because of its fast growth and the ease of genetic manipulation) could generate D-lactate (utilized in biodegradable plastic productions). The research team demonstrated that malic enzyme accelerates the production of D-lactate via genetically modifying D-lactate synthesis pathways using cyanobacteria. They eventually succeeded in producing the world's maximum amount of D-lactate (26.6 g/L) directly from light and carbon dioxide [129,130].

The process of generating petroleum products (such as jet fuel, heating oil, propane, gasoline, and diesel) could be a serious potential environmental threat (pollution), and the release of toxic substances such as greenhouse gasses (methane, ozone, carbon dioxide (CO_2) and nitrous oxide) into the atmosphere could be hazardous to humans, plants and animals [131,132]. Cyanobacteria provide excellent assurance as suppliers of renewable biofuels for the energy sector [133]. Biofuels such as 2-methyl-1-butanol, isobutanol, 2,3-butanediol ethanol, isobutanol and ethylene have been produced from engineered cyanobacteria (*Synechocystis* sp.) [134]. Biofuels including gasoline, jet fuel, biodiesel and ethanol are being generated from genetically modified cyanobacteria by some US-based companies such as Joule Unlimited and Algenol [15,135–137].

3.4. Cyanobacteria in Bioremediation

Cyanobacteria are characterized by high adaptability to various stress conditions, being at the same time quite resistant to toxic compounds of different origins [31]. Therefore, such photosynthetic bacteria seem to be relevant for numerous approaches in the field of bioremediation, including soil remediation, wastewater treatment and degradation of organic pollutants. Water pollution represents a real environmental problem as a consequence of anthropogenic activity both in the context of urbanization and industrialization of the environment as well agricultural practices. Several research groups have successfully explored the potential application of cyanobacteria for wastewater treatment, demonstrating that polluted water from different sources can be treated effectively with the help of such microorganisms, and the system based on photosynthetic procaryotic cells can be considered as a promising alternative to conventional biological processes such as activated sludge.

Extensive aquaculture generates a huge amount of polluted nitrogen-rich water released into the costal seas, directly improving the nitrogen pool in marine ecosystems, altering the balance of species [138]. The marine cyanobacterium *Synechococcus* sp. has been shown to effectively remove ammonium from brackish aquaculture wastewater [139]. The tested microorganism assimilated ammonium through the actions of glutamine synthetase (GS, EC 6.3.1.2) and glutamate synthase (GOGAT, EC 1.4.1.13) cooperated in the GS-GOGAT cycle, which is closely related to the adaptive strategy of the *Synechococcus* species to changing nutrient conditions [140].

An interesting approach to biological wastewater treatment is the use of a cyanobacterial-bacterial consortium that operates on the synergistic action between photosynthetic microorganisms and heterotrophic bacteria. It should be stressed that such photosynthetic microbes are known from exopolysaccharides production that are useful in establishing a symbiotic association of cyanobacteria with other organisms [141]. Brewery wastewater containing high concentrations of organic pollutants and significant amounts of nitrogen and phosphorus has been treated using a cyanobacterial-bacterial consortium dominated by the filamentous cyanobacterium *Leptolyngbya* sp. [142] Cyanobacterial-bacterial aggregates grown under optimal pH and temperature conditions were found to be highly effective and successfully reduced the levels of nitrogenous compounds, including nitrate (up to 80%), ammonium (up to 90%) and phosphorus compounds (up to 70%) in crude wastewater. The introduction of an additional wastewater pretreatment step involving electrocoagulation followed by the use of an electrochemically treated supernatant as a medium for the cultivation of microorganisms increased the level of removal of pollutants [143]. The bioremediation potential of the studied consortium was successfully verified under stressed conditions in a flat-plate photobioreactor filled with hydrophilic support [144]. The proposed solution is very important from a technological point of view as it brings laboratory ideas closer to the practical disposal of brewery wastewater. Other examples illustrating the applicability of bacterial consortia for wastewater treatment are as follows: *Dinophysis acuminata* and *Dinophysis caudata* living in a consortium were reported to effectively remove phosphate, phenol and cyanide from coke-oven wastewater [145], the effective mixing of nitrogen-fixing soil cyanobacterial culture was municipal wastewater treatment [146] and mixed cyanobacteria formed mats degraded pesticide lindane in pesticide-contaminated effluents, showing high resistance against its toxicity [147].

The most important demonstration of the degradation potential of cyanobacterial cells is their ability to remove heavy metals from sewage of various origins. The ability of cyanobacteria to tolerate and interact with metal ions makes them an attractive tool for environmental biotechnology [148]. Cyanobacteria employ a variety of mechanisms such as biosorption, bioaccumulation, activation of metal transporters, biotransformation, and induction of detoxifying enzymes to sequester and minimize the toxic effects of heavy metals [149]. Hexavalent chromium can be present in some aquatic systems as a result of textile, paint, metal cleaning, plating, electroplating, and mining industries [150]. As chromium (Cr(VI)) is potentially toxic and carcinogenic to humans, its removal from water and wastewater is required to avoid serious health and environmental problems. The living cyanobacterial consortium consisting of *Limnococcus limneticus* and *Leptolyngbya subtilis* has been found to be efficient in the removal of Cr (VI) from wastewater [151], whereas the mat-forming cyanobacterial consortium consisting of *Chlorella* sp., *Phormidium* sp. and *Oscillatoria* sp. efficiently removed hexavalent chromium from the sewage in the tannery industry [152]. It is worth noting that cultivation of microbes under conditions that force cells to organize themselves into a mat or application of naturally forming mats can be applied on a large scale and used practically in bioremediation. Cadmium is a toxic metal and its exposure remains a global concern [153]. An interesting strategy for the sequestration of Cd (II) from aqueous solutions was proposed, including axenic cultures of *Nostoc muscorum* immobilized on the glass surface through the formation of biofilms [154]. A microorganism growing as a biofilm expressed the ability to adsorb Cd(II) in a wide concentration range of ~24 ppb to 100 ppm and a pH range of 5 to 9. Strains belonging to *Nostoc* genera that produce EPS are known for their ability to remove metals, and the application of mixotrophic cultivation conditions increased the uptake capacity of heavy metal ions by the *Nostoc* species [155]. The self-flocculating *Oscillatoria* sp. was shown to possess metabolic properties to eliminate Cd from metal-contaminated water [156]. Authors analyzed the mechanism of bacterial activity against Cd and they found that metal adsorption by negatively charged functional groups in cyanobacterial biomasses was the major mechanism used by *Oscillatoria* sp. to remove metals from the aqueous medium followed by Cd bioaccumulation in living cells. The potential in municipal sewage remediation has also been shown for *Anabaena oryzae,* characterized by a high removal efficiency for cadmium, lead, zinc, iron, copper and manganese [157].

Cyanobacteria are present very often in polluted environments [158], and due to their naturally evolved resistance and selectivity against environmental pollutants, they exude a significant metabolic potential for xenobiotic degradation. For example, *Spirulina* spp. demonstrated the ability to metabolize the phosphonate xenobiotic Dequest 2054® [159], *Leptolyngbya* sp. has the ability to degrade phenol, significantly decreasing its concentration in the cultivation medium [160], *Aphanothece conferta* demonstrated high degradation efficiency against aliphatic hydrocarbons, whereas aromatic hydrocarbons were degraded by *Synechocystis aquatilis* [161]; cyanobacteria have also been reported to have the relevant enzymatic system to participate in the degradation of textile dyes [162,163].

3.5. Applications of Cyanobacteria Species in Biocatalytic Processes

Cyanobacteria, as organisms capable of photosynthesis, are a group of biocatalysts with unusual activities that can find application in biocatalytic processes for obtaining crucial derivatives used in various industries. One of the derivatives with extensive use in the pharmaceutical and fragrance industry are unsaturated alcohols, e.g., cinnamyl alcohol, essential both for the synthesis of Taxol and Chloramycin and for the production of aromatic compounds [164,165]. Traditionally, this compound is obtained from cinnamaldehyde of natural origin, but the chemical reduction of the aldehyde group to a primary hydroxyl group with the double bond intact is problematic, so a mixture of products (Scheme 1) is usually obtained. The solution to this problem may be the use of selected cyanobacterial strains that can chemoselectively reduce aldehyde to alcohol without breaking the double bond in the side chain [165].

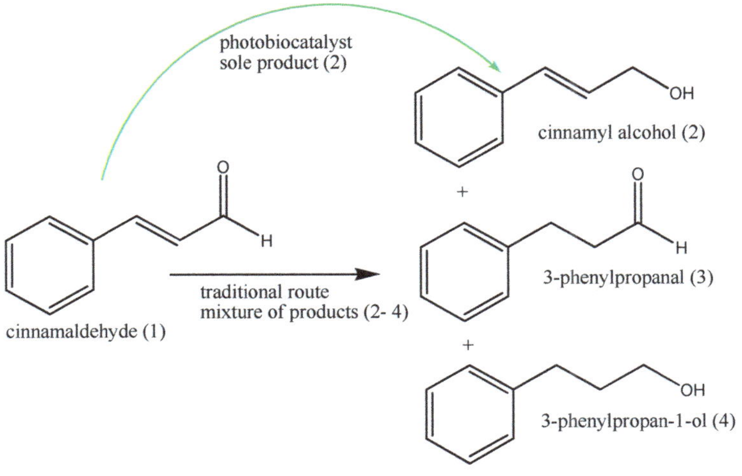

Scheme 1. Possible reduction of cinnamaldehyde.

Yamanaka et al. [165] have selected the strain *Synechocystis* sp. PCC 6803 as the most efficient, capable of converting cinnamaldehyde to cinnamyl alcohol with an efficiency of up to 98%. In addition, they demonstrated that the reaction is strictly light-dependent, and the coenzymes (NADPH) needed for reduction are regenerated in photosynthetic electron transfer reactions; thus, there is no need for additional supporting substrates, which greatly simplifies the whole process.

Another group of compounds of industrial importance are optically pure secondary alcohols used as chiral building blocks for the synthesis of optically active products. The whole cell biocatalytic reduction of the corresponding ketones is a favorable alternative to traditional chemical processes. Acetophenone is often used as a model substrate for screening in case of bioreduction and it can undergo enantioselective bioreduction to *R*- or *S*-phenyl ethanol (**1**, Scheme 2). Several methods have been developed using whole cells of heterotrophic microorganisms or plant tissues [166–168], but typically these processes require co-substrates to support coenzyme regeneration systems. The intracellular oxidoreductase that is responsible for the asymmetric acetophenone reduction reaction is not itself dependent on light but requires the presence of reduced NAD(P)H, which is one of the products of the light phase of photosynthesis (Scheme 2). Lighting is therefore essential for the regeneration of the cofactor. According to the literature data, cyanobacterial enzymatic systems could be effectively applied for the reduction of acetophenone to *S*- phenyl ethanol (e.g., *Spirulina platensis*–45% yield, 97% e.e [169]; *Arthrospira maxima*- 45,8% yield, 98,8% e.e. [170]; however, the efficiency and enantioselectivity of processes are dependent on the growth rate of biocatalysts, substrate concentration and light regime [169,170].

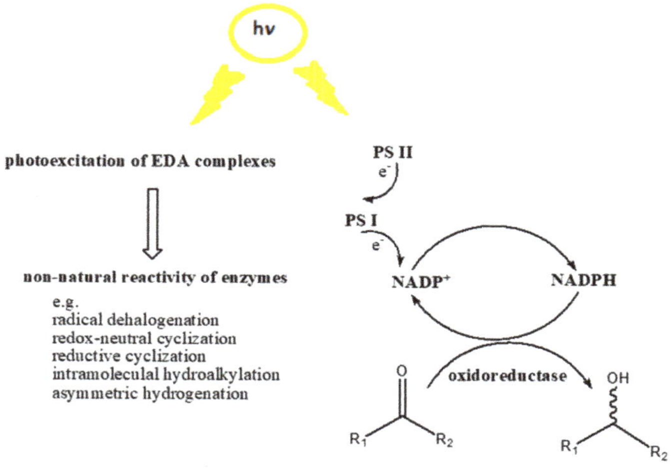

(1) R_1= $-C_6H_5$; R_2= $-CH_3$
(2) R_1= $-CH_3$; R_2= $-C_5H_{12}PO_3$
(3) R_1= $-C_6H_5$; R_2= $-C_5H_{12}PO_3$
(4) R_1= $-CH_2CH_3$; R_2= $-C_5H_{12}PO_3$

Scheme 2. Possible effects of light on enzyme activity in phototrophic organisms.

Similar relationships have been observed in the reduction of diethyl 2-oxophosphonates to corresponding *S*-2-hydroxyalkylphosphonates (Scheme 2), which in an optically pure form can also be used as a synthon for the synthesis of derivatives used in medicine to agriculture [171–173]. The biological activity of phosphonates is based, among other things, on enzyme inhibition and therefore they are difficult substrates for biocatalytic processes, especially in the scaling-up context. Additionally, phosphonates with a carbonyl group that is located right next to the aromatic ring were inefficiently transformed by fungal biocatalysts [174,175]; therefore, special attention was turned to the strain of *Nodularia sphaerocarpa* capable of reducing the substrate (**3**) (Scheme 2), with a yield of 99% and an enantiomeric excess of 93%. As in the case of acetophenone, the efficiency of the reaction was closely correlated with the concentration of the substrate used and the best results were obtained using 1 mM of diethyl 2-oxo-2-phenylethylphosphonate (**3**) [176].

Other substrates tested in the discussed study were not reduced very effectively (diethyl 2-oxopropylphosphonate (**2**)—transformed with *Arthrospira maxima* with 20% yield and 99% e.e, diethyl 2-oxobutylphosphonate (**3**)—transformed with *Nodularia sphaerocarpa* with 27% yield and 80% e.e.) [176]. In the next stage of the research mentioned above, immobilization in calcium alginate was used to increase the resistance of the biocatalyst to the toxic effects of substrates, and the influence of shaking (better contact of the substrate with the biocatalyst) on the reduction efficiency was checked. Also, in this case *Nodularia sphaerocarpa* turned out to be the most effective strain. The use of mixing or immobilization made it possible to increase the scale of the process. Packing the immobilized photobiocatalyst in a simplified flow reactor allowed for increasing the substrate (**3**) (2-oxo-2-phenylethylphosphonate) concentration to the value of 10 mM, although in the 500 mL batch culture conducted with stirring, a better efficiency (44% of yield, 91% of e.e.) was obtained compared to 38% of the yield and 86% of e.e. obtained in the column reactor [177].

A quite recent finding was that NAD(P)H or flavin-dependent enzymes involved in many light-independent metabolic processes may have unnatural activities after exposure to light. Cofactors of these enzymes can form electron-donor-acceptor (EDA) complexes with unnatural substrates (Scheme 2). EDA can be excited by visible light which allows for the flow of electrons and consequently the formation of the reduced product [178]. The first enzyme of this type to be described was NADPH-dependent carbonyl ketoreductase, which catalyzed the radical dehalogenation of halolactone upon exposure to light [179]. Later, in a similar manner, asymmetric reductive cyclization [180,181], intermolecular hydroalkylation [178] or asymmetric hydrogenation [182] using ene-reductase was achieved.

Another meaningful application of cyanobacteria is related to the activity of the fatty acid photodecarboxylase (FAP) of microalgae *Chlorella variabilis* origin. This enzyme is inside the cell, and is involved in the lipids metabolism and driven by blue light and the presence of FAD. This mechanism was deeply analysed and described by Damine Sourigue et al. [183]. Scientists studied the microalgal strain *Chlorella variabilis* NC64A and discovered that phototrophs produced by photodecarboxylase belong to the oxidoreductases and catalyses decarboxylation of saturated and unsaturated fatty acids, with the releasing of the corresponding alkanes or alkenes. This enzyme interacts with substrates by binding them in a tunnel-like site, which leads directly to the flavin dinucleotide, which is crucial as a moiety sensitive to light excitement, which is followed by the electron transfer from the substrate. Such a sequence initiates the reaction. This discovery has not gone unnoticed by researchers involved in the application of biocatalysts of different origins for the synthesis of variable chemical compounds according to green chemistry rules. Photodecarboxylase from *Chlorella variabilis* NC64A (Cv FAP) was considered a part of the cascade of reactions finally leading to biofuel production. The very first approaches were focused on the evaluation of the activity of the enzyme (Cv FAP) towards structurally different fatty acids (saturated and unsaturated) [184,185]. Usually, reactions were carried out at pH 8.5 under illuminations by blue light of intensity = 13.7 $\mu EL^{-1}:s^{-1}$ and the proportions of the substrate and enzyme were as follows: [substrate] = 30 mM, [decarboxylase] = 6.0 μM. According to the general scheme (Scheme 3) below, a number of fatty acids were tested: lauric, myristic, palmitic, margaric, stearic, oleic, linoleic and arachidic.

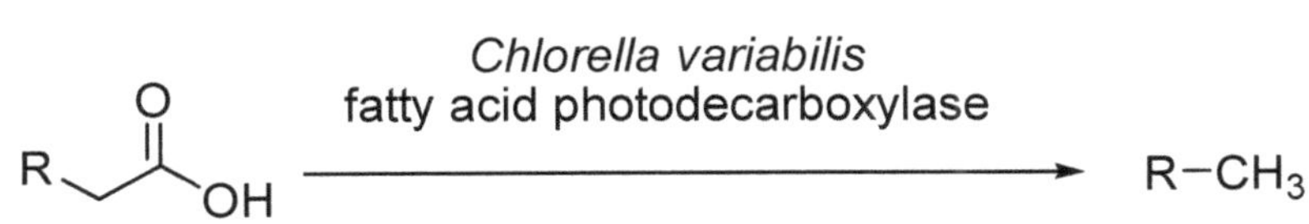

Scheme 3. Photodecarboxylase action.

The best conversion degrees were obtained for the four substrates: palmitic, margaric, stearic and arachidic acids (above 90%). The differences in the results were correlated to the differences in the substrates biding, which in turn is due to differences in matching between the tunnel site in enzyme structures and the substrate moiety. However, further experiments were meant for the scaling of the selected reactions and applying the palmitic acid as a starting compound; this approach succeeds and leads to the scale elevation up to 155 mg of pentadecane production. These results were important in relation to the necessity of green chemistry solution implementations in the chemical industry. The practical side of the mentioned results appeared in the cascade of the reactions, starting from triglycerides hydrolysis via decarboxylation and leading to the release of glycerol, hydrocarbon and carbon dioxide (Scheme 4) [184–186]. Previous approaches were based on the two-step process design as a sequence starting from the hydrolysis conducted by the lipase from *Candida rugosa,* which delivers fatty acids for the next reaction catalysed by photocarboxylase (Scheme 4).

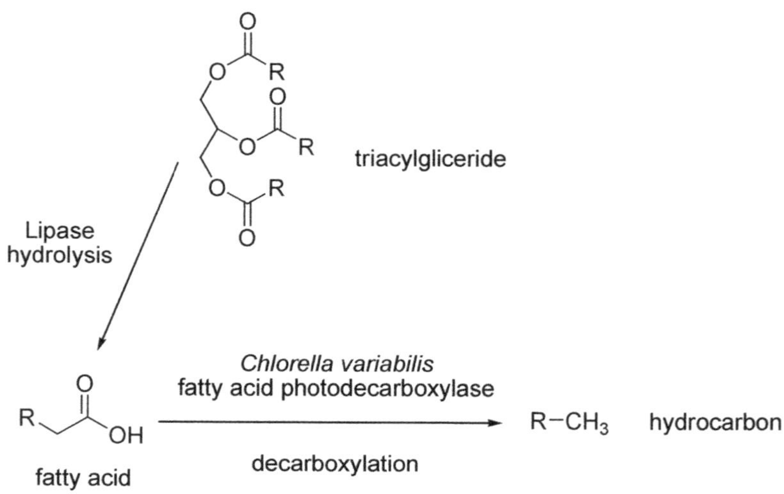

Scheme 4. Cascade of the reactions.

Further developed protocols were performed as one-step ones. In addition, the mode of the biocatalysts applied for photodecarboxylations was modified. Instead of using the purified enzyme, whole-cell biocatalysts were involved in the reaction. Genetically engineered cells of *E. coli* were able to produce photodecarboxylase at ae high level and with great activity towards studied substrates, which is why for the next set of experiments such a biocatalyst was applied. The limiting factor was finding a compatible reaction–environment lipase, which is at the same time insensitive to blue light, which is essential for decarboxylase activity. These boundary conditions were met by immobilized lipase from the fungus *Rhizopus oryzae* [186] and finally the one pot sequence of reactions was set. The effectiveness of this protocol was checked against plants oils (e.g., soybean oil) and waste cooking oil and for different physical chemical parameters (e.g., temperature, biocatalysts and substrates concentrations, reaction duration). This allowed for selecting the best protocol for scaling (from 1 mL to 15 mL of the final volume of reaction mixture), which was conducted with soybean oil with the receiving of almost 1 g of hydrocarbon (21.2% of isolated yield).

The above achievements are the base for further discoveries as they proved the practical and biotechnological function of photobiocatalysts; nevertheless, it should be emphasized that they require the individual protocols for almost every designed biocatalytic process.

4. Conclusions and Future Perspective

This review has highlighted that cyanobacteria are promising sources of several primary and secondary metabolites that are known for their biological activities. These bioactive compounds were reviewed to possess several impacts in various biotechnological industries. Nonetheless, to improve the usefulness of cyanobacteria for human and animals exploitation, extensive research is essential for profiling cyanobacterial secondary metabolites, known to be generally employed in different countries in the areas of nutraceuticals and pharmaceuticals applications, including in antimicrobial and antitumor activities. Furthermore, cyanobacterial enzymatic systems are currently applied as biocatalysts for conducting chemically different reactions to achieve various synthetic goals. This field is still not fully explored and offers biological tools of different features compared with light independent organisms (e.g., fungi and heterotrophic bacteria). There is a need for further research on how to effectively exploit cyanobacteria using green technological advancements as well as efficient insights on cyanobacteria applications on larger scales. Future investigation that is also promising is essential to explore more beneficial roles of cyanobacteria in biotechnology.

Author Contributions: Conceptualization, E.Ż.-D., Data source preparation, E.Ż.-D., S.O.S., M.B.-R., M.K.-O. All authors have read and agreed to the published version of the manuscript.

Funding: This research was funded by a subsidy from the Ministry of Education and Science for 471 the Faculty of Chemistry of Wrocław University of Science and Technology (subsidy nr: 472 8211104160).

Informed Consent Statement: Not applicable.

Data Availability Statement: Study does not report any data.

Conflicts of Interest: The authors declare no conflict of interest.

References

1. Huisman, J.; Codd, G.A.; Paerl, H.W.; Ibelings, B.W.; Verspagen, J.M.H.; Visser, P.M. Cyanobacterial Blooms. *Nat. Rev. Microbiol.* **2018**, *16*, 471–483. [CrossRef] [PubMed]
2. Stomp, M.; Huisman, J.; Vörös, L.; Pick, F.R.; Laamanen, M.; Haverkamp, T.; Stal, L.J. Colourful Coexistence of Red and Green Picocyanobacteria in Lakes and Seas. *Ecol. Lett.* **2007**, *10*, 290–298. [CrossRef] [PubMed]
3. Whitton, B.A.; Potts, M. Introduction to the Cyanobacteria. In *Ecology of Cyanobacteria II: Their Diversity in Space and Time*; Whitton, B.A., Ed.; Springer: Dordrecht, The Netherlands; Heidelberg, Germany; New York, NY, USA; London, UK, 2012; pp. 1–13.
4. Rasmussen, B.; Fletcher, I.R.; Brocks, J.J.; Kilburn, M.R. Reassessing the First Appearance of Eukaryotes and Cyanobacteria. *Nature* **2008**, *455*, 1101–1104. [CrossRef] [PubMed]
5. Catherine, Q.; Susanna, W.; Isidora, E.S.; Mark, H.; Aurélie, V.; Jean-François, H. A Review of Current Knowledge on Toxic Benthic Freshwater Cyanobacteria–Ecology, Toxin Production and Risk Management. *Water Res.* **2013**, *47*, 5464–5479. [CrossRef] [PubMed]
6. Xiao, M.; Li, M.; Reynolds, C.S. Colony Formation in the Cyanobacterium Microcystis. *Biol. Rev. Camb. Philos. Soc.* **2018**, *93*, 1399–1420. [CrossRef]
7. Amadu, A.A.; de Graft-Johnson, K.A.A.; Ameka, G.K. Industrial Applications of Cyanobacteria. *Cyanobacteria Recent Adv. Taxon. Appl.* **2021**, *106*, 1–12.
8. Costa, J.A.V.; Moreira, J.B.; Lucas, B.F.; Da Silva Braga, V.; Cassuriaga, A.P.A.; De Morais, M.G. Recent Advances and Future Perspectives of PHB Production by Cyanobacteria. *Ind. Biotechnol.* **2018**, *14*, 249–256. [CrossRef]
9. Sukenik, A.; Hadas, O.; Kaplan, A.; Quesada, A. Invasion of Nostocales (Cyanobacteria) to Subtropical and Temperate Freshwater Lakes—Physiological, Regional, and Global Driving Forces. *Front. Microbiol.* **2012**, *3*, 86. [CrossRef]
10. Paerl, H.W.; Paul, V.J. Climate Change: Links to Global Expansion of Harmful Cyanobacteria. *Water Res.* **2012**, *46*, 1349–1363. [CrossRef]
11. Berg, M.; Sutula, M. Factors Affecting Growth of Cyanobacteria With Special Emphasis on the Sacramento-San Joaquin Delta. *South. Calif. Coast. Water Res. Proj. Tech. Rep.* **2015**, *869*, 100.
12. Koller, M. Advances in Polyhydroxyalkanoate (PHA) Production, Volume 3. *Bioengineering* **2022**, *9*, 328. [CrossRef] [PubMed]
13. Afreen, R.; Tyagi, S.; Singh, G.P.; Singh, M. Challenges and Perspectives of Polyhydroxyalkanoate Production From Microalgae/Cyanobacteria and Bacteria as Microbial Factories: An Assessment of Hybrid Biological System. *Front. Bioeng. Biotechnol.* **2021**, *9*, 109. [CrossRef] [PubMed]
14. Kurmayer, R.; Deng, L.; Entfellner, E. Role of Toxic and Bioactive Secondary Metabolites in Colonization and Bloom Formation by Filamentous Cyanobacteria Planktothrix. *Harmful Algae* **2016**, *54*, 69–86. [CrossRef]
15. Mazard, S.; Penesyan, A.; Ostrowski, M.; Paulsen, I.T.; Egan, S. Tiny Microbes with a Big Impact: The Role of Cyanobacteria and Their Metabolites in Shaping Our Future. *Mar. Drugs* **2016**, *14*, 97. [CrossRef] [PubMed]
16. Ng, H.-S.; Chew, L.-L. Valuable Compounds Produced by Microalgae. *Handb. Biorefinery Res. Technol.* **2022**, 1–23. [CrossRef]

17. Orona-Navar, A.; Aguilar-Hernández, I.; Nigam, K.D.P.; Cerdán-Pasarán, A.; Ornelas-Soto, N. Alternative Sources of Natural Pigments for Dye-Sensitized Solar Cells: Algae, Cyanobacteria, Bacteria, Archaea and Fungi. *J. Biotechnol.* **2021**, *332*, 29–53. [CrossRef]
18. Chittora, D.; Meena, M.; Barupal, T.; Swapnil, P. Cyanobacteria as a Source of Biofertilizers for Sustainable Agriculture. *Biochem. Biophys. Rep.* **2020**, *22*, 100737. [CrossRef]
19. Nicoletti, M. The Nutraceutical Potential of Cyanobacteria. *Pharmacol. Potential Cyanobacteria* **2022**, 287–330. [CrossRef]
20. Agarwal, P.; Soni, R.; Kaur, P.; Madan, A.; Mishra, R.; Pandey, J.; Singh, S.; Singh, G. Cyanobacteria as a Promising Alternative for Sustainable Environment: Synthesis of Biofuel and Biodegradable Plastics. *Front. Microbiol.* **2022**, *13*, 939347. [CrossRef]
21. Böttger, A.; Vothknecht, U.; Bolle, C.; Wolf, A. Plant Secondary Metabolites and Their General Function in Plants. In *Lessons on Caffeine, Cannabis & Co*; Springer: Cham, Switzerland, 2018; pp. 3–17.
22. Chomel, M.; Guittonny-Larchevêque, M.; Fernandez, C.; Gallet, C.; DesRochers, A.; Paré, D.; Jackson, B.G.; Baldy, V. Plant Secondary Metabolites: A Key Driver of Litter Decomposition and Soil Nutrient Cycling. *J. Ecol.* **2016**, *104*, 1527–1541. [CrossRef]
23. Baselga-Cervera, B.; Balboa, C.G.; Costas, E.; Lopez-Rodas, V. Why Cyanobacteria Produce Toxins? Evolutionary Game Theory Suggests the Key. *Int. J. Biol.* **2014**, *7*, 64. [CrossRef]
24. Singh, D.P.; Prabha, R.; Verma, S.; Meena, K.K.; Yandigeri, M. Antioxidant Properties and Polyphenolic Content in Terrestrial Cyanobacteria. *3 Biotech* **2017**, *7*, 134. [CrossRef] [PubMed]
25. Formighieri, C.; Melis, A. Cyanobacterial Production of Plant Essential Oils. *Planta* **2018**, *248*, 933–946. [CrossRef] [PubMed]
26. Knoot, C.J.; Khatri, Y.; Hohlman, R.M.; Sherman, D.H.; Pakrasi, H.B. Engineered Production of Hapalindole Alkaloids in the Cyanobacterium *Synechococcus* sp. UTEX 2973. *ACS Synth. Biol.* **2019**, *8*, 1941–1951. [CrossRef] [PubMed]
27. Khalifa, S.A.M.; Shedid, E.S.; Saied, E.M.; Jassbi, A.R.; Jamebozorgi, F.H.; Rateb, M.E.; Du, M.; Abdel-Daim, M.M.; Kai, G.Y.; Al-Hammady, M.A.M.; et al. Cyanobacteria—From the Oceans to the Potential Biotechnological and Biomedical Applications. *Mar. Drugs* **2021**, *19*, 241. [CrossRef]
28. Dong, Q.; Dong, R.; Xing, X.; Li, Y. A new antibiotic produced by the cyanobacterium-symbiotic fungus *Simplicillium lanosoniveum*. *Nat. Prod. Res.* **2018**, *32*, 1348–1352. [CrossRef]
29. Sunday, O.S.; Emmanuel, E.; Cornelius, A.O. In-Vitro Assessment of Antibacterial Activity of Crude Methanolic and Aqueous Extracts of Mitracarpus Villosus. *Afr. J. Microbiol. Res.* **2021**, *15*, 62–68. [CrossRef]
30. Zahra, Z.; Choo, D.H.; Lee, H.; Parveen, A. Cyanobacteria: Review of Current Potentials and Applications. *Environments* **2020**, *7*, 13. [CrossRef]
31. Rachedi, R.; Foglino, M.; Latifi, A. Stress Signaling in Cyanobacteria: A Mechanistic Overview. *Life* **2020**, *10*, 312. [CrossRef]
32. Chen, B.; Wan, C.; Mehmood, M.A.; Chang, J.S.; Bai, F.; Zhao, X. Manipulating Environmental Stresses and Stress Tolerance of Microalgae for Enhanced Production of Lipids and Value-Added Products—A Review. *Bioresour. Technol.* **2017**, *244*, 1198–1206. [CrossRef]
33. Ge, S.; Champagne, P.; Plaxton, W.C.; Leite, G.B.; Marazzi, F. Microalgal Cultivation with Waste Streams and Metabolic Constraints to Triacylglycerides Accumulation for Biofuel Production. *Biofuels Bioprod. Biorefining* **2017**, *11*, 325–343. [CrossRef]
34. Singh, R.; Parihar, P.; Singh, M.; Bajguz, A.; Kumar, J.; Singh, S.; Singh, V.P.; Prasad, S.M. Uncovering Potential Applications of Cyanobacteria and Algal Metabolites in Biology, Agriculture and Medicine: Current Status and Future Prospects. *Front. Microbiol.* **2017**, *8*, 515. [PubMed]
35. El-Baz, F.K.; El-Senousy, W.M.; El-Sayed, A.B.; Kamel, M.M. In vitro antiviral and antimicrobial activities of algal extract of Spirulina platensis. *J. Appl. Pharm. Sci.* **2013**, *3*, 52–56.
36. Vatansever, F.; de Melo, W.C.M.A.; Avci, P.; Vecchio, D.; Sadasivam, M.; Gupta, A.; Chandran, R.; Karimi, M.; Parizotto, N.A.; Yin, R.; et al. Antimicrobial Strategies Centered around Reactive Oxygen Species–Bactericidal Antibiotics, Photodynamic Therapy and Beyond. *FEMS Microbiol. Rev.* **2013**, *37*, 955. [CrossRef]
37. Rastogi, R.P.; Sinha, R.P.; Moh, S.H.; Lee, T.K.; Kottuparambil, S.; Kim, Y.J.; Rhee, J.S.; Choi, E.M.; Brown, M.T.; Häder, D.P.; et al. Ultraviolet Radiation and Cyanobacteria. *J. Photochem. Photobiol. B Biol.* **2014**, *141*, 154–169. [CrossRef]
38. Mahfooz, S.; Shamim, A.; Husain, A.; Hasan, Z.; Farooqui, A. Oxidative Stress Biomarkers in Cyanobacteria Exposed to Heavy Metals. *Contam. Water Health Risk Assess. Treat. Strateg.* **2021**, 385–403. [CrossRef]
39. Badri, H.; Monsieurs, P.; Coninx, I.; Wattiez, R.; Leys, N. Molecular investigation of the radiation resistance of edible cyanobacterium *Arthrospira* sp. PCC 8005. *Microbiologyopen* **2015**, *4*, 187–207. [CrossRef]
40. Singh, D.P.; Prabha, R.; Meena, K.K.; Sharma, L.; Sharma, A.K.; Singh, D.P.; Prabha, R.; Meena, K.K.; Sharma, L.; Sharma, A.K. Induced Accumulation of Polyphenolics and Flavonoids in Cyanobacteria under Salt Stress Protects Organisms through Enhanced Antioxidant Activity. *Am. J. Plant Sci.* **2014**, *5*, 726–735. [CrossRef]
41. Patipong, T.; Hibino, T.; Waditee-Sirisattha, R.; Kageyama, H. Induction of Antioxidative Activity and Antioxidant Molecules in the Halotolerant Cyanobacterium *Halothece* sp. PCC7418 by Temperature Shift. *Nat. Prod. Commun.* **2019**, *14*, 1934578X19865680.
42. Park, H.J.; Cho, J.H.; Hong, S.H.; Kim, D.H.; Jung, H.Y.; Kang, I.K.; Cho, Y.J. Whitening and Anti-Wrinkle Activities of Ferulic Acid Isolated from Tetragonia Tetragonioides in B16F10 Melanoma and CCD-986sk Fibroblast Cells. *J. Nat. Med.* **2018**, *72*, 127–135. [CrossRef]
43. Monteiro e Silva, S.A.; Calixto, G.M.F.; Cajado, J.; de Carvalho, P.C.A.; Rodero, C.F.; Chorilli, M.; Leonardi, G.R. Gallic Acid-Loaded Gel Formulation Combats Skin Oxidative Stress: Development, Characterization and Ex Vivo Biological Assays. *Polymers* **2017**, *9*, 391. [CrossRef] [PubMed]

44. Sun, S.; Kee, H.J.; Ryu, Y.; Choi, S.Y.; Kim, G.R.; Kim, H.S.; Kee, S.J.; Jeong, M.H. Gentisic Acid Prevents the Transition from Pressure Overload-Induced Cardiac Hypertrophy to Heart Failure. *Sci. Rep.* **2019**, *9*, 3018. [CrossRef] [PubMed]
45. Frazzini, S.; Scaglia, E.; Dell'anno, M.; Reggi, S.; Panseri, S.; Giromini, C.; Lanzoni, D.; Rossi, C.A.S.; Rossi, L. Antioxidant and Antimicrobial Activity of Algal and Cyanobacterial Extracts: An In Vitro Study. *Antioxidants* **2022**, *11*, 992. [CrossRef] [PubMed]
46. Monroe, M.B.B.; Easley, A.D.; Grant, K.; Fletcher, G.K.; Boyer, C.; Maitland, D.J. Multifunctional Shape-Memory Polymer Foams with Bio-Inspired Antimicrobials. *ChemPhysChem* **2018**, *19*, 1999–2008. [CrossRef] [PubMed]
47. Li, R.; Narita, R.; Nishimura, H.; Marumoto, S.; Yamamoto, S.P.; Ouda, R.; Yatagai, M.; Fujita, T.; Watanabe, T. Antiviral Activity of Phenolic Derivatives in Pyroligneous Acid from Hardwood, Softwood, and Bamboo. *ACS Sustain. Chem. Eng.* **2018**, *6*, 119–126. [CrossRef]
48. Del Mondo, A.; Smerilli, A.; Sané, E.; Sansone, C.; Brunet, C. Challenging microalgal vitamins for human health. *Microb Cell Fact.* **2020**, *19*, 201. [CrossRef] [PubMed]
49. Helliwell, K.E.; Lawrence, A.D.; Holzer, A.; Kudahl, U.J.; Sasso, S.; Kräutler, B.; Scanlan, D.J.; Warren, M.J.; Smith, A.G. Cyanobacteria and Eukaryotic Algae Use Different Chemical Variants of Vitamin B12. *Curr. Biol.* **2016**, *26*, 999–1008. [CrossRef] [PubMed]
50. Asensi-Fabado, M.A.; Munné-Bosch, S. Vitamins in Plants: Occurrence, Biosynthesis and Antioxidant Function. *Trends Plant Sci.* **2010**, *15*, 582–592. [CrossRef] [PubMed]
51. Youness, R.A.; Dawoud, A.; ElTahtawy, O.; Farag, M.A. Fat-souluble vitamins: Updated review of their role and orchestration in human nutrition throughout life cycle with sex differences. *Nutr. Metab.* **2022**, *19*, 60–70. [CrossRef] [PubMed]
52. Golriz, F.; Donnelly, L.F.; Devaraj, S.; Krishnamurthy, R. Modern American Scurvy—Experience with Vitamin C Deficiency at a Large Children's Hospital. *Pediatr. Radiol.* **2017**, *47*, 214–220. [CrossRef] [PubMed]
53. Fenech, M.; Amaya, I.; Valpuesta, V.; Botella, M.A. Vitamin C Content in Fruits: Biosynthesis and Regulation. *Front. Plant Sci.* **2018**, *9*, 2006. [CrossRef] [PubMed]
54. Al Saleh, Y.; Beshyah, S.A.; Hussein, W.; Almadani, A.; Hassoun, A.; Al Mamari, A.; Ba-Essa, E.; Al-Dhafiri, E.; Hassanein, M.; Fouda, M.A.; et al. Diagnosis and Management of Vitamin D Deficiency in the Gulf Cooperative Council (GCC) Countries: An Expert Consensus Summary Statement from the GCC Vitamin D Advisory Board. *Arch. Osteoporos.* **2020**, *15*, 35. [CrossRef] [PubMed]
55. Merchant, R.E.; Phillips, T.W.; Udani, J. Nutritional Supplementation with Chlorella Pyrenoidosa Lowers Serum Methylmalonic Acid in Vegans and Vegetarians with a Suspected Vitamin B12 Deficiency. *J. Med. Food* **2015**, *18*, 1357–1362. [CrossRef] [PubMed]
56. Grosshagauer, S.; Kraemer, K.; Somoza, V. The True Value of *Spirulina*. *J. Agric. Food Chem.* **2020**, *68*, 4109–4115. [CrossRef]
57. Watanabe, F.; Katsura, H.; Takenaka, S.; Fujita, T.; Abe, K.; Tamura, Y.; Nakatsuka, T.; Nakano, Y. Pseudovitamin B12 Is the Predominant Cobamide of an Algal Health Food, *Spirulina* Tablets. *J. Agric. Food Chem.* **1999**, *47*, 4736–4741. [CrossRef]
58. Agostoni, C.; Canani, R.B.; Fairweather-Tait, S.; Heinonen, M.; Korhonen, H.; La Vieille, S.; Marchelli, R.; Martin, A.; Naska, A.; Neuhäuser-Berthold, M.; et al. Scientific Opinion on Dietary Reference Values for Cobalamin (Vitamin B12). *EFSA J.* **2015**, *13*, 4150.
59. Muys, M.; Sui, Y.; Schwaiger, B.; Lesueur, C.; Vandenheuvel, D.; Vermeir, P.; Vlaeminck, S.E. High Variability in Nutritional Value and Safety of Commercially Available Chlorella and *Spirulina* Biomass Indicates the Need for Smart Production Strategies. *Bioresour. Technol.* **2019**, *275*, 247–257. [CrossRef]
60. Anantharajappa, K.; Dharmesh, S.M.; Ravi, S. Gastro-Protective Potentials of *Spirulina*: Role of Vitamin B12. *J. Food Sci. Technol.* **2020**, *57*, 745–753. [CrossRef]
61. Ekeuku, S.O.; Chong, P.N.; Chan, H.K.; Mohamed, N.; Froemming, G.R.A.; Okechukwu, P.N. *Spirulina* Supplementation Improves Bone Structural Strength and Stiffness in Streptozocin-Induced Diabetic Rats. *J. Tradit. Complement. Med.* **2022**, *12*, 225–234. [CrossRef]
62. Carcea, M.; Sorto, M.; Batello, C.; Narducci, V.; Aguzzi, A.; Azzini, E.; Fantauzzi, P.; Finotti, E.; Gabrielli, P.; Galli, V.; et al. Nutritional Characterization of Traditional and Improved Dihé, Alimentary Blue-Green Algae from the Lake Chad Region in Africa. *LWT Food Sci. Technol.* **2015**, *62*, 753–763. [CrossRef]
63. Santiago-Morales, I.S.; Trujillo-Valle, L.; Márquez-Rocha, F.J.; Hernández, J.F.L. Tocopherols, Phycocyanin and Superoxide Dismutase from Microalgae: As Potential Food Antioxidants. *Appl. Food Biotechnol.* **2018**, *5*, 19–27.
64. Aaronson, S.; Dhawale, S.W.; Patni, N.J.; Deangelis, B.; Frank, O.; Baker, H. The Cell Content and Secretion of Water-Soluble Vitamins by Several Freshwater Algae. *Arch. Microbiol.* **1977**, *112*, 57–59. [CrossRef] [PubMed]
65. Ljubic, A.; Jacobsen, C.; Holdt, S.L.; Jakobsen, J. Microalgae Nannochloropsis Oceanica as a Future New Natural Source of Vitamin D3. *Food Chem.* **2020**, *320*, 126627. [CrossRef] [PubMed]
66. Fais, G.; Manca, A.; Bolognesi, F.; Borselli, M.; Concas, A.; Busutti, M.; Broggi, G.; Sanna, P.; Castillo-Aleman, Y.M.; Rivero-Jiménez, R.A.; et al. Wide Range Applications of *Spirulina*: From Earth to Space Missions. *Mar. Drugs* **2022**, *20*, 299. [CrossRef] [PubMed]
67. Tarento, T.D.C.; McClure, D.D.; Vasiljevski, E.; Schindeler, A.; Dehghani, F.; Kavanagh, J.M. Microalgae as a Source of Vitamin K1. *Algal Res.* **2018**, *36*, 77–87. [CrossRef]
68. Scheidler, C.M.; Kick, L.M.; Schneider, S. Ribosomal Peptides and Small Proteins on the Rise. *ChemBioChem* **2019**, *20*, 1479–1486. [CrossRef]
69. Wei, Y.; Zhang, L.; Zhou, Z.; Yan, X. Diversity of Gene Clusters for Polyketide and Nonribosomal Peptide Biosynthesis Revealed by Metagenomic Analysis of the Yellow Sea Sediment. *Front. Microbiol.* **2018**, *9*, 295. [CrossRef]

70. Katz, L.; Baltz, R.H. Natural Product Discovery: Past, Present, and Future. *J. Ind. Microbiol. Biotechnol.* **2016**, *43*, 155–176. [CrossRef]
71. Welker, M.; Von Döhren, H. Cyanobacterial Peptides—Nature's Own Combinatorial Biosynthesis. *FEMS Microbiol. Rev.* **2006**, *30*, 530–563. [CrossRef]
72. Maurya, S.K.; Niveshika; Mishra, R. Importance of Bioinformatics in Genome Mining of Cyanobacteria for Production of Bioactive Compounds. *Cyanobacteria Basic Sci. Appl.* **2019**, 477–506. [CrossRef]
73. Wang, H.; Fewer, D.P.; Holm, L.; Rouhiainen, L.; Sivonen, K. Atlas of Nonribosomal Peptide and Polyketide Biosynthetic Pathways Reveals Common Occurrence of Nonmodular Enzymes. *Proc. Natl. Acad. Sci. USA* **2014**, *111*, 9259–9264. [CrossRef] [PubMed]
74. Kehr, J.C.; Picchi, D.G.; Dittmann, E. Natural Product Biosyntheses in Cyanobacteria: A Treasure Trove of Unique Enzymes. *Beilstein J. Org. Chem.* **2011**, *7*, 1622–1635. [CrossRef] [PubMed]
75. Li, B.; Sher, D.; Kelly, L.; Shi, Y.; Huang, K.; Knerr, P.J.; Joewono, I.; Rusch, D.; Chisholm, S.W.; Van Der Donk, W.A. Catalytic Promiscuity in the Biosynthesis of Cyclic Peptide Secondary Metabolites in Planktonic Marine Cyanobacteria. *Proc. Natl. Acad. Sci. USA* **2010**, *107*, 10430–10435. [CrossRef] [PubMed]
76. Knerr, P.J.; Van Der Donk, W.A. Discovery, Biosynthesis, and Engineering of Lantipeptides. *Annu. Rev. Biochem.* **2012**, *81*, 479–505. [CrossRef] [PubMed]
77. Cubillos-Ruiz, A.; Berta-Thompson, J.W.; Becker, J.W.; Van Der Donk, W.A.; Chisholm, S.W. Evolutionary Radiation of Lanthipeptides in Marine Cyanobacteria. *Proc. Natl. Acad. Sci. USA* **2017**, *114*, E5424–E5433. [CrossRef]
78. Breitmaier: Terpenes: Importance, General Structure. Available online: https://scholar.google.com/scholar_lookup?title=Terpenes:+Importance,+general+structure,+and+biosynthesis&author=Breitmaier,+E.&publication_year=2006&journal=Terpenes+Flavors+Fragr.+Pharmaca+Pheromones&volume=1&pages=1--3 (accessed on 22 September 2022).
79. Kandi, S.; Godishala, V.; Rao, P.; Biomedicine, K.R. Biomedical Significance of Terpenes: An Insight. *Citeseer* **2015**, *3*, 8–10.
80. Jahnke, L.L.; Embaye, T.; Hope, J.; Turk, K.A.; van Zuilen, M.; des Marais, D.J.; Farmer, J.D.; Summons, R.E. Lipid Biomarker and Carbon Isotopic Signatures for Stromatolite-Forming, Microbial Mat Communities and Phormidium Cultures from Yellowstone National Park. *Geobiology* **2004**, *2*, 31–47. [CrossRef]
81. Summons, R.E.; Jahnke, L.L.; Hope, J.M.; Logan, G.A. 2-Methylhopanoids as Biomarkers for Cyanobacterial Oxygenic Photosynthesis. *Nature* **1999**, *400*, 554–557. [CrossRef]
82. Agger, S.A.; Lopez-Gallego, F.; Hoye, T.R.; Schmidt-Dannert, C. Identification of Sesquiterpene Synthases from Nostoc Punctiforme PCC 73102 and *Nostoc* sp. Strain PCC 7120. *J. Bacteriol.* **2008**, *190*, 6084–6096. [CrossRef]
83. Vítek, P.; Ascaso, C.; Artieda, O.; Casero, M.C.; Wierzchos, J. Discovery of Carotenoid Red-Shift in Endolithic Cyanobacteria from the Atacama Desert. *Sci. Rep.* **2017**, *7*, 11116. [CrossRef]
84. Boucar, M.C.M.; Shen, L.Q.; Wang, K.; Zhang, Z.C.; Qiu, B.S. UV-B Irradiation Enhances the Production of Unique Mycosporine-like Amino Acids and Carotenoids in the Subaerial Cyanobacterium *Pseudanabaena* sp. CCNU1. *Eur. J. Phycol.* **2021**, *56*, 316–323. [CrossRef]
85. Kusama, Y.; Inoue, S.; Jimbo, H.; Takaichi, S.; Sonoike, K.; Hihara, Y.; Nishiyama, Y. Zeaxanthin and Echinenone Protect the Repair of Photosystem II from Inhibition by Singlet Oxygen in *Synechocystis* sp. PCC 6803. *Plant Cell Physiol.* **2015**, *56*, 906–916. [CrossRef] [PubMed]
86. Dienst, D.; Wichmann, J.; Mantovani, O.; Rodrigues, J.S.; Lindberg, P. High Density Cultivation for Efficient Sesquiterpenoid Biosynthesis in *Synechocystis* sp. PCC 6803. *Sci. Rep.* **2020**, *10*, 5932. [CrossRef] [PubMed]
87. Höckelmann, C.; Becher, P.G.; Von Reuß, S.H.; Jüttner, F. Sesquiterpenes of the Geosmin-Producing Cyanobacterium Calothrix PCC 7507 and Their Toxicity to Invertebrates. *Z. Fur Naturforsch. Sect. C J. Biosci.* **2009**, *64*, 49–55. [CrossRef]
88. Lopes, G.; Clarinha, D.; Vasconcelos, V. Carotenoids from Cyanobacteria: A Biotechnological Approach for the Topical Treatment of Psoriasis. *Microorganisms* **2020**, *8*, 302. [CrossRef]
89. Cabanillas, A.H.; Tena Pérez, V.; Maderuelo Corral, S.; Rosero Valencia, D.F.; Martel Quintana, A.; Ortega Doménech, M.; Rumbero Sánchez, Á. Cybastacines A and B: Antibiotic Sesterterpenes from a *Nostoc* sp. Cyanobacterium. *J. Nat. Prod.* **2018**, *81*, 410–413. [CrossRef]
90. Mo, S.; Krunic, A.; Pegan, S.D.; Franzblau, S.G.; Orjala, J. An Antimicrobial Guanidine-Bearing Sesterterpene from the Cultured Cyanobacterium *Scytonema* sp. *J. Nat. Prod.* **2009**, *72*, 2043–2045. [CrossRef]
91. Kulasooriya, S.A.; Magana-Arachchi, D.N. Nitrogen Fixing Cyanobacteria: Their Diversity, Ecology and Utilization with Special Reference to Rice Cultivation. *J. Natl. Sci. Found. Sri Lanka* **2016**, *44*, 111–128. [CrossRef]
92. Singh, J.S.; Kumar, A.; Rai, A.N.; Singh, D.P. Cyanobacteria: A Precious Bio-Resource in Agriculture, Ecosystem, and Environmental Sustainability. *Front. Microbiol.* **2016**, *7*, 529. [CrossRef]
93. Bajpai, V.K.; Shukla, S.; Kang, S.M.; Hwang, S.K.; Song, X.; Huh, Y.S.; Han, Y.K. Developments of Cyanobacteria for Nano-Marine Drugs: Relevance of Nanoformulations in Cancer Therapies. *Mar. Drugs* **2018**, *16*, 179. [CrossRef]
94. Deviram, G.; Mathimani, T.; Anto, S.; Ahamed, T.S.; Ananth, D.A.; Pugazhendhi, A. Applications of Microalgal and Cyanobacterial Biomass on a Way to Safe, Cleaner and a Sustainable Environment. *J. Clean. Prod.* **2020**, *253*, 119770. [CrossRef]
95. de Farias Silva, C.E.; Bertucco, A. Bioethanol from Microalgae and Cyanobacteria: A Review and Technological Outlook. *Process Biochem.* **2016**, *51*, 1833–1842. [CrossRef]

96. Pulz, O.; Gross, W. Valuable Products from Biotechnology of Microalgae. *Appl. Microbiol. Biotechnol.* **2004**, *65*, 635–648. [CrossRef] [PubMed]
97. Mourelle, M.L.; Gómez, C.P.; Legido, J.L. The Potential Use of Marine Microalgae and Cyanobacteria in Cosmetics and Thalassotherapy. *Cosmetics* **2017**, *4*, 46. [CrossRef]
98. Angermayr, S.A.; Hellingwerf, K.J.; Lindblad, P.; Teixeira de Mattos, M.J. Energy Biotechnology with Cyanobacteria. *Curr. Opin. Biotechnol.* **2009**, *20*, 257–263. [CrossRef]
99. Abed, R.M.M.; Dobretsov, S.; Sudesh, K. Applications of Cyanobacteria in Biotechnology. *J. Appl. Microbiol.* **2009**, *106*, 1–12. [CrossRef]
100. Markou, G.; Chentir, I.; Tzovenis, I. Microalgae and Cyanobacteria as Food: Legislative and Safety Aspects. *Cult. Microalgae Food Ind. Curr. Potential Appl.* **2021**, 249–264. [CrossRef]
101. Khan, Z.; Bhadouria, P.; Bisen, P. Nutritional and Therapeutic Potential of *Spirulina*. *Curr. Pharm. Biotechnol.* **2005**, *6*, 373–379. [CrossRef]
102. Campanella, L.; Russo, M.; Chim, P.A.-A. Free and Total Amino Acid Composition in Blue-Green Algae. *Ann. Chim* **2002**, *92*, 343–352.
103. Marsan, D.W.; Conrad, S.M.; Stutts, W.L.; Parker, C.H.; Deeds, J.R. Evaluation of Microcystin Contamination in Blue-Green Algal Dietary Supplements Using a Protein Phosphatase Inhibition-Based Test Kit. *Heliyon* **2018**, *4*, e00573. [CrossRef]
104. Omatola, C.A.; Okolo, M.L.O.; Adaji, D.M.; Mofolorunsho, C.K.; Oyiguh, J.A.; Zige, D.V.; Akpala, N.S.; Samson, S.O. Coinfection of Human Immunodeficiency Virus-Infected Patients with Hepatitis B Virus in Lokoja, North Central Nigeria. *Viral Immunol.* **2020**, *33*, 391–395. [CrossRef] [PubMed]
105. Gregoire, E.; Pirotte, B.F.; Moerman, F.; Altdorfer, A.; Gaspard, L.; Firre, E.; Moonen, M.; Darcis, G. Mycobacterium Avium Complex and Cryptococcus Neoformans Co-Infection in a Patient with Acquired Immunodeficiency Syndrome: A Case Report. *Acta Clin. Belgica Int. J. Clin. Lab. Med.* **2022**, *77*, 679–684. [CrossRef] [PubMed]
106. Qamar, H.; Hussain, K.; Soni, A.; Khan, A.; Hussain, T.; Chénais, B. Cyanobacteria as Natural Therapeutics and Pharmaceutical Potential: Role in Antitumor Activity and as Nanovectors. *Molecules* **2021**, *26*, 247. [CrossRef] [PubMed]
107. Mayer, A.M.S.; Rodríguez, A.D.; Berlinck, R.G.S.; Hamann, M.T. Marine Pharmacology in 2005-6: Marine Compounds with Anthelmintic, Antibacterial, Anticoagulant, Antifungal, Anti-Inflammatory, Antimalarial, Antiprotozoal, Antituberculosis, and Antiviral Activities; Affecting the Cardiovascular, Immune and Nervous Systems, and Other Miscellaneous Mechanisms of Action. *Biochim. Biophys. Acta* **2009**, *1790*, 283–308. [PubMed]
108. Wase, N.V.; Wright, P.C. Systems Biology of Cyanobacterial Secondary Metabolite Production and Its Role in Drug Discovery. *Expert Opin. Drug Discov.* **2008**, *3*, 903–929. [CrossRef] [PubMed]
109. Dittmann, E.; Neilan, B.; Börner, T. Molecular Biology of Peptide and Polyketide Biosynthesis in Cyanobacteria. *Appl. Microbiol. Biotechnol.* **2001**, *57*, 467–473.
110. Volk, R.B. Screening of Microalgae for Species Excreting Norharmane, a Manifold Biologically Active Indole Alkaloid. *Microbiol. Res.* **2008**, *163*, 307–313. [CrossRef]
111. Sato, S.I.; Murata, A.; Orihara, T.; Shirakawa, T.; Suenaga, K.; Kigoshi, H.; Uesugi, M. Marine Natural Product Aurilide Activates the OPA1-Mediated Apoptosis by Binding to Prohibitin. *Chem. Biol.* **2011**, *18*, 131–139. [CrossRef]
112. Han, B.; Gross, H.; Goeger, D.E.; Mooberry, S.L.; Gerwick, W.H. Aurilides B and C, Cancer Cell Toxins from a Papua New Guinea Collection of the Marine Cyanobacterium Lyngbya Majuscula. *J. Nat. Prod.* **2006**, *69*, 572–575. [CrossRef]
113. Watanabe, A.; Ohno, O.; Morita, M.; Inuzuka, T.; Suenaga, K. Structures and Biological Activities of Novel Biselyngbyaside Analogs Isolated from the Marine Cyanobacterium *Lyngbya* sp. *Bull. Chem. Soc. Jpn.* **2015**, *88*, 1256–1264. [CrossRef]
114. Teruya, T.; Sasaki, H.; Kitamura, K.; Nakayama, T.; Suenaga, K. Biselyngbyaside, a Macrolide Glycoside from the Marine Cyanobacterium *Lyngbya* sp. *Org. Lett.* **2009**, *11*, 2421–2424. [CrossRef] [PubMed]
115. Kang, H.K.; Choi, M.C.; Seo, C.H.; Park, Y. Therapeutic Properties and Biological Benefits of Marine-Derived Anticancer Peptides. *Int. J. Mol. Sci.* **2018**, *19*, 919. [CrossRef] [PubMed]
116. Borbély, A.; Figueras, E.; Martins, A.; Esposito, S.; Auciello, G.; Monteagudo, E.; Di Marco, A.; Summa, V.; Cordella, P.; Perego, R.; et al. Synthesis and Biological Evaluation of RGD−Cryptophycin Conjugates for Targeted Drug Delivery. *Pharmaceutics* **2019**, *11*, 151. [CrossRef] [PubMed]
117. Hermann, P.; Armant, M.; Brown, E.; Rubio, M.; Ishihara, H.; Ulrich, D.; Caspary, R.G.; Lindberg, F.P.; Armitage, R.; Maliszewski, C.; et al. The Vitronectin Receptor and Its Associated CD47 Molecule Mediates Proinflammatory Cytokine Synthesis in Human Monocytes by Interaction with Soluble CD23. *J. Cell Biol.* **1999**, *144*, 767–775. [CrossRef]
118. Robles-Bañuelos, B.; Durán-Riveroll, L.M.; Rangel-López, E.; Pérez-López, H.I.; González-Maya, L. Marine Cyanobacteria as Sources of Lead Anticancer Compounds: A Review of Families of Metabolites with Cytotoxic, Antiproliferative, and Antineoplastic Effects. *Molecules* **2022**, *27*, 4814. [CrossRef]
119. Thornburg, C.C.; Cowley, E.S.; Sikorska, J.; Shaala, L.A.; Ishmael, J.E.; Youssef, D.T.A.; McPhail, K.L. Apratoxin H and Apratoxin A Sulfoxide from the Red Sea Cyanobacterium Moorea Producens. *J. Nat. Prod.* **2013**, *76*, 1781. [CrossRef]
120. Tiwari, A.K.; Tiwari, B.S. Cyanotherapeutics: An emerging field for future drug discovery. *Appl. Phycol.* **2020**, *1*, 44–57. [CrossRef]
121. Vijayakumar, S.; Menakha, M. Pharmaceutical Applications of Cyanobacteria—A Review. *J. Acute Med.* **2015**, *5*, 15–23. [CrossRef]

122. Favas, R.; Morone, J.; Martins, R.; Vasconcelos, V.; Lopes, G. Cyanobacteria Secondary Metabolites as Biotechnological Ingredients in Natural Anti-Aging Cosmetics: Potential to Overcome Hyperpigmentation, Loss of Skin Density and UV Radiation-Deleterious Effects. *Mar. Drugs* **2022**, *20*, 183. [CrossRef]
123. Morone, J.; Lopes, G.; Oliveira, B.; Vasconcelos, V.; Martins, R. Cyanobacteria in Cosmetics: A Natural Alternative for Anti-Aging Ingredients. *Pharmacol. Potential Cyanobacteria* **2022**, 257–286. [CrossRef]
124. Samadhiya, K.; Sangtani, R.; Nogueira, R.; Bala, K. Insightful Advancement and Opportunities for Microbial Bioplastic Production. *Front. Microbiol.* **2022**, *12*, 3755. [CrossRef] [PubMed]
125. Savakis, P.; Hellingwerf, K.J. Engineering Cyanobacteria for Direct Biofuel Production from CO_2. *Curr. Opin. Biotechnol.* **2015**, *33*, 8–14. [CrossRef] [PubMed]
126. Klanchui, A.; Raethong, N.; Prommeenate, P.; Vongsangnak, W.; Meechai, A. Cyanobacterial Biofuels: Strategies and Developments on Network and Modeling. *Adv. Biochem. Eng. Biotechnol.* **2017**, *160*, 75–102. [PubMed]
127. Kiran, B.; Kumar, R.; Deshmukh, D. Perspectives of Microalgal Biofuels as a Renewable Source of Energy. *Energy Convers. Manag.* **2014**, *88*, 1228–1244. [CrossRef]
128. Durall, C.; Kukil, K.; Hawkes, J.A.; Albergati, A.; Lindblad, P.; Lindberg, P. Production of Succinate by Engineered Strains of Synechocystis PCC 6803 Overexpressing Phosphoenolpyruvate Carboxylase and a Glyoxylate Shunt. *Microb. Cell Fact.* **2021**, 20–39. [CrossRef] [PubMed]
129. Successful Method Yielding High Rate of D-Lactate Using Cyanobacteria Could Revolutionize Bioplastic Production | Research at Kobe. Available online: https://www.kobe-u.ac.jp/research_at_kobe_en/NEWS/news/2020_03_10_02.html (accessed on 26 September 2022).
130. Hidese, R.; Matsuda, M.; Osanai, T.; Hasunuma, T.; Kondo, A. Malic Enzyme Facilitates d -Lactate Production through Increased Pyruvate Supply during Anoxic Dark Fermentation in *Synechocystis* sp. PCC 6803. *ACS Synth. Biol.* **2020**, *9*, 260–268. [CrossRef]
131. Coverage of Greenhouse Gas Emissions from Petroleum Use under Climate Policy—Center for Climate and Energy SolutionsCenter for Climate and Energy Solutions. Available online: https://www.c2es.org/document/coverage-of-greenhouse-gas-emissions-from-petroleum-use-under-climate-policy/ (accessed on 26 September 2022).
132. What Are Greenhouse Gases | GHG Emissions & Canada's Carbon Footprint. Available online: https://www.capp.ca/explore/greenhouse-gas-emissions/ (accessed on 26 September 2022).
133. Doamekpor, L.K.; Onwona-Agyeman, R.; Ameka, G.K. Bioenergy: Biodiesel from Freshwater Green Microalgae and a Cyanobacterium Occurring in Ghana. *West Afr. J. Appl. Ecol.* **2020**, *27*, 51–60.
134. Nozzi, N.E.; Oliver, J.W.K.; Atsumi, S. Cyanobacteria as a Platform for Biofuel Production. *Front. Bioeng. Biotechnol.* **2013**, *1*, 7. [CrossRef] [PubMed]
135. Cyanobacteria: Recent Advances in Taxonomy and Applications–Google Books. Available online: https://books.google.pl/books?id=sbZaEAAAQBAJ&pg=PA30&lpg=PA30&dq=cyanobacteria+capable+of+growing+in+brackish+or+saltwater+to+produce+a+range+of+biofuels+such+as+ethanol,+biodiesel,+gasoline+and+jet+fuel+in+addition+to+other+valuable+chemicals&source=bl&ots=ZrzLnyti-S&sig=ACfU3U0LletVZ4G5VWJpux2zCP8AojcWOQ&hl=en&sa=X&ved=2ahUKEwirhNDE27L6AhXzhP0HHc-uA7kQ6AF6BAgCEAM#v=onepage&q=cyanobacteriacapableofgrowinginbrackishorsaltwatertoproducearangeofbiofuelssuchasethanol%2Cbiodiesel%2Cgasolineandjetfuelinadditiontoothervaluablechemicals&f=false (accessed on 26 September 2022).
136. Joule Unlimited | Solving the Energy Crisis with Affordable, Renewable Clean Fuel. Available online: https://www.jouleunlimited.com/ (accessed on 26 September 2022).
137. The Algenol Advantage | Algenol Biotech. Available online: https://www.algenol.com/ (accessed on 26 September 2022).
138. Tom, A.P.; Jayakumar, J.S.; Biju, M.; Somarajan, J. Aquaculture wastewater treatment technologies and their sustainability: A review. *Energy Nexus* **2021**, *4*, 100022. [CrossRef]
139. Srimongkola, P.; Thongchul, N.; Phunpruchd, S.; Karnchanatat, A. Ability of marine cyanobacterium *Synechococcus* sp. VDW to remove ammonium from brackish aquaculture wastewater. *Agric. Water Manag.* **2019**, *212*, 155–161. [CrossRef]
140. Callieri, C. *Synechococcus* plasticity under environmental changes. *FEMS Microbiol. Lett.* **2017**, *364*, fnx229. [CrossRef] [PubMed]
141. Potnis, A.A.; Raghavan, P.S.; Rajaram, H. Overview on cyanobacterial exopolysaccharides and biofilms: Role in bioremediation. *Rev. Environ. Sci. Biotechnol.* **2021**, *20*, 781–794. [CrossRef]
142. Papadopoulos, K.P.; Economou, C.N.; Dailianis, S.; Charalampous, N.; Stefanidou, N.; Moustaka-Gouni, M.; Tekerlekopoulou, A.G.; Vayenasa, D.V. Brewery wastewater treatment using cyanobacterial-bacterial settleable aggregates. *Algal. Res.* **2020**, *49*, 101957. [CrossRef]
143. Papadopoulos, K.P.; Economou, C.N.; Tekerlekopoulou, A.G.; Vayenasa, D.V. Two-step treatment of brewery wastewater using electrocoagulation and cyanobacteria-based cultivation aggregates. *J. Environ. Manag.* **2020**, *265*, 110543. [CrossRef] [PubMed]
144. Papadopoulos, K.P.; Economou, C.N.; Tekerlekopoulou, A.G.; Vayenasa, D.V. A Cyanobacteria-Based Biofilm System for Advanced Brewery Wastewater Treatment. *Appl. Sci.* **2021**, *11*, 174. [CrossRef]
145. Sen, S.; Bhardwaj, K.; Thakurta, S.G.; Chakrabarty, J.; Ghanta, K.C.; Dutta, S. Phycoremediation of cyanide from coke–oven wastewater using cyanobacterial consortium. *Int. J. Environ. Sci. Technol.* **2018**, *15*, 2151–2164. [CrossRef]
146. Arias, D.M.; Uggetti, E.; García, J. Assessing the potential of soil cyanobacteria for simultaneous wastewater treatment and carbohydrate-enriched biomass production. *Algal. Res.* **2020**, *51*, 102042. [CrossRef]

147. El-Bestawy, E.A.; Abd El-Salam, A.Z.; Mansy, A.E. Potential use of environmental cyanobacterial species in bioremediation of lindane-contaminated effluents. *Int. Biodeterior. Biodegrad.* **2007**, *59*, 180–192. [CrossRef]
148. Baptista, M.; Vasconcelos, T. Cyanobacteria metal interactions: Requirements, toxicity, and ecological implications. *Crit. Rev. Microbiol.* **2006**, *32*, 127–137. [CrossRef] [PubMed]
149. Cui, J.; Xie, Y.; Sun, T.; Chen, L.; Zhang, W. Deciphering and engineering photosynthetic cyanobacteria for heavy metal bioremediation. *Sci. Total Environ.* **2021**, *761*, 144111. [CrossRef] [PubMed]
150. Kumar, V.; Dwivedi, S.K. A review on accessible techniques for removal of hexavalent Chromium and divalent Nickel from industrial wastewater: Recent research and future outlook. *J. Clean. Prod.* **2021**, *295*, 126229. [CrossRef]
151. Sena, S.; Dutta, S.; Guhathakurata, S.; Chakrabarty, J.; Nandi, S.; Dutta, A. Removal of Cr(VI) using a cyanobacterial consortium and assessment of biofuel production. *Int. Biodeterior. Biodegrad.* **2017**, *119*, 211–224. [CrossRef]
152. Shukla, D.; Vankar, P.S.; Srivastava, S.K. Bioremediation of hexavalent chromium by a cyanobacterial mat. *Appl. Water Sci.* **2012**, *2*, 245–251. [CrossRef]
153. Fatima, G.; Raza, A.M.; Hadi, N.; Nigam, N.; Mahdi, A.A. Cadmium in Human Diseases: It's More than Just a Mere Metal. *Indian J. Clin. Biochem.* **2019**, *34*, 371–378. [CrossRef] [PubMed]
154. Raghavana, P.S.; Potnis, A.A.; Bhattacharyya, K.; Salaskar, D.A.; Rajaram, H. Axenic cyanobacterial (*Nostoc muscorum*) biofilm as a platform for Cd(II) sequestration from aqueous solutions. *Algal. Res.* **2020**, *46*, 101778. [CrossRef]
155. Ghorbani, E.; Nowruzi, B.; Nezhadali, M.; Hekmat, M. Metal removal capability of two cyanobacterial species in autotrophic and mixotrophic mode of nutrition. *BMC Microbiol.* **2022**, *22*, 58. [CrossRef]
156. Bon, I.C.; Salvatierra, L.C.; Lario, L.D.; Morató, J.; Pérez, L.M. Prospects in Cadmium-ContaminatedWater Management Using Free-Living Cyanobacteria (*Oscillatoria* sp.). *Water* **2021**, *13*, 542. [CrossRef]
157. Fawzy, M.A.; Mohamed, A.K.S.H. Bioremediation of heavy metals from municipal sewage by cyanobacteria and its effects on growth and some metabolites of *Beta vulgaris*. *J. Plant. Nutr.* **2017**, *40*, 2550–2561. [CrossRef]
158. Akoijam, C.; Langpoklakpam, J.S.; Chettri, B.; Singh, K.M. Cyanobacterial diversity in hydrocarbon-polluted sediments and their possible role in bioremediation. *Int. Biodeterior. Biodegrad.* **2015**, *103*, 97–104. [CrossRef]
159. Forlani, G.; Prearo, V.; Wieczorek, D.; Kafarski, P.; Lipok, J. Phosphonate degradation by *Spirulina* strains: Cyanobacterial biofilters for the removal of anticorrosive polyphosphonates from wastewater. *Enzyme Microb. Technol.* **2011**, *48*, 299–305. [CrossRef]
160. Thakurta, S.G.; Aakula, M.; Chakrabarty, J.; Dutta, S. Bioremediation of phenol from synthetic and real wastewater using *Leptolyngbya* sp.: A comparison and assessment of lipid production. *3 Biotech* **2018**, *8*, 206. [CrossRef] [PubMed]
161. Ibraheem, I.B.M. Biodegradability of hydrocarbons by cyanobacteria. *J. Phycol.* **2010**, *46*, 818–824. [CrossRef]
162. El-Sheekh, M.M.; Gharieb, M.M.; Abou-El-Souod, G.W. Biodegradation of dyes by some green algae and cyanobacteria. *Int. Biodeterior. Biodegrad.* **2009**, *63*, 699–704. [CrossRef]
163. Dellamatrice, P.M.; Silva-Stenico, M.E.; Beraldo de Moraes, L.A.; Fiore, M.F.; Rosim Monteiro, R.T. Degradation of textile dyes by cyanobacteria. *Braz. J. Microbiol.* **2017**, *48*, 25–31. [CrossRef]
164. Bonini, C.; Righi, G. Enantio-and stereo-selective route to the taxol side chain via asymmetric epoxidation of trans-cinnamyl alcohol and subsequent expoxide ring opening. *J. Chem. Soc. Chem. Commun.* **1994**, *24*, 2767–2768. [CrossRef]
165. Yamanaka, R.; Nakamura, K.; Murakami, M.; Murakami, A. Selective synthesis of cinnamyl alcohol by cyanobacterial photobiocatalysts. *Tetrahedron Lett.* **2015**, *56*, 1089–1091. [CrossRef]
166. Molinari, F.; Gandolfi, R.; Villa, R.; Occhiato, E.G. Lyophilised yeasts: Easy-to-handle biocatalysts for stereoselective reduction of ketones. *Tetrahedron Asymmetr.* **1999**, *10*, 3515–3520. [CrossRef]
167. Głąb, A.; Szmigiel-Merena, B.; Brzezińska-Rodak, M.; Żymańczyk-Duda, E. Biotransformation of 1- and 2-phenylethanol to products of high value via redox reactions. *BioTechnologia* **2016**, *97*, 203–210. [CrossRef]
168. Chang, X.; Yang, Z.; Zeng, R.; Yang, G.; Yan, J. Production of chiral aromatic alcohol by asymmetric reduction with vegetable catalyst. *Chin. J. Chem. Eng.* **2010**, *18*, 1029–1033. [CrossRef]
169. Yang, Z.-H.; Luo, L.; Chang, X.; Zhou, W.; Chen, G.-H.; Zhao, Y.; Wang, Y.-J. Production of chiral alcohols from prochiral ketones by microalgal photo-biocatalytic asymmetric reduction reaction. *J. Ind. Microbiol. Biotechnol.* **2012**, *39*, 835–841. [CrossRef]
170. Żymańczyk-Duda, E.; Głąb, A.; Górak, M.; Klimek-Ochab, M.; Brzezińska-Rodak, M.; Strub, D.; Śliżewska, A. Reductive capabilities of different cyanobacterial strains towards acetophenone as a model substrate—Prospect of applications for chiral building blocks synthesis. *Bioorg. Chem.* **2019**, *93*, 102810. [CrossRef] [PubMed]
171. Patel, D.V.; Rielly-Gauvin, K.; Ryono, D.E.; Free, C.A.; Rogers, W.L.; Smith, S.A.; DeForrest, J.M.; Oehl, R.S.; Petrillo, E.W., Jr. Alpha-Hydroxy phosphinyl-based inhibitors of human renin. *J. Med. Chem.* **1995**, *38*, 4557–4569. [CrossRef]
172. Stowasser, B.; Budt, K.-H.; Jian-Qi, L.; Peyman, A. New hybrid transition state analog inhibitors of HIV protease with peripheric C2-symmetry. *Tetrahedron Lett.* **1992**, *33*, 6625. [CrossRef]
173. Yokomatsu, T.; Murano, T.; Akiyama, T.; Koizumi, J.; Shimeno, H.; Tsuji, Y.; Soeda, S.; Shimeno, H. Improved Synthesis of 1,3-Propanediol Derivatives Having a Diethoxyphosphoryldifluoroethyl Functional Group at the 2-Position: Application to Chemoenzymatic Synthesis of Novel Acyclic Nucleotide Analogues of Adenosine Bisphosphates. *Bioorg. Med. Chem. Lett.* **2003**, *13*, 229. [CrossRef] [PubMed]
174. Żymańczyk-Duda, E.; Kafarski, P.; Lejczak, B. Reductive biotransformation of diethyl beta-, gamma- and delta-oxoalkylphosphonates by cells of baker's yeast. *Enzyme Microb. Technol.* **2000**, *26*, 265. [CrossRef]

175. Żymańczyk-Duda, E.; Brzezińska-Rodak, M.; Klimek-Ochab, M.; Latajka, R.; Kafarski, P.; Lejczak, B. Chiral O-phosphorylated derivative of 2-hydroxy-phenylethylphosphonate as a valuable product of microbial biotransformation of diethyl 2-oxo-2-phenylethylphosphonate. *J. Mol. Catal. B Enzym.* **2008**, *52*, 74. [CrossRef]
176. Górak, M.; Żymańczyk-Duda, E. Application of cyanobacteria for chiral phosphonate synthesis. *Green Chem.* **2015**, *17*, 4570. [CrossRef]
177. Górak, M.; Żymańczyk-Duda, E. Reductive activity of free and immobilized cells of cyanobacteria towar oxophosphonates—Comparative study. *J. Appl. Phycol.* **2017**, *29*, 245–253. [CrossRef]
178. Harrison, W.; Huang, X.; Zhao, H. Photobiocatalysis for abiological transformations. *Acc. Chem. Res.* **2022**, *55*, 1087–1096. [CrossRef]
179. Emmanuel, M.A.; Greenberg, N.R.; Oblinsky, D.G.; Hyster, T.K. Accessing non-natural reactivity by irradiating nicotinamide dependent enzymes with light. *Nature* **2016**, *540*, 414–417. [CrossRef]
180. Biegasiewicz, K.F.; Cooper, S.J.; Gao, X.; Oblinsky, D.G.; Kim, J.H.; Garfinkle, S.E.; Joyce, L.A.; Sandoval, B.A.; Scholes, G.D.; Hyster, T.K. Photoexcitation of flavoenzymes enables a stereoselective radical cyclization. *Science* **2019**, *364*, 1166–1169. [CrossRef] [PubMed]
181. Gao, X.; Turek-Herman, J.R.; Joo Choi, Y.; Cohen, R.; Hyster, T. Photoenzymatic synthesis of α-tertiary amines by engineered flavin-dependent 'ene'-reductases. *J. Am. Chem. Soc.* **2021**, *143*, 19643–19647. [CrossRef] [PubMed]
182. Sandoval, B.A.; Clayman, P.D.; Oblinsky, D.G.; Oh, S.; Nakano, Y.; Bird, M.; Scholes, G.D.; Hyster, T.K. Photoenzymatic reductions enabled by direct excitation of flavin-dependent "ene"- reductases. *J. Am. Chem. Soc.* **2021**, *143*, 1735–1739. [CrossRef] [PubMed]
183. Sorigue, D.; Legeret, B.; Cuine, S.; Blangy, S.; Moulin, S.; Billon, E.; Richaud, P.; Brugiere, S.; Coute, Y.; Nurizzo, D.; et al. An algal photoenzyme converts fatty acids to hydrocarbons. *Science* **2017**, *357*, 903–907. [CrossRef] [PubMed]
184. Huijbers, M.M.E.; Zhang, W.; Tonin, F.; Hollman, F. Light-Driven Enzymatic Decarboxylation of Fatty Acids. *Angew. Chem. Int. Ed.* **2018**, *57*, 13648–13651. [CrossRef] [PubMed]
185. Zhang, W.; Ma, M.; Huijbers, M.M.E.; Filonenko, G.A.; Pidko, E.A.; van Schie, M.; de Boer, S.; Burek, B.O.; Bloh, J.Z.; van Berkel, W.J.H.; et al. Hydrocarbon Synthesis via Photoenzymatic Decarboxylation of Carboxylic Acids. *J. Am. Chem. Soc.* **2019**, *141*, 3116–3120. [CrossRef] [PubMed]
186. Ma, M.; Zhang, X.; Zhang, W.; Li, P.; Hollmann, F.; Wang, Y. Photoenzymatic production of next generation biofuels from natural triglycerides combining a hydrolase and a photodecarboxylase. *ChemPhotoChem* **2020**, *4*, 39–44. [CrossRef]

microorganisms

MDPI

Article

Bioactive Potential of Two Marine Picocyanobacteria Belonging to *Cyanobium* and *Synechococcus* Genera

Patrizia Pagliara [1,*], Giuseppe Egidio De Benedetto [2,3], Matteo Francavilla [4], Amilcare Barca [1] and Carmela Caroppo [5,*]

1 Department of Biological and Environmental Sciences and Technologies, University of Salento, Via Provin-Ciale Lecce-Monteroni, 73100 Lecce, Italy; amilcare.barca@unisalento.it
2 Laboratory of Analytical and Isotopic Mass Spectrometry, Department of Cultural Heritage, University of Salento, 73100 Lecce, Italy; giuseppe.debenedetto@unisalento.it
3 National Research Council, Institute of Heritage Sciences (CNR-ISPC), 73100 Lecce, Italy
4 STAR*Facility Centre, Department of Agriculture, Foods, Natural Resources and Engineering, University of Foggia, 71122 Foggia, Italy; matteo.francavilla@unifg.it
5 National Research Council, Water Research Institute (CNR-IRSA), 74123 Taranto, Italy
* Correspondence: patrizia.pagliaral@unisalento.it (P.P.); carmela.caroppo@irsa.cnr.it (C.C.)

Abstract: Coccoid cyanobacteria produce a great variety of secondary metabolites, which may have useful properties, such as antibacterial, antiviral, anticoagulant or anticancer activities. These cyanobacterial metabolites have high ecological significance, and they could be considered responsible for the widespread occurrence of these microorganisms. Considering the great benefit derived from the identification of competent cyanobacteria for the extraction of bioactive compounds, two strains of picocyanobacteria (coccoid cyanobacteria < 3 μm) (*Cyanobium* sp. ITAC108 and *Synechococcus* sp. ITAC107) isolated from the Mediterranean sponge *Petrosia ficiformis* were analyzed. The biological effects of organic and aqueous extracts from these picocyanobacteria toward the nauplii of *Artemia salina*, sea urchin embryos and human cancer lines (HeLa cells) were evaluated. Methanolic and aqueous extracts from the two strains strongly inhibited larval development; on the contrary, in ethyl acetate and hexane extracts, the percentage of anomalous embryos was low. Moreover, all the extracts of the two strains inhibited HeLa cell proliferation, but methanol extracts exerted the highest activity. Gas chromatography–mass spectrometry analysis evidenced for the first time the presence of β-*N*-methylamino-L-alanine and microcystin in these picocyanobacteria. The strong cytotoxic activity observed for aqueous and methanolic extracts of these two cyanobacteria laid the foundation for the production of bioactive compounds of pharmacological interest.

Keywords: picocyanobacteria toxicity; *Cyanobium*; *Synechococcus*; *Artemia salina*; sea urchin embryos; HeLa cell line; microcystin; BMAA

Citation: Pagliara, P.; De Benedetto, G.E.; Francavilla, M.; Barca, A.; Caroppo, C. Bioactive Potential of Two Marine Picocyanobacteria Belonging to *Cyanobium* and *Synechococcus* Genera. *Microorganisms* **2021**, *9*, 2048. https://doi.org/10.3390/microorganisms9102048

Academic Editor: Johannes F. Imhoff

Received: 8 September 2021
Accepted: 22 September 2021
Published: 28 September 2021

1. Introduction

Picocyanobacteria (PCCs) (0.2–3 μm) [1] are a conspicuous component of the phytoplankton communities. Unicellular coccoid forms [2–4] and prochlorophytes [5,6] represent them. In marine environments, the most represented genera are *Synechococcus*, *Synechocystis*, *Cyanobium*, *Cyanobacterium* [7] and *Prochlorococcus* [8]. PCCs can be detected both in the free-living population in the water column and associated with other organisms (foraminifers, corals, sponges, mollusks) or substrates [9].

PCCs may contribute significantly (i.e., up to 50%) to phytoplankton productivity and biomass in marine waters [10] and they can be responsible for up to 98% of the total biomass in brackish systems [11,12]. PCCs play an important role in the functioning of the microbial loop [13,14], and they actively participate in the modulation of energy and matter flows as well as in the sustenance and development of higher trophic levels [15,16].

However, interest in these microorganisms has recently increased due to the property they have of producing a great variety of secondary metabolites, mainly isolated until now from the filamentous species [17–19]. Indeed, studies on PCC biomolecules have developed in more recent years [20–24]. The identified secondary metabolites may have useful properties, such as antibacterial, antiviral, anticoagulant or antitumor activities [25–28]. These cyanobacterial metabolites have high ecological significance and could be considered responsible for the widespread occurrence of these microorganisms [29]. Their ability to produce bioactive molecules could be associated with ultraviolet radiation protection, anti-predatory defense, allelopathy, resource competition and signaling [22,30]. As an example of anti-predatory defense, *Synechococcus* blooms are known to inhibit zooplankton grazing by producing extracellular polysaccharides [31]. Moreover, *Synechococcus* and *Synechocystis* marine strains are responsible for the production of the microcystin [32–34], a toxin considered highly toxic to potential grazers of cyanobacteria [35] and previously detected only in freshwater strains.

The secondary metabolites produced by PCCs comprise oligopeptides, cyanobactines, 2-methylisoborneol (MIB) and geosmin (1,2,7,7-tetramethyl-2-norborneol) (GSM) [20,22,24], some of which are cyanotoxins. The most frequently detected cyanotoxins are hepatotoxins (microcystins, nodularins), neurotoxins (β-methylamino-L-alanine, β-*N*-methylamino-L-alanine, i.e., BMAA) [36,37] and dermotoxins (lipopolysaccharide, LPS) [38].

To date, a limited number of studies explored these small forms of cyanobacteria to identify new bioactive compounds [24,27] also because their isolation is time consuming and costly [20]. However, some PPC species isolated from water samples and substrate collected on beaches were tested to evaluate their ability to produce secondary metabolites with promising applications in cosmetics [39] and pharmaceutics [22,27,40]. Despite the enormous range of biochemicals potentially available from numerous species, currently commercial production of cyanobacteria relies on few genera/species as *Arthrospira, Spirulina, Nostoc* and *Aphanizomenon flos-aquae*, primarily used as "health food" or added while manufacturing food supplements and food additives [41,42]. Furthermore, a potential commercial development of cyanobacterial compounds for non-biomedical applications, as herbicides, algicides and insecticides, must be considered [43].

Among the various investigations, the assessment of the cyanotoxin potential of marine PCCs living in association with sponges is particularly interesting, as these invertebrates are considered a huge source of bioactive molecules [44].

Previous investigations on *Cyanobium* and *Synechococcus* strains isolated from the marine sponge *P. ficiformis* demonstrated that these two strains are toxic to *Artemia salina* nauplii and can interfere with sea urchin (*Paracentrotus lividus*) embryonic development [28,45]. Furthermore, toxic effects of these PCC strains have also been evidenced on mussel hemocytes, in which chromatin condensation and fragmentation, typical signs of apoptosis, occurred after cyanobacteria challenge [46].

With the aim to investigate more in depth the bioactivity and chemical nature of secondary metabolites from *Cyanobium* sp. ITAC108 and *Synechococcus* sp. ITAC107, fractionated extracts with different solvent polarity were tested on *A. salina*, sea urchin embryos and human cancer cell lines. Moreover, chemical analyses (GC-MS and LC-MS) were performed to try to identify the compounds responsible for the extract's bioactivity.

2. Materials and Methods

2.1. *Cyanobacterial Biomass: Culture Conditions, Growth Curve and Harvest*

Two cyanobacterial strains, *Synechococcus* sp. ITAC107 and *Cyanobium* sp. ITAC108, previously isolated from the Mediterranean marine sponge *Petrosia ficiformis* [45] were investigated in this study as target species for biological activity screening. Cyanobacterial strain identification was based on both morphological [7,47,48] and molecular criteria as previously described [49].

Cyanobacterial cultures were grown in MN medium enriched with B12 vitamin (5 μg/L) [50]. Culture media was prepared with natural seawater filtered through glass fiber filters (Whatman GF/C, GE Healthcare Company, Maidstone, England) and autoclaved. The cultures were performed in 6 L flasks with 4 L of medium with constant aeration. They were incubated at 24.0 ± 1.0 °C under white fluorescent light at a photosynthetic photon flux of 20 µmol photon^{-2} s^{-1} [51] and an illumination cycle of 18D:6N.

Picocyanobacteria cells were counted under an epifluorescence microscope (AXIOSKOP ZEISS, Oberkochen, Germany). According to the Guillard and Sieracki [52] method, OD was measured spectrophotometrically at 750 nm with a Varian Cary UV–Visible Spectrophotometer. These data were used to fit a linear regression model between the variables N and OD. The linear correlation between N and OD for *Synechococcus* was y = 8E + 07x − 21,943 (R^2 = 0.99) and for *Cyanobium*, y = 4E + 06x − 35,670 (R^2 = 0.99), where y = N (mL^{-1}) and x = OD.

Cyanobacteria cells were harvested after seven days of growth (exponential phase) by centrifugation (5100 rpm for 20 min at 5 °C). Concentrated biomass was washed with distilled water to remove NaCl. The solid residue was separated from the supernatant, freeze-dried and stored at −20 °C under N_2 atmosphere in amber vials until the extraction process.

2.2. Fractionated Extraction of Cyanobacterial Biomass

A fractionated extraction was performed on cyanobacterial freeze-dried samples. The extraction procedure was performed according to Francavilla et al. [53]. The idea was to sequentially extract organic compounds from cyanobacterial biomass playing on their difference in polarity. Therefore, four solvents with different polarity indexes (P) were used to perform the extractions [54]. n-Hexane (P = 0.1), ethyl acetate (P = 4.4), methanol (P = 5.1) and water (P = 10.2) were selected as solvents. Briefly, 0.5 g of ground freeze-dried cyanobacterial biomass was homogenized at 17,000 rpm, using an UltraTurrax T18 IKA homogenizer (IKA®-WERKE GMBH & CO. KG, Staufen, Germany), twice each time with 5 mL of n-hexane at room temperature for 1 min followed by centrifugation. Further extractions in the similar manner were performed sequentially on the pellet using ethyl acetate, methanol and then deionized water (at 80 °C). The isolated supernatants for each solvent (hexane, ethyl acetate, methanol and aqueous) were combined and evaporated under vacuum. The dried extracts were weighed and stored at −20 °C until analysis. To evaluate the effect of the extracts on *A. salina* vitality and on sea urchin development, the dried extracts were dissolved in filtered (0.22 µm) sea water (FSW) to obtain a stock solution of 100 µg/mL.

2.3. Acute Toxicity Assay Using Nauplii of Artemia salina

Dried cysts of *A. salina* (JBL NovoTemia, Germany) were hatched in FSW (1 g cysts/L) at 25 °C under continuous illumination and aeration to obtain the nauplii. The *A. salina* nauplii were collected and transferred to 96-well microplates after 24 h of incubation. Ten individuals were transferred to each microplate well containing 100 µL of total volume. A two-fold serial dilution of extracts over 10 wells was performed. Cyanobacteria extracts, prepared as described in the previous paragraphs, were used starting with a dilution of stock solution (50 µg/mL). After 24 h of exposure at 25 °C in darkness, the number of dead larvae in each well was counted. Filtered seawater, methanol, hexane and ethyl acetate were used as negative control. Results are presented as percentage of mortality ± standard deviation (SD) and LC_{50} values was estimated using the Probits statistical method [55].

2.4. Sea Urchins and Embryo Toxicity Assay

Adult *P. lividus* sea urchins were collected during the breeding season by SCUBA diving in the Ionian Sea (E 18°00′47″, N 39°58′14″). The sea urchins were acclimated at least for 24 h in natural FSW at 18.0 ± 1 °C (salinity 38.0 ± 0.2‰, pH 8.0 ± 0.2). The field studies did not involve endangered or protected species. All animal procedures were in compliance with the guidelines of the European Union (directive 2010/63/U.E.).

P. lividus embryos were obtained as previously described [56]. Briefly, sperm solution was added to an egg suspension, and in vitro fertilization was assured by the observation of the vitelline membrane under a light microscope.

Twenty microliters of the cyanobacterial extracts were added to 20 mL of embryo culture immediately after the vitelline membrane elevation (within 5 min). After 24 and 48 h, aliquots of embryos suspension were fixed with cold methanol and observed under a light microscope (Nikon Eclipse 50i, Tokyo, Japan). One hundred embryos from each treatment were counted to obtain the percentage of normal embryos. Pictures were taken using a camera connected to the microscope.

Three replicates for each extract were performed. The results of the sea urchin embryo toxicity test are reported as the mean of the percentage of deformed embryos ± SD.

The plutei dimensions were evaluated by using the ImageJ program, and differences were obtained by the Student's *t*-test. Differences were indicated as statistically significant with P-values < 0.01.

2.5. *Cytotoxicity*

2.5.1. HeLa Cell Culture Conditions and Treatments

HeLa cells (human cervix adenocarcinoma; cell line ATCC® CCL 2™) were cultured in sterile conditions in 25 or 75 cm^2 plastic flasks with DMEM, supplemented with 10% (v/v) FBS, 2 mM L-glutamine and 100 μg/mL penicillin/streptomycin and maintained in 5 % CO_2 at 37 °C in an incubator (Thermo Fisher Scientific, Waltham, MA, USA). Cells were detached and harvested with a 0.3% (v/v) trypsin solution and then transferred to new flasks every 2–3 days (70–90% confluence) for propagation. The culture medium was changed every 2 days. All experiments were performed between passage 3 and 10 of propagation. Concentrated extracts were dissolved in DMSO, and the experiments were carried out incubating HeLa cells with DMEM containing 10 μg/mL of different solvent-free extracts. The final concentrations of DMSO did not exceed 0.05%. Cells incubated with fresh medium were used as control.

2.5.2. MTT Test

MTT assay was performed using different extracts (ethyl acetate, methanol and aqueous) from the two cyanobacteria on the HeLa cells.

Cells were seeded in 96-well plates (20 × 10^3 cells/well) and incubated for 24 h at 37 °C. After incubation, the medium was removed and replaced with a medium containing aqueous, methanol or acetate extracts (diluted 1:100 in the medium). After treatment, MTT solution (5 mg/mL in sterile filtered PBS, pH 7.4) was added to each well to reach a final concentration of 0.5 mg MTT/mL, and plates were incubated at 37 °C for 3 h. The dark-blue formazan crystals were then solubilized by cell lysis with 200 μL/well 2-propanol/HCl 4N, and absorbance was measured at 550 nm with a Multiskan Fc Microplate Photometer (Thermo Fisher Scientific, Waltham, MA, USA). Data were reported as percentage of control (mean ± SEM) of eight sample replicates per treatment. Three independent experiments were performed.

2.6. *Secondary Metabolite Determination*

Secondary metabolite analysis was performed by using a gas-chromatograph coupled with a single quadrupole mass spectrometer (GC-MS) apparatus (6890 GC coupled to a 5973 inert MSD, Agilent Technologies, Santa Clara, CA, USA) as already described [57]. Methanolic and aqueous extracts were first methoximated with 10 μL of a 20 mg/mL MeOX solution in pyridine at 40 °C for 90 min and then silylated with 90 μL of MSTFA at 70 °C for 60 min. After cooling at room temperature for 5 min, 1 μL of the solution was injected into the GC-MS, which operated in splitless mode.

The injector temperature was set at 280 °C. Metabolites were separated on a DB-1ht capillary column 30 m × i.d. 250 μm × 0.1 μm using a continuous flow rate of 1 mL/min of ultrapure helium. The column oven temperature was set at 50 °C for 2 min; then the temperature was increased from 50 to 350 °C with a ramp of 10 °C/min and a hold time of 10 min. The total run time was 42 min. The mass detector was operated at 70 eV in the electron impact (EI) ionization mode. The ion source and transfer line temperatures were set at 250 and 300 °C, respectively, whereas nominal mass scan spectra were acquired with a mass scan range of 50–550 m/z.

Fractionated extracts were also analyzed by gas chromatography coupled to high-resolution and high-accuracy mass spectrometry (Thermo Scientific™ Q Exactive™ GC Orbitrap™ GC-MS/MS system, GC-HRMS, Waltham, MA USA) using the same column and gas-chromatograph conditions. Ion source and transfer line temperatures were set at 330 and 280 °C, respectively, HRMS spectra were acquired with a mass scan range of 50–750 m/z, resolution (FWHM at m/z 200) was 60,000, mass accuracy 1 ppm using internal lock mass correction and 207.03235 m/z as lock mass. Low-resolution data were analyzed by Agilent ChemStation, whereas for HRMS measurements, deconvolution, feature identification and putative metabolite identification were carried out using Thermo TraceFinder 4.1 software. Peak deconvolution was automatically performed with TraceFinder, where NIST 2014 and GC-Orbitrap Metabolomics mass spectral libraries were used to annotate the peaks with a search index threshold of >700. Compound identification was made using a total confidence score that considers the NIST spectral match as well as the percentage of fragment ions that can be explained from the elemental composition of the molecular ion assigned by NIST.

Extracts were also analyzed by liquid chromatography coupled with mass spectrometry (LC-MS). A Surveyor MS Pump coupled with an LCQ DECA XP Plus (Thermo Finnigan, Thermo Fisher Scientific, Waltham, MA, USA) ion trap mass spectrometer, equipped with an ESI source, was used for our purposes [58]. A reverse-phase Thermo Scientific Biobasic-C18 column (100 mm × 2.1 mm i.d., particle size 5 µm) with a C18-Security Guard cartridge precolumn (10 mm × 2.1 mm i.d., particle size 5 µm) was used for chromatographic separation at room temperature (25 °C). The injection volume was set at 2 µL. A gradient binary elution was performed using solvent A (0.1% formic acid in water) and solvent B (0.1% formic acid in acetonitrile) at a constant flow of 0.2 mL/min.

The gradient program was set as 5% B (0–5 min), 5–95% B (5–40 min), 95% B (40–45 min), 5% B (45–50 min), 5% B (50–60 min). The mass spectrometer was run in the positive ion mode, and the capillary voltage was set at 3.5 kV. The ion trap scanned the 200–1400 m/z range during the separation and detection. Automatic gain control was employed using three microscans and a maximum injection time of 140 ms. Electrospray operating conditions were as follows: heated capillary temperature, 280 °C; sheath gas, 55 L min^{-1}; auxiliary gas 28 L min^{-1}. The spectrometer was calibrated externally with a mixture of caffeine, MRFA and Ultramark according to the manufacturer's instruction.

Data were acquired using Thermo Xcalibur software, whereas untargeted analysis was carried out using MS-DIAL [59].

3. Results

3.1. Fractionated Extraction Yields

The yield of fractionated extractions using solvents with different polarity index is showed in Figure 1. The increase in polarity from hexane to methanol was associated with an increase in the extraction yield. Therefore, the highest yields were found using methanol, but different values were observed between the strain *Synechococcus* sp. ITAC107 (87.6% d.w.) and *Cyanobium* sp. ITAC108 (54.4% d.w.). Interestingly, a further increase in polarity achieved by switching from methanol (P = 5.1) to water at 80 °C (P = 10.2) did not correspond to an increase in extraction yield. Moreover, the water extract was higher for *Cyanobium* sp. ITAC108 (44.1% d.w.) compared to *Synechococcus* sp. ITAC107 (10.6% d.w.). Hexane and ethyl acetate fractions showed lower extraction yields that ranged between 0.02% d.w. (*Cyanobium* sp. ITAC108) and 0.04% d.w. (*Synechococcus* sp. ITAC107) in hexane and 0.08% d.w. (*Cyanobium* sp. ITAC108) and 0.06% d.w. (*Synechococcus* sp. ITAC107) in ethyl acetate, respectively.

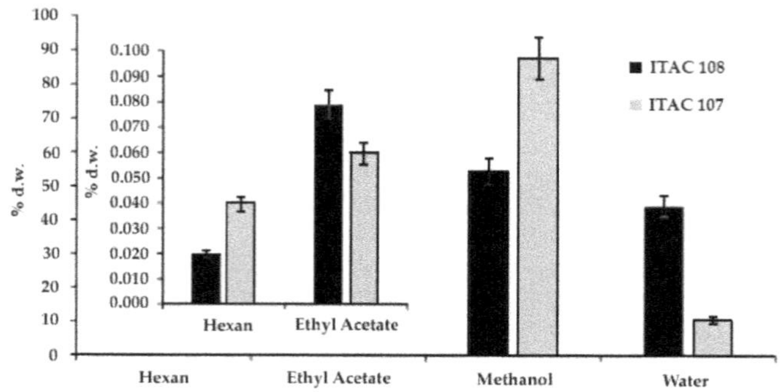

Figure 1. Fractionated extraction yield (expressed as % dry weight) of *Cyanobium* sp. ITAC108 and *Synechococcus* sp. ITAC107 biomass. Different solvents with different polarity indexes (hexane, ethyl acetate, methanol and water) were used.

3.2. Toxicity Test

3.2.1. *Artemia salina*

The toxicity test showed that the extracts from *Synechococcus* sp. ITAC107 and *Cyanobium* sp. ITAC108 were toxic to *Artemia* nauplii after 24 h in a dose-dependent manner. Most of the extracts were found to be toxic, inducing the maximum percentage (100%) of death at the maximum administered dose. As an exception, the ethyl acetate extract of *Cyanobium* sp. ITAC108 did not affect brine shrimp vitality. On the contrary, the most toxic sample was the water extract from *Cyanobium* sp. ITAC108 since its LC_{50} was 3.13 mg/mL (Table 1). For *Synechococcus* sp. ITAC107, we observed that the methanolic extract exerted a stronger activity (6.25 μg/mL) than polar and water fractions. In Table 1, data about the LC_{50} of all the fractions are reported.

Table 1. The 24 h LC_{50} values (with 95% confidence limits) for *A. salina* exposed to cyanobacterial extracts.

Strains	LC_{50} 24 h μg/mL	95% Confidence Limits
Cyanobium sp. ITAC108 MetOH	6.25	4.71–8.29
Cyanobium sp. ITAC108 Water	3.13	1.87–4.76
Cyanobium sp. ITAC108 Hexane	9	7–11
Cyanobium sp. ITAC108 Ethyl Acetate	-	-
Synechococcus sp. ITAC107 MetOH	6.25	3.96–9.86
Synechococcus sp. ITAC107 Water	10.19	5.93–17.55
Synechococcus sp. ITAC107 Hexane	7	4–11
Synechococcus sp. ITAC107 Ethyl Acetate	16.3	13–20.67

3.2.2. Effects on Sea Urchin Embryo Development

Twenty-four hours after fertilization, the control sample presented a prismatic shape. All the cyanobacterial extracts affected the development of sea urchin larvae. Toxicity assays with water and methanolic extracts from *Cyanobium* sp. ITAC108 and *Synechococcus* sp. ITAC107 displayed an evident toxic effect by completely inhibiting embryogenesis (Figure 2): after 24 h, in a sample treated with water and methanolic extracts, most embryos showed, respect to control (Figure 3A), an anomalous development, blocked at the morula stage (Figure 3B,C,E,F). Morulae exhibited anomalous cell division in most cases. On the contrary, in ethyl acetate (Figure 3D) and hexane extracts the percentage of anomalous embryos was low: 21.8 ± 10.6 and 19.4 ± 3.2 for *Cyanobium* sp. ITAC108 and 21.1 ± 15.1 and 32.8 ± 19.5 for *Synechococcus* sp. ITAC107. There were no significant differences ($P > 0.01$) with control, which showed that only 7.3% ± 1.3% were anomalous embryos.

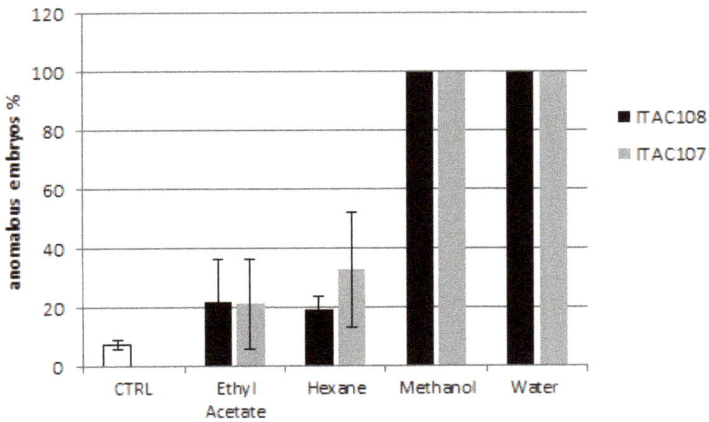

Figure 2. Percentage of anomalous sea urchin embryos after 24 h of treatment with cyanobacterial extracts.

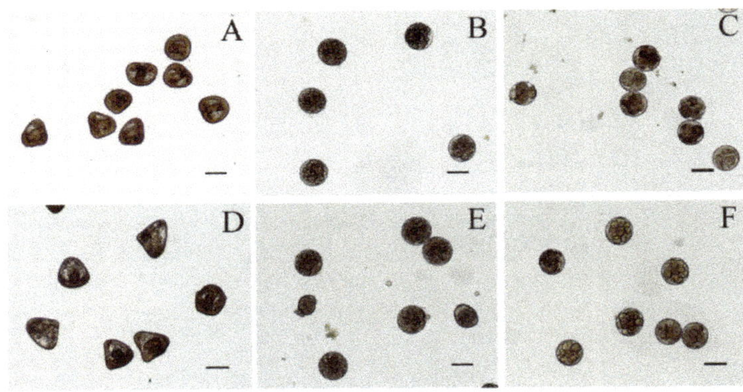

Figure 3. Micrographs of *P. lividus* embryos after 24 h of treatment with cyanobacteria extracts. (**A**) Control sample; (**B**,**C**) sea urchin embryos treated with aqueous extracts from *Cyanobium* sp. ITAC108 and *Synechococcus* sp. ITAC107, respectively; (**D**) sea urchin embryos treated with hexane extracts from *Synechococcus* sp. ITAC107; (**E**,**F**) sea urchin embryos treated with methanolic extracts from *Cyanobium* sp. ITAC108 and *Synechococcus* sp. ITAC107, respectively. Bar represents 100 μm.

Forty-eight hours after treatment, in control 97.3% ± 1.6 % of sea urchin embryos developed into normal pluteus larvae (Figure 4A,E) with an average length of 518 ± 35 μm. No effects were observed for embryos incubated with hexane and ethyl acetate extracts from *Cyanobium* sp. ITAC108 (Figure 4B). Hexane extract from *Synechococcus* sp. ITAC107 exerted a bland effect on sea urchin embryos that reached the pluteus stage (Figure 4C), but in this case the larval dimensions had a significantly ($P < 0.01$) reduced size (Figure 4F) compared to controls (Figure 4A) with an average length of 338 ± 33 μm. In addition, skeletal deformations (Figure 4F) were recorded in this sample treatment. On the contrary, a very dangerous effect was detected with inhibition of embryonic development after incubation with ethyl acetate extract from *Synechococcus* sp. ITAC107. In the last case, embryos did not survive more than 24 h (Figure 4D), presenting in most cases a completely deformed skeleton (Figure 4G).

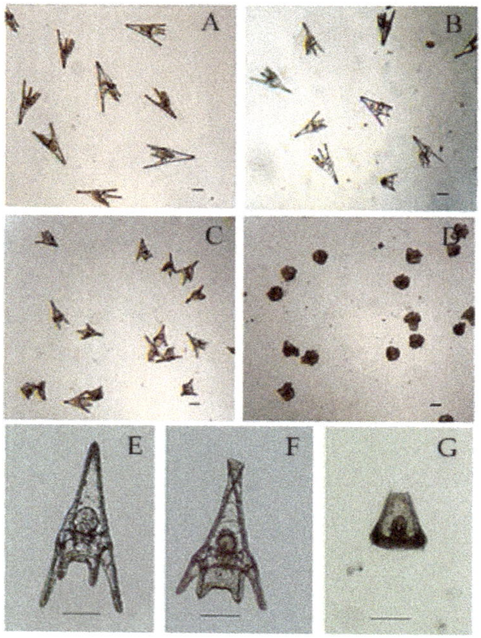

Figure 4. Micrographs of *P. lividus* embryos after 48 h of treatment with cyanobacteria extracts. (**A**) Control sample. (**B**) Sea urchin embryos treated with ethyl acetate extracts from *Cyanobium* sp. ITAC108. (**C**) Sea urchin embryos treated with hexane extracts from *Synechococcus* sp. ITAC107. (**D**) Sea urchin embryos treated with ethyl acetate extracts from *Synechococcus* sp. ITAC107. (**E**) Untreated *P. lividus* embryo. (**F**) Embryo treated with ethyl acetate extracts from *Synechococcus* sp. ITAC107. (**G**) Embryo treated with ethyl acetate extracts from *Synechococcus* sp. ITAC107. Bar represents 100 μm.

3.2.3. Effects on HeLa Cells

In this study, we have explored the effect of various extracts of two cyanobacterial strains on human epithelial-like HeLa cells.

The cytotoxicity of hexane, ethyl acetate, methanol and water extracts from *Cyanobium* sp. ITAC108 and *Synechococcus* sp. ITAC107 was assessed using the MTT assay. As shown in Figure 5, all the extracts inhibited HeLa cell proliferation, but methanol extracts from both cyanobacteria strains exerted the highest activity with 52% and 53% of cell growth inhibition, respectively. Aqueous extracts from the two cyanobacteria also inhibited HeLa growth with a percentage of 48% and 36%, respectively. Cells treated with ethyl acetate extracts were metabolically more active, showing 36% and 32% of inhibition, while for hexane extracts from strain *Cyanobium* sp. ITAC108 and *Synechococcus* sp. ITAC107, 24% and 35% of cell growth inhibition were observed, respectively.

3.3. Secondary Metabolite Determination

Both the methanolic and water extracts from *Cyanobium* sp. ITAC108 and *Synechococcus* sp. ITAC107 were analyzed by GC-MS to identify those metabolites responsible for the toxicity of extracts. The β-*N*-methylamino-L-alanine, BMAA, was identified in the chromatograms of the methanolic extract at 16.71 min as tms derivates using the accurate mass 116.0889 and 291.1263. The relevant extracted ion chromatograms of all the samples are shown in Figure 6, where it is possible to observe, along with BMAA, the presence of 2,4-diaminobutyric acid (2,4-DAB), which also has neurotoxic properties [60].

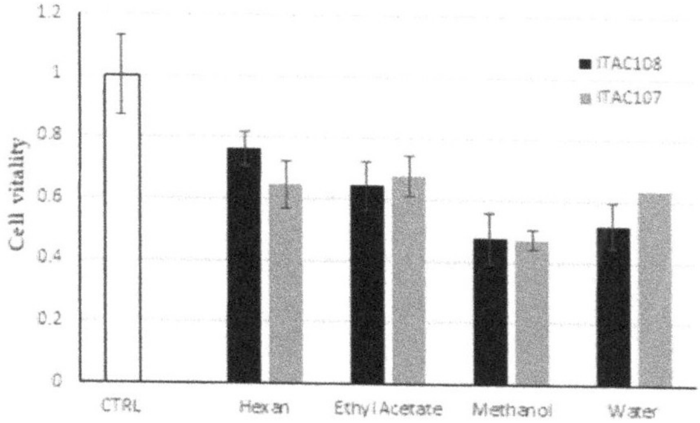

Figure 5. Effect of cyanobacterial fractions on cultured cells viability. The MTT assays was performed on human HeLa cells exposed for 6 h to cyanobacterial fractions from *Synechococcus* sp. ITAC107 and *Cyanobium* sp. ITAC108 strains.

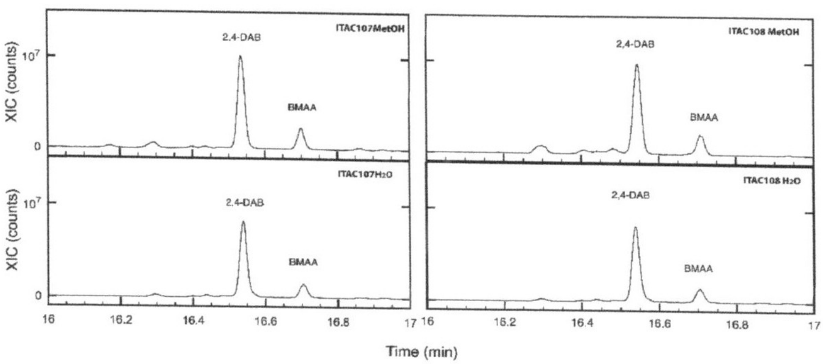

Figure 6. Extracted chromatograms of m/z 291.1263 ion for methanolic and aqueous extracts for *Cyanobium* sp. ITAC108 and *Synechococcus* sp. ITAC107 in the region comprised between 16 and 17 min. In the chromatograms, it is possible to identify 2,4-DAB and BMAA.

Table S1 shows all the other annotated metabolites identified in the extracts using GC-HRMS.

The microcystin-VF was identified in LC-ESI(+)-MS at m/z 972.51 ([M + H]+) in the methanolic extract of both PCCs. As it was not possible to identify other secondary metabolites among those already identified in PCCs [61], an untargeted approach was used to analyze LC-ESI(+)-MS data. As methanolic extracts showed a greater activity toward HeLa cells than aqueous extracts did, during the untargeted analysis carried out using MS-DIAL, methanolic and aqueous extracts were considered as different classes. Figure S1 reports the PCA score plot of the samples, where it is possible to see that methanolic extracts are separated from aqueous along PC1, whereas PC2 permits the separation of *Cyanobium* sp. ITAC108 from *Synechococcus* sp. ITAC107 aqueous extracts.

Table S2 lists all the significant (P-values < 0.05) characteristics that were identified and putatively annotated. Data curation is beyond the scope of the present paper; however, it is evident that there are different classes of compounds characterizing the extracts and possibly having a role in the recorded PCC activity.

4. Discussion

In the present investigation, we focused our attention on two marine strains of PCCs: *Cyanobium* sp. ITAC108 and *Synechococcus* sp. ITAC107, already characterized by morphological and molecular criteria [49]. Previous investigation on these two PCC strains, suggested better characterization of their potential as bioproducers given their ability to affect HeLa cell vitality [49] and to gain more insight into the nature of the more active molecules. The potential as producers of bioactive compounds of some strains of *Synechococcus* and *Cyanobium* genera was previously reported by some authors [27,62–64] who emphasize the importance of also studying marine PCCs [22,24,27]. As unicellular coccoid forms, these two PCCs were easily cultured in the laboratory, requiring less care and exhibiting a faster growth rate than the filamentous forms. These properties make them good candidates as bioproducers.

The first step toward the identification of the chemical nature of these compounds was the fractionation of the crude extracts, which suggested that both samples contain a very low amount of apolar (lipophilic) compounds (aldehydes and ketones, hydrocarbons, esters, short chain alcohols, etc.). However, these types of compounds from both cyanobacteria strains were found to be toxic to *A. salina* at the highest concentration used (50 μg/mL) except for the *Cyanobium* sp. ITAC108 ethyl acetate sample. On the other hand, a very high content of polar compounds likely phenols, ammines, amides, carboxylic, etc. (87.6% d.w.) was found in *Synechococcus* sp. ITAC107, while *Cyanobium* sp. ITAC108 contained a significantly lower amount (54.4%) of them. All these fractions were toxic to brine shrimp. Methanolic extracts from the two strains showed comparable activity.

During the fractionated extraction, hot water was used to solubilize very polar and protic compounds including most likely carbohydrates, polymers and proteins. *Cyanobium* sp. ITAC108 seems to contain a significantly higher amount of this latter class of compounds compared to the sample from *Synechococcus* sp. ITAC107. The amount of protic and polar compounds reflects the stronger toxicity of *Cyanobium* sp. ITAC108 toward *Artemia* nauplii compared to that of *Synechococcus* sp. ITAC107. Although in some cases *Cyanobium* strains did not demonstrate toxicity for *A. salina* [27], other studies [63,65] showed that one strain of *Cyanobium* and one of *Synechococcus* caused acute toxicity in the brine shrimp nauplii, inducing 100% mortality with aqueous and methanolic extracts.

The toxic effect of the different extracts of the cyanobacterial strains was also evident for the embryos of the sea urchin, *P. lividus*. After treatment with water and methanolic extracts from both cyanobacteria strains, embryogenesis stopped at the first stages (segmentation) of development (within 24 h), providing strong evidence for the presence of polar compounds interfering with growth factors. Affecting the segmentation phase of the embryo, we may suppose that both extracts interfere with cell mitotic activity.

For some extracts (i.e., hexane), embryogenesis of the sea urchin embryos was not inhibited. However, after a longer treatment with hexane extract from *Synechococcus* sp. ITAC107, the resulting larvae were significantly smaller than the control larvae. This suggests that a chronic effect may occur with this hexane extract and that the reduced size of the larvae could be due to the action of compounds present in low concentrations. Sellem and coworkers [66] also suggested this possibility when they tested the effects of different concentrations of unsaturated fatty acids of a dinoflagellate species to embryos of the sea urchin *P. lividus*.

Besides invertebrates, toxicity is often evaluated on mammalian cells by using cancer cell lines [67–69]. Interestingly, some study reported a good correlation between the toxic activity evidenced by the *A. salina* test and that observed with tumor cell lines [70]. Here, we report about the effect of the various fractionated extracts on the human cervical cancer HeLa cell line that was already affected by the total crude extracts from both PCCs [49]. The toxicity was checked using the MTT assay, which showed that the cells were metabolically less active in the presence of all the extracts in comparison to control (without extracts). Although the response of cells to various extracts and the percentage change in metabolic activity was different depending on the extract type, methanolic fractions from both PCCs exerted the highest negative effect. The different responses may be due to the selective cellular response and affinity to compounds present in the various extracts.

Methanolic extracts from other cyanobacteria strains are known to induce an antiproliferative effect on HeLa cells [71], but to our knowledge, this is the first time that different fractionated extracts from these two strains were tested on Hela cells.

All these results highlighting a strong bioactivity of the methanolic and aqueous extracts of both PCCs suggested analyzing them by GC-MS and identifying those metabolites responsible for the toxicity of the extracts.

All four extracts of the two PCCs had various bioactive components possessing numerous activities. BMAA is one of the important compounds that we found in methanolic extracts along with 2,4-DAB, whereas it was not possible to identify the other BMAA constitutional isomer, i.e., N-(2-aminoethyl)-glycine (AEG). BMAA is a neurotoxin previously identified by other researchers in symbionts and free-living cyanobacteria [68]. *Prochlorococcus marinus*, *Synechoccous* spp. and *Synechocystis* spp. [22,72,73] represent the PCC strains until now identified as BMAA producers

Here, we report for the first time the presence of BMAA and one of its constitutional isomers, 2,4-DAB, in a *Cyanobium* and in a *Synechoccoccus* strain isolated from a marine invertebrate (i.e., the sponge *Petrosia ficiformis*) that could be the main compound responsible for the observed toxicity.

LC-MS also revealed the presence of the hepatotoxic microcystins (MCs) in our samples. Microcystins, representing the main secondary metabolites here identified, are cyclic heptapeptides typically produced by cyanobacteria that could be the main products influencing the survival of different model organisms we used. These cyanotoxins have been already found in PCCs as reported in Śliwińska-Wilczewska and coauthors [22]. According to their review, the most common MC isomers are MC-LR and MC-YR, while the MC-VF identified in our strains was already detected only in the filamentous *Microcystis aeruginosa* [74]. Other secondary metabolites produced by PCCs are 2-MIB and GSM [20]. However, both compounds were not identified either in the methanolic or in the aqueous extracts.

Finally, considering that the two strains used here have shown high growth rates in laboratory cultures, the metabolites of their active fractions could be considered as easily accessible by-products from biomass that can increase the value of their large-scale culture. However, it is known that the genomic organization of biosynthetic gene clusters, complex gene expression patterns and low yields of compounds synthesized by native producers restrict access to many of these valuable molecules for detailed studies [75]. Recently some investigations have highlighted the importance of genetic engineering in this field. An example of this is represented by the production of key compounds of the hapalindole family of indole-isonitrile alkaloids obtained by engineering the fast-growing cyanobacterium *Synechococcus elongatus* UTEX 2973 [75].

5. Conclusions

This work provides further useful data to expand the range of cyanobacteria from which new compounds with significant bioactivity could be identified. Particularly intriguing was the bioactivity of polar extracts (especially the methanolic ones) of two PCC strains belonging to *Cyanobium* and *Synechococcus* genera. To our knowledge, this is the first time in which BMAA, 2,4-DAB and microcystin have been found in cyanobacteria isolated from a marine sponge. We consider these substances to be responsible for cytotoxicity and antimitotic activity, but further investigations are needed to isolate and better characterize these compounds. Furthermore, it will be important to deepen the observations on the long-term effects as the extracts of ethyl acetate, which, although not having antimitotic activity, block the development of the sea urchin by preventing the gastrulation phase. Furthermore, future investigations could help to know the cause of the skeletal deformation induced by the hexane extract from *Synechococcus* sp. ITAC 107. According to our results, we consider the active fractions to be promising for the isolation of compounds for potential biotechnological applications, including the field of pharmacological agents. In addition, considering that the two strains used here have shown high growth rates in laboratory cultures, the metabolites of their active fractions can be considered as easily accessible by-products from biomass that can increase the value of their large-scale culture. However, although a high growth rate has been shown in the laboratory, large-scale production of bioproducts is likely to be best achieved by engaging in the engineering of these cyanobacteria. Another challenge to be faced in the near future.

Supplementary Materials: The following are available online at https://www.mdpi.com/article/10.3390/microorganisms9102048/s1, Figure S1: PCA score plot of the samples where it is possible to see that methanolic (square) extracts are separated from aqueous ones (dot) along PC1 whereas PC2 permit to separate *Cyanobium* sp. ITAC108 from *Synechococcus* sp. ITAC107 aqueous extracts, Table S1: Metabolites identified in the extracts using GC-HRMS, Table S2: lists of all the significant (*P*-values < 0.05) characteristics distinguishing methanolic from aqueous extracts obtained and annotated by MS-DIAL.

Author Contributions: Conceptualization, P.P. and C.C.; methodology, P.P. and C.C.; formal analysis, P.P., G.E.D.B., M.F., A.B. and C.C.; investigation, P.P., G.E.D.B., M.F., A.B. and C.C.; data curation, P.P. and C.C.; methodology, P.P. and C.C.; validation, P.P., G.E.D.B., M.F., A.B. and C.C.; writing—original draft preparation, P.P. and C.C.; writing—review and editing, P.P. and C.C.; visualization, P.P. and C.C.; supervision, P.P. All authors have read and agreed to the published version of the manuscript.

Funding: This research received no external funding.

Data Availability Statement: Most of the data presented in this study are contained within the article or Supplementary Materials; data not present are available on request from the corresponding author.

Conflicts of Interest: The authors declare no conflict of interest.

References

1. Six, C.; Finkel, Z.V.; Irwin, A.J.; Campbell, D.A. Light Variability Illuminates Niche-Partitioning among Marine Picocyanobacteria. *PLoS ONE* **2007**, *2*, e1341. [CrossRef] [PubMed]
2. Johnson, P.W.; Sieburth, J.M.N. Chroococcoid cyanobacteria in the diverse phototrophic biomass. *Limnol. Oceanogr.* **1979**, *24*, 928–935. [CrossRef]
3. Waterbury, J.B.; Watson, S.W.; Guillard, R.R.L.; Brand, L.E.E. Widespread occurrence of a unicellular, marine, planktonic cyanobacterium. *Nature* **1979**, *227*, 293–294. [CrossRef]
4. Li, W.K.W.; Rao, D.V.S.; Harrison, W.G.; Smith, J.C.; Cullen, J.J.; Irwin, B.; Platt, T. Autotrophic picoplankton in the tropical ocean. *Science* **1983**, *219*, 292–295. [CrossRef] [PubMed]
5. Chisholm, S.W.; Olson, R.J.; Zettler, E.R.; Goericke, R.; Waterbury, J.B.; Welschmeyer, N.A. A novel free-living prochlorophyte abundant in the oceanic euphotic zone. *Nature* **1988**, *334*, 340–343. [CrossRef]
6. Chisholm, S.W.; Frankel, S.L.; Goericke, R.; Olson, R.J.; Palenik, B.; Waterbury, J.B.; West-Johnsrud, L.; Zettler, E.R. *Prochlorococcus marinus* nov. gen. nov. sp.: An oxyphototrophic marine prokaryote containing divinyl chlorophyll a and b. *Arch. Microbiol.* **1992**, *157*, 297–300. [CrossRef]
7. Komárek, J.; Anagnostidis, K. *Cyanoprokaryota, Part. 1: Chroococcales, Süsswasserflora von Mitteleuropa*; Elsevier/Spektrum Akademischer Verlag: Heidelberg, Germany, 2008; pp. 1–548.
8. Johnson, Z.I.; Zinser, E.R.; Coe, A.; McNulty, N.P.; Woodward, E.M.S.; Chisholm, S.W. Niche Partitioning Among *Prochlorococcus* Ecotypes Along Ocean-Scale Environmental Gradients. *Science* **2006**, *311*, 1737–1740. [CrossRef]
9. Golubic, S.; Le Campion-Alsumard, T.; Campbell, S.E. Diversity of marine cyanobacteria. In *Marine Cyanobacteria*; Charpy, L., Larkum, A.W.D., Eds.; Institute of Oceanography: Le Rocher, Monaco, 1999; pp. 53–76.
10. Pittera, J.; Humily, F.; Thorel, M.; Grulois, D.; Garczarek, L.; Six, C. Connecting thermal physiology and latitudinal niche partitioning in marine *Synechococcus*. *ISME J.* **2014**, *8*, 1221–1236. [CrossRef]
11. Sorokin, P.Y.; Sorokin, Y.I.; Boscolo, R.; Giovanardi, O. Bloom of picocyanobacteria in the Venice lagoon during summer–autumn 2001: Ecological sequences. *Hydrobiologia* **2004**, *523*, 71–85. [CrossRef]
12. Sorokin, Y.I.; Zakuskina, O.Y. Features of the Comacchio Ecosystem Transformed during Persistent Bloom of Picocyanobacteria. *J. Oceanogr.* **2010**, *66*, 373–387. [CrossRef]
13. Azam, F.; Fenchel, T.; Field, J.G.; Gray, J.S.; Mayer-Reil, L.A.; Thingstad, F. The ecological role of water column microbes in the sea. *Mar. Ecol. Prog. Ser.* **1983**, *10*, 257–263. [CrossRef]
14. Worden, A.Z.; Nolan, J.K.; Palenik, B. Assessing the dynamics and ecology of marine picophytoplankton: The importance of the eukaryotic component. *Limnol. Oceanogr.* **2004**, *49*, 168–179. [CrossRef]
15. Pomeroy, L.R.; Williams, P.J.L.; Azam, F.; Hobbie, J.E. The microbal loop. *Oceanography* **2007**, *2*, 28–33. [CrossRef]
16. Karuza, A.; Caroppo, C.; Camatti, E.; Di Poi, E.; Monti, M.; Stabili, L.; Auriemma, R.; Pansera, M.; Cibic, T.; Del Negro, P. 'End to end' planktonic trophic web and its implications for the mussel farms in the Mar Piccolo of Taranto (Ionian Sea, Italy). *Environ. Sci. Pollut. Res.* **2016**, *23*, 12707–12724. [CrossRef] [PubMed]

17. Costa, M.; Garcia, M.; Costa-Rodrigues, J.; Costa, M.S.; Ribeiro, M.J.; Fernandes, M.H.; Barros, P.; Barreiro, A.; Vasconcelos, V.; Martins, R. Exploring Bioactive Properties of Marine Cyanobacteria Isolated from the Portuguese Coast: High Potential as a Source of Anticancer Compounds. *Mar. Drugs* **2014**, *12*, 98–114. [CrossRef] [PubMed]
18. Mazard, S.; Penesyan, A.; Ostrowski, M.; Paulsen, I.T.; Egan, S. Tiny Microbes with a Big Impact: The Role of Cyanobacteria and Their Metabolites in Shaping Our Future. *Mar. Drugs* **2016**, *14*, 97. [CrossRef]
19. Lopes, G.; Clarinha, D.; Vasconcelos, V. Carotenoids from Cyanobacteria: A Biotechnological Approach for the Topical Treatment of Psoriasis. *Microorganisms* **2020**, *8*, 302. [CrossRef]
20. Jakubowska, N.; Szelag-Wasielewska, E. Toxic Picoplanktonic Cyanobacteria—Review. *Mar. Drugs* **2015**, *13*, 1497–1518. [CrossRef] [PubMed]
21. Jasser, I.; Callieri, C. Picocyanobacteria: The smallest cell-size cyanobacteria. In *Handbook on Cyanobacterial Monitoring and Cyanotoxin Analysis*; Meriluoto, J., Spoof, L., Codd, G.A., Eds.; Wiley: Hoboken, NJ, USA, 2017; pp. 19–27.
22. Śliwińska-Wilczewska, S.; Maculewicz, J.; Felpeto, A.B.; Latała, A. Allelopathic and bloom-forming picocyanobacteria in a changing world. *Toxins* **2018**, *10*, 48. [CrossRef]
23. Huang, I.-S.; Zimba, P.V. Cyanobacterial bioactive metabolites-A review of their chemistry and Biology. *Harmful Algae* **2019**, *83*, 42–94. [CrossRef]
24. Konarzewska, Z.; Śliwińska-Wilczewska, S.; Felpeto, A.B.; Vasconcelos, V.; Latała, A. Assessment of the Allelochemical Activity and Biochemical Profile of Different Phenotypes of Picocyanobacteria from the Genus *Synechococcus*. *Mar. Drugs* **2020**, *18*, 179. [CrossRef] [PubMed]
25. Leão, P.N.; Costa, M.; Ramos, V.; Pereira, A.R.; Fernandes, V.C.; Domingues, V.F.; Gerwick, W.H.; Vasconcelos, V.M.; Martins, R. Antitumor activity of hierridin B, a cyanobacterial secondary metabolite found in both filamentous and unicellular marine strains. *PLoS ONE* **2013**, *8*, e69562. [CrossRef] [PubMed]
26. Agha, R.; Quesada, A. Oligopeptides as biomarkers of cyanobacterial subpopulations. Toward an understanding of their biological role. *Toxins* **2014**, *6*, 1929–1950. [CrossRef] [PubMed]
27. Costa, M.S.; Costa, M.; Ramos, V.; Leao, P.N.; Barreiro, A.; Vasconcelos, V.; Martins, R. Picocyanobacteria from a clade of marine *Cyanobium* revealed bioactive potential against microalgae, bacteria, and marine invertebrates. *J. Toxicol. Environ. Health Part A* **2015**, *78*, 432–442. [CrossRef]
28. Regueiras, A.; Pereira, S.; Costa, M.S.; Vasconcelos, V. Differential Toxicity of Cyanobacteria Isolated from Marine Sponges towards Echinoderms and Crustaceans. *Toxins* **2018**, *10*, 297. [CrossRef]
29. Caroppo, C. Ecology and biodiversity of picoplanktonic cyanobacteria in coastal and brackish environments. *Biodivers. Conserv.* **2015**, *24*, 949–971. [CrossRef]
30. Leão, P.N.; Engene, N.; Antunes, A.; Gerwick, W.H.; Vasconcelos, V. The chemical ecology of cyanobacteria. *Nat. Prod. Rep.* **2012**, *29*, 372–391. [CrossRef]
31. Goleski, J.A.; Koch, F.; Marcoval, M.A.; Wall, C.C.; Jochem, F.J.; Peterson, B.J.; Gobler, C.J. The role of zooplankton grazing and nutrient loading in the occurrence of harmful marine cyanobacterial blooms in Florida Bay, USA. *Estuar. Coast.* **2010**, *33*, 1202–1215. [CrossRef]
32. Carmichael, W.W.; Li, R. Cyanobacteria toxins in the Salton Sea. *Saline Syst.* **2006**, *2*, 5–17. [CrossRef]
33. Vareli, K.; Zarali, E.; Zacharioudakis, G.S.A.; Vagenas, G.; Varelis, V.; Pilidis, G.; Briasoulis, E.; Sainis, I. Microcystin producing cyanobacterial communities in Amvrakikos Gulf (Mediterranean Sea, NW Greece) and toxin accumulation in mussels (*Mytilus galloprovincialis*). *Harmful Algae* **2012**, *15*, 109–118. [CrossRef]
34. Vareli, K.; Jaeger, W.; Touka, A.; Frillingos, S.; Briasoulis, E.; Sainis, I. Hepatotoxic seafood poisoning (HSP) due to microcystins: A threat from the ocean? *Mar. Drugs* **2013**, *11*, 2751–2768. [CrossRef]
35. Blom, J.F.; Baumann, H.I.; Codd, G.A.; Jüttner, F. Sensitivity and adaptation of aquatic organisms to oscillapeptin J and [D-Asp3,(E)-Dhb7] microcystin-RR. *Arch. Hydrobiol.* **2006**, *167*, 547–559. [CrossRef]
36. Błaszczyk, A.; Mazur-Marzec, H. BMAA and other cyanobacterial neurotoxins. *Pol. Hyperb. Res.* **2006**, *4*, 7–14.
37. Wiese, M.; D'Agostino, P.M.; Mihali, T.K.; Moffitt, M.C.; Neilan, B.A. Neurotoxic alkaloids: Saxitoxin and its analogs. *Mar. Drugs* **2010**, *8*, 2185–2211. [CrossRef] [PubMed]
38. Stewart, I.; Schluter, P.J.; Shaw, G.R. Cyanobacterial lipopolysaccharides and human health—A review. *Environ. Health* **2006**, *5*, 7. [CrossRef]
39. Morone, J.; Lopes, G.; Preto, M.; Vasconcelos, V.; Martins, R. Exploitation of Filamentous and Picoplanktonic Cyanobacteria for Cosmetic Applications: Potential to Improve Skin Structure and Preserve Derma Matrix Components. *Mar. Drugs* **2020**, *18*, 486. [CrossRef]
40. Freitas, S.; Martins, R.; Campos, A.; Azevedo, J.; Osorio, H.; Costa, M.; Barros, P.; Vasconcelos, V.; Urbatzka, R. Insights into the potential of picoplanktonic marine cyanobacteria strains for cancer therapies–Cytotoxic mechanisms against the RKO colon cancer cell line. *Toxicon* **2016**, *119*, 140–151. [CrossRef]
41. Milledge, J.J. Commercial application of microalgae other than as biofuels: A brief review. *Rev. Environ. Sci. Bioethanol.* **2011**, *10*, 31–41. [CrossRef]
42. Massey, I.Y.; Yanga, F.; Dingd, Z.; Yanga, S.; Guoa, J.; Tezia, C.; Al-Osmana, M.; Kamegnia, R.B.; Zeng, W. Exposure routes and health effects of microcystins on animals and humans: A mini-review. *Toxicon* **2018**, *151*, 156–162. [CrossRef]

43. Uzair, B.; Tabassum, S.; Rasheed, M.; Rehman, S.F. Exploring marine cyanobacteria for lead compounds of pharmaceutical importance. *Sci. World J.* **2012**, *2012*, 179782. [CrossRef]
44. Blunt, J.W.; Copp, B.R.; Munro, M.H.; Northcote, P.T.; Prinsep, M.R. Marine natural products. *Nat. Prod.* **2010**, *27*, 165–237. [CrossRef]
45. Pagliara, P.; Caroppo, C. Cytotoxic and antimitotic activities in aqueous extracts of eight cyanobacterial strains isolated from the marine sponge *Petrosia ficiformis*. *Toxicon* **2011**, *57*, 889–896. [CrossRef] [PubMed]
46. Pagliara, P.; Caroppo, C. Toxicity of two cyanobacterial strains on *Mytilus galloprovincialis* hemocytes. *Rapp. Comm. Int. Mer. Médit.* **2013**, *40*, 389.
47. Komárek, J.; Anagnostidis, K. *Cyanoprokaryota, Part. 2: Oscillatoriales, Süsswasserflora von Mitteleuropa, Bd19/2*; Elsevier/Spektrum Akademischer Verlag: Heidelberg, Germany, 2005; pp. 1–759.
48. Boone, D.R.; Castenholz, R.W.; Garrity, G.M. *Bergey's Manual of Systematic Bacteriology*; Springer: New York, NY, USA, 2001; pp. 1–357.
49. Pagliara, P.; Barca, A.; Verri, T.; Caroppo, C. The marine sponge *Petrosia ficiformis* harbors different cyanobacteria strains with potential biotechnological application. *J. Mar. Sci. Eng.* **2020**, *8*, 638. [CrossRef]
50. Rippka, R.; Deruelles, J.; Waterbury, J.B.; Herdman, M.; Stanier, R.Y. Generic assignments, strain histories and properties of pure cultures of cyanobacteria. *J. Gen. Microbiol.* **1979**, *111*, 1–61. [CrossRef]
51. Kana, T.M.; Glibert, P.M.G. Effect of irradiances up to 2000 mE m-2s-1 on marine Synechococcus WH7803-I. Growth, pigmentation and cell composition. *Deep-Sea Res.* **1987**, *34*, 479–495. [CrossRef]
52. Guillard, R.; Sieracki, M. Counting Cells in Cultures with the Light Microscope. In *Algal Culturing Techniques*; Andersen, R.A., Ed.; Academic Press: Cambridge, MA, USA, 2005; pp. 239–252. [CrossRef]
53. Francavilla, M.; Franchi, M.; Monteleone, M.; Caroppo, C. The red seaweed *Gracilaria gracilis* as a multi products source. *Mar. Drugs* **2013**, *11*, 3754–3776. [CrossRef]
54. Broekaert, J.A.C.; Harris, D.C. Quantitative chemical analysis. *Anal. Bioanal. Chem.* **2015**, *407*, 8943–8944. [CrossRef]
55. Finney, D.J. *Probits Analysis*; Cambridge University Press: New York, NY, USA, 1971; pp. 1–333.
56. Pagliara, P.; Caroppo, C. Toxicity assessment of *Amphidinium carterae*, *Coolia* cfr. *monotis* and *Ostreopsis* cfr. ovata (Dinophyta) isolated from the northern Ionian Sea (Mediterranean Sea). *Toxicon* **2012**, *60*, 1203–1214. [CrossRef]
57. De Benedetto, E.; Savino, A.; Fico, D.; Rizzo, D.; Pennetta, A.; Cassiano, A.; Minerva, B. A multi-analytical approach for the characterisation of the oldest pictorial cycle in the 12th century monastery Santa Maria delle Cerrate. *Open J. Archaeom.* **2013**, *1*, 12. [CrossRef]
58. Pizzolante, G.; Durante, M.; Rizzo, D.; Di salvo, M.; Tredici, S.M.; Tufariello, M.; De Paolis, A.; Talà, A.; Mita, G.; Alifano, P.; et al. Characterization of two *Pantoea* strains isolated from extra-virgin olive oil. *AMB Expr.* **2018**, *8*, 113. [CrossRef]
59. Tsugawa, H.; Ikeda, K.; Takahashi, M.; Satoh, A.; Mori, Y.; Uchino, H.; Okahashi, N.; Yamada, Y.; Tada, Y.; Bonini, P.; et al. A lipidome atlas in MS-DIAL 4. *Nat. Biotechnol.* **2020**, *38*, 1159–1163. [CrossRef]
60. Violi, J.P.; Facey, J.A.; Mitrovic, S.M.; Colville, A.; Rodgers, K.J. Production of β-methylamino-L-alanine (BMAA) and Its Isomers by Freshwater Diatoms. *Toxins* **2019**, *11*, 512. [CrossRef] [PubMed]
61. Le Manach, S.; Charlotte, D.; Arul, M.; Chakib, D.; Arnaud, C.; Marc, E.; Cécile, B.; Benjamin, M. Global Metabolomic Characterizations of Microcystis spp. Highlights Clonal Diversity in Natural Bloom-Forming Populations and Expands Metabolite Structural Diversity. *Front. Microbiol.* **2019**, *10*, 791. [CrossRef]
62. Hamilton, T.J.; Paz-Yepes, J.; Morrison, R.A.; Palenik, B.; Tresguerres, M. Exposure to bloom-like concentrations of two marine *Synechococcus* cyanobacteria (Strains Cc9311 and Cc9902) differentially alters fish behaviour. *Conserv. Physiol.* **2014**, *2*, 1–9. [CrossRef] [PubMed]
63. Martins, R.; Fernandez, N.; Beiras, R.; Vasconcelos, V. Toxicity assessment of crude and partially purified extracts of marine *Synechocystis* and *Synechococcus* cyanobacterial strains in marine invertebrates. *Toxicon* **2007**, *50*, 791–799. [CrossRef] [PubMed]
64. Martins, R.; Ramos, M.; Herfindal, L.; Sousa, J.; Skaerven, K.; Vasconcelos, V. Antimicrobial and cytotoxic assessment of marine cyanobacteria *Synechocystis* and *Synechococcus*. *Mar. Drugs* **2008**, *6*, 1. [CrossRef] [PubMed]
65. Frazão, B.; Martins, R.; Vasconcelos, V. Are Known Cyanotoxins Involved in the Toxicity of Picoplanktonic and Filamentous North Atlantic Marine Cyanobacteria? *Mar. Drugs* **2010**, *8*, 1908–1919. [CrossRef]
66. Sellem, F.; Pesando, D.; Bodennec, G.; Abed, A.E.; Girard, J.P. Toxic effects of *Gymnodinium* cf. *mikimotoi* unsaturated fatty acids to gametes and embryos of the sea urchin *Paracentrotus lividus*. *Water Res.* **2000**, *34*, 550–556. [CrossRef]
67. Jaki, B.; Orjala, J.; Bürgi, H.-R.; Sticher, O. Biological Screening of Cyanobacteria for Antimicrobial and Molluscicidal Activity, Brine Shrimp Lethality, and Cytotoxicity. *Pharm. Biol.* **1999**, *37*, 138–143. [CrossRef]
68. Carballo, J.L.; Hernández-Inda, Z.L.; Pérez, P.; García-Grávalos, M.D. A comparison between two brine shrimp assays to detect *in vitro* cytotoxicity in marine natural products. *BMC Biotechnol.* **2002**, *2*, 17. [CrossRef] [PubMed]
69. Hisem, D.; Hrouzek, P.; Tomek, P.; Tomsícková, J.; Zapomelová, E.; Skácelová, K.; Lukesová, A.; Kopecký, J. Cyanobacterial cytotoxicity versus toxicity to brine shrimp *Artemia salina*. *Toxicon* **2011**, *57*, 76–83. [CrossRef] [PubMed]
70. Anderson, J.E.; Goetz, C.M.; McLaughlin, J.L.; Suffness, M. A blind comparison of simple bench-top bioassays and human tumour cell cytotoxicities as antitumor prescreens. *Phytochem. Anal.* **1991**, *2*, 107–111. [CrossRef]
71. Andeden, E.E.; Ozturk, S.; Aslim, B. Antiproliferative, neurotoxic, genotoxic and mutagenic effects of toxic cyanobacterial extracts. *Interdiscip. Toxicol.* **2018**, *11*, 267–274. [CrossRef] [PubMed]

72. Cox, P.A.; Banack, S.A.; Murch, S.J.; Rasmussen, U.; Tien, G.; Bidigare, R.R.; Metcalf, J.S.; Morrison, L.F.; Codd, G.A.; Bergman, B. Diverse taxa of cyanobacteria produce β-N-methylamino-L-alanine, a neurotoxic amino acid. *Proc. Natl. Acad. Sci. USA* **2005**, *102*, 5074–5078. [CrossRef]
73. Cianca, R.C.C.; Baptista, M.S.; Lopes, V.R.; Vasconcelos, V.M. The non-protein amino acid-N-methylamino-L-alanine in Portuguese cyanobacterial isolates. *Amino Acids* **2012**, *42*, 2473–2479. [CrossRef]
74. Massey, I.Y.; Wu, P.; Wei, J.; Luo, J.; Ding, P.; Wei, H.; Yang, F. A Mini-Review on Detection Methods of Microcystins. *Toxins* **2020**, *12*, 641. [CrossRef]
75. Knoot, C.J.; Khatri, Y.; Hohlman, R.M.; Sherman, D.H.; Pakrasi, H.B. Engineered Production of Hapalindole Alkaloids in the Cyanobacterium *Synechococcus* sp. UTEX 2973. *ACS Synth. Biol* **2019**, *8*, 1941–1951. [CrossRef]

microorganisms MDPI

Article

A Combination of Aqueous Extraction and Ultrafiltration for the Purification of Phycocyanin from *Arthrospira maxima*

Dante Matteo Nisticò [1], Amalia Piro [1], Daniela Oliva [1], Vincenzo Osso [1], Silvia Mazzuca [1,*], Francesco Antonio Fagà [2], Rosanna Morelli [3], Carmela Conidi [3], Alberto Figoli [3] and Alfredo Cassano [3,*]

1 Laboratorio di Biologia e Proteomica Vegetale, Dipartimento di Chimica e Tecnologie Chimiche, Università della Calabria, Via P. Bucci 12/C, 87036 Rende, Italy; dante.nistico@unical.it (D.M.N.); amalia.piro@unical.it (A.P.); daniela.oliva@unical.it (D.O.); vincent0593@hotmail.it (V.O.)
2 BIORISI S.r.l.—Oil Fox Europe, Via G. Pinna 78, 88046 Lamezia Terme, Italy; direzione@biorisi.it
3 Istituto per la Tecnologia delle Membrane (ITM-CNR), Università della Calabria, Via P. Bucci 17/C, 87036 Rende, Italy; r.morelli@itm.cnr.it (R.M.); c.conidi@itm.cnr.it (C.C.); a.figoli@itm.cnr.it (A.F.)
* Correspondence: silvia.mazzuca@unical.it (S.M.); a.cassano@itm.cnr.it (A.C.)

Abstract: The purification of phycocyanin (PC) from *Spirulina* generally involves a combination of different techniques. Here, we report the results on PC yields from a combined aqueous extraction-ultrafiltration (UF) process of a strain of *Arthrospira maxima* cultivated in a farm devoted to producing PC with food-grade purity. Samples optimized from different biomass/solvent ratios were purified by using a polyethersulphone (PES) membrane with a molecular weight cut-off (MWCO) of 20 kDa. The UF system was operated at 2.0 ± 0.1 bar and at 24 ± 2 °C up to a volume concentration factor (VCF) of 5. A diafiltration (DF) process was conducted after UF in order to increase the PC recovery in the retentate. Samples were collected during both UF and DF processes in order to evaluate membrane productivity and PC purity. The average permeate fluxes of about 14.4 L/m^{2}h were measured in the selected operating conditions and more than 96% of PC was rejected by the UF membrane independently ofthe extraction yields and times. The concentration of PC in the final retentate was 1.17 mg/mL; this confirmed the observed rejection and the final VCF of the process (about 5-fold when compared to the concentration of PC in the crude extract). In addition, the combination of UF and diafiltration allowed the removal of about 91.7% of the DNA from the crude extract, thereby improving the purity of the phycocyanin in the retentate fraction.

Keywords: *Spirulina*; phycocyanin; extraction; ultrafiltration; diafiltration

Citation: Nisticò, D.M.; Piro, A.; Oliva, D.; Osso, V.; Mazzuca, S.; Fagà, F.A.; Morelli, R.; Conidi, C.; Figoli, A.; Cassano, A. A Combination of Aqueous Extraction and Ultrafiltration for the Purification of Phycocyanin from *Arthrospira maxima*. *Microorganisms* **2022**, *10*, 308. https://doi.org/10.3390/microorganisms10020308

Academic Editor: Assaf Sukenik

Received: 15 December 2021
Accepted: 26 January 2022
Published: 28 January 2022

1. Introduction

Photosynthetic organisms, such as microalgae and cyanobacteriarepresent promising renewable sources of healthy food ingredients and functional food products due to their high contents of bioactive compounds, such as essential amino acids, antioxidant molecules, minerals and fibers. Compared to other natural sources of bioactive ingredients, these organisms have many advantages, including a wide biodiversity, the possibility to grow under conditions of low water utilization and the plasticity of their metabolism, which can be induced to produce specific molecules [1,2].

The production of microalgae-based protein products, for example, involve microalgae cultivation followed by harvesting, drying, cell disruption, protein extraction, hydrolysis and separation [3]. Among cyanobacteria, *Arthrospira platensis* (traditionally known as *Spirulina*), a blue-green coil shaped species, has received increased attention in recent years due to its high content of proteins, vitamins, minerals and many essential amino-acids and fatty acids [4]. It is an important source of phycocyanin (PC) and allophycocyanin (APC) water soluble proteins belonging to the phycobiliprotein family [5]. PC is a natural blue colorant, with an estimated molecular weight of 100–200 kDa; thanks to its therapeutic properties as an antioxidant, anti-inflammatory molecule, as well as its anti-cancer activities [6–9], the PC has great potential for industrial and commercial development. Its market value is estimated to be around 10–50 million US$ per annum [10].

The purification of PC from *Spirulina* and other microalgae has been investigated and optimized by several authors [11–16]. Purification generally involves different techniques that include extraction, centrifugation and separation by chromatography or ion exchange and dialysis. These procedures are time consuming and directly increase facilities and equipment-related expenses, thereby leading to an increase in production costs. In this context, membrane-based processes, such as microfiltration (MF) and ultrafiltration (UF), offer a useful approach as fractionation, purification and concentration steps asalternatives to expensive sequential purification techniques such as ammonium sulphate precipitation and gel filtration chromatography. These processes, thanks to their mild operating conditions, prevent possible thermal denaturation and deactivation of the PC molecules. Additional advantages over conventional technologies includean easy scale-up combined with high selectivity, modularity, low energy consumption, no phase change and no use of chemical additives [17,18].

UF membranes with a molecular weight cut-off (MWCO) of 50 kDa were used by Herrera et al. [19] to concentrate a spirulina extract up to a VCF of 1.9. A food grade phycocyanin powder with a purity ratio of 0.74 was obtained after adsorption on activated charcoal and spray drying. Jaouen et al. [20] investigated the use of MF and UF tubular inorganic membranes for the clarification of raw extracts after sonication of a spirulina culture, while UF, nanofiltration (NF) and reverse osmosis (RO) organic membranes were used for the concentration of the clarified extract. NF membranes exhibited the best performance in the concentration of the clarified extract: in selected conditions of operating pressure and tangential velocity (30 bar and 1.5 m/s, respectively), the recovery of PC and permeation flux resulted in 100% and 85 L/m^{2}h, respectively.

A combination of an aqueous two-phase system, UF and precipitation was developed by Rito-Palomares et al. [15] in order to reduce the number of unit operations and increased the yield of the protein. In this approach, the use of a 30 kDa UF membrane followed by precipitation with ammonium sulfate led to a protein purity of 3.8 $\pm$ 0.1% and an overall product yield of 29.5% (w/w).

Figueira et al. [21] obtained a C-PC extract with purity of 0.95 suitable for use as a food dye (purity between 0.75 and 1.5) by using a combination of UF and 6 diafiltration cycles. In this approach, a 50 kDa polyethersulfone UF membrane in flat-sheet configuration was used.

Recently, Brião et al. [22] simplified the purification step by using the phosphate buffer extraction followed by UF and one-step of diafiltration (DF) membranes; throughthis procedure, a food-grade PC from *A. platensis* has been produced. Similarly, UF in diafiltration mode was assessed by Balti et al. [23] for protein fractionation of *A. platensis* in order to produce different protein extracts more or less purified, notably in salt and chlorophyll.

This study aimed at developing a sustainable process for the recovery, fractionation and purification of PC from a strain of *A. maxima* extracts. It was based on a 'green' aqueous extraction of PC from the *Arthrosphira* biomass followed by a purification/fractionation step of the centrifuged extract through the use of a 20 kDa UF membrane aimed at removing mainly non-protein molecules, such as DNA, co-extracted with phycobiliproteins. To enhance the PC purification, UF was combined with a diafiltration (DF) step based on the addition of purified water to the UF retentate.

The effect of the biomass-water ratio on the PC concentration was investigated. Membrane productivity, cleaning efficiency and membrane retention towards PC and DNA were also assessed.

2. Materials and Methods

2.1. Chemicals

Bradford reagent, water CHROMASOLV®Plus for HPLC, bromophenol blue ACS reagent, ammonium bicarbonate BioUltra (≥99.5%) and acetonitrile anhydrous (99.8%) wereall acquired from Sigma-Aldrich (Milan, Italy). A 40% acrylamide/bis solution, ammonium persulfate, sodium dodecyl sulfate (SDS), trimethamine and glycine were purchased from Bio-Rad Laboratories (Hercules, CA, USA). Boric acid and ethylenediaminetetraacetic acid (EDTA) were acquired from Carlo Erba (Milan, Italy). Glycerol 99.0–101.0% was purchased from Honeywell Research Chemicals (Seelze, Germany). Coomassie R-250 solution was purchased fromAmersham Pharmacia Biotech (Uppsala, Sweden). Agarose CSL-AG5 for gel electrophoresis was purchased from Cleaver Scientific (Rugby, UK). Gel red nucleic acid gel stain 10,000× in water was purchased from Biotium (Fremont, CA, USA). Tetramethylethylendiamine (TEMED) was purchased from Chem-Lab (Zedelgem, Belgium). GeneRuler DNA Ladders 10Kpb was purchased from Thermo Fisher Scientific Inc. (Swindon, UK). All other reagents were of analytical grade.

2.2. Biological Materials

Dried biomass samples of the *Arthrospira maxima* strain cultivated according to the Oil Fox® technology and registered trademark Spirulina-fox® and covered by an international patent [24] were used.

2.3. Aqueous Extraction of Phycocyanin (PC) from A. maxima Biomass

The PC extraction from *A. maxima* samples was carried out at different biomass-solvent ratios in order to investigate the amount of solvent needed to maximize the yield of extraction. Ultrapure distilled water was added to dried *A. maxima* biomass to obtain suspensions of 200 mL at concentrations of 0.005, 0.01, 0.015 and 0.02 g/mL, respectively. All suspensions were stirred on a magnetic plate at 500 rpm for 24 h at RT. Three independent extractions were carried out and results of the PC concentration were expressed as mean value ± SD.

2.4. Feed Solution

After incubation time, the crude extract of *A. maxima* was centrifuged for 15 min at 6000 rpm, the supernatant removed and then frozen at −20 °C. The supernatant was used as a feed solution in the ultrafiltration experiments.

2.5. Ultrafiltration (UF): Experimental Set-Up and Procedure

Dead-end UF experiments were performed using the Sterlitech™ HP 4750 high-pressure stirred cell (Sterlitech, Kent, WA, USA) with a filter area of 13.85 cm^2 and a processing capacity of 300 mL. The cell filtration system was equipped with a flat-sheet polyethersulphone (PES) membrane (Nadir® UP020 P, from Microdyn-Nadir GmbH, Wiesbaden, Germany) with a thickness of 210–250 μm and a nominal MWCO of 20 kDa. Filtration experiments were carried out according to a batch concentration configuration at a transmembrane pressure (TMP) of 2 bar and an operating temperature of 24 ± 2 °C. Stirring inside the cell was accomplished by using a magnetic stirrer. An initial volume of crude extract (UF feed) of 100 mL was used and the permeate was collected separately up to a final volume of 80 mL corresponding to a volume concentration factor (VCF) of 5.

VCF is dimensionless and defined as:

$$\mathrm{VCF} = \frac{V_f}{V_r} \quad (1)$$

where V_f and V_r are the feed and retentate volumes, respectively.

Diafiltration experiments were performed by adding distilled water to the UF retentate at the same flowrate of the permeate so as to keep the retentate volume constant (15 mL) during the process. Six diafiltration volumes (addition of 90 mL of distilled water) were added before stopping filtration.

The diafiltration volume (ratio of the solvent volume added per volume of feed solution) was determined as follows [25]:

$$DV = \frac{V_p}{V_0} = \frac{V_w}{V_0} \quad (2)$$

where V_p and V_0 are the volumes of permeate and feed solution, respectively, and V_w is the volume of water added during the diafiltration process.

The diafiltration was operated at the same parameters of temperature, stirring conditions and transmembrane pressure asthe UF.

The schematic flowchart of the combined extraction–filtration process investigated in this study is depicted in Figure 1.

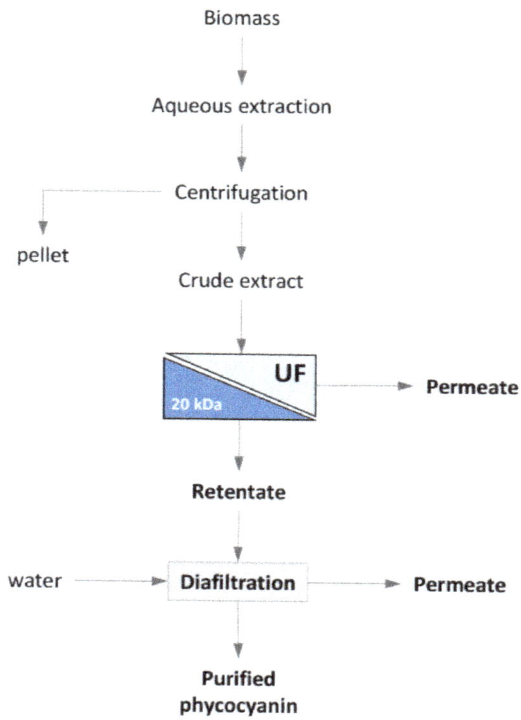

Figure 1. Schematic flowchart of the combined extraction/UF process investigated.

2.6. Data Processing

The permeate flux (J_p), expressed in L/m^{2}h, was determined by measuring the permeate volume collected in a given time according to Equation (3):

$$J_p = \frac{V_p}{t\,A} \quad (3)$$

where V_p (L) is the permeate volume, t (h) is the permeation time and A (m^2) is the membrane area.

The hydraulic permeability of the UF membrane was determined by measuring the water permeate flux at different TMP values (at an operating temperature of 25 °C). The fouling index was determined by measuring the hydraulic permeability before and after filtration experiments, according to the following equation:

$$FI = \left(\frac{W_{p1}}{W_{p0}}\right) \cdot 100 \quad (4)$$

where W_{p0} and W_{p1} are the pure water permeability before and after UF experiments, respectively.

After the experiments with the crude extract, the UF membrane was cleaned in two steps. The first cleaning step was performed with distilled water for 30 min at a temperature of 40 °C in order to remove the reversible polarized layer. The hydraulic permeability measured afterwards was W_{p2}.

In the second step, the membrane was submitted to an enzymatic cleaning with a solution of Ultrasil P3 (Henkel-Ecolab, Dusseldorf, Germany) at 0.5% *w*/*w* for 30 min at 40 °C. At the end of the cleaning procedure, the membrane was rinsed with distilled water for 10 min and the hydraulic permeability, indicated as W_{p3}, was measured.

The cleaning efficiency (*CE*) was evaluated by using the flux recovery method according to the following equation:

$$CE = \left(\frac{W_{p3}}{W_{p0}}\right) \cdot 100 \quad (5)$$

where W_{p3} is the water permeability measured after the enzymatic cleaning.

The rejection rate (*R*) of PC was calculated as described in Equation (6):

$$R = \left(1 - \frac{C_p}{C_f}\right) \cdot 100 \quad (6)$$

where C_p and C_f refer to the concentration of PC within the permeate and feed, respectively.

2.7. Analytical Assays

2.7.1. Phycocyanin (PC) Concentration

The PC concentration was measured by using a UV-VIS spectrophotometer (Shimadzu UV-160A, Kyoto, Japan) at wavelengths of 615 nm and 652 nm. The total amount of PC was calculated by using the following equation [26,27]:

$$\mathrm{PC} = \frac{\mathrm{A}_{615} - 0.474(\mathrm{A}_{652})}{5.34} \quad (7)$$

where PC is the phycocyanin concentration (mg/mL), A_{615} is the optical density of the sample at 615 nm and A_{652} is the optical density of the sample at 652 nm.

The PC purity was monitored spectrophotometrically by the A_{615}/A_{280} ratio [28] according to the Equation (8):

$$\mathrm{PP} = \frac{\mathrm{A}_{615}}{\mathrm{A}_{280}} \quad (8)$$

where A_{280} is the optical density of the sample at 280 nm.

The purification factor (*PF*) was calculated from the ratio of PC concentration in purified (PC_p) and crude extract (PC_c) by Equation (9) [29]:

$$PF = \frac{PC_p}{PC_c} \quad (9)$$

2.7.2. Protein Analysis by SDS-PAGE of the Feed, Retentate and Permeate after UF

The protein concentration was measured by the Bradford method [30] and the protein sample was separated by the acrylammide gel electrophoresis under denaturant conditions (SDS-PAGE) [31]. The protein extract, solubilized in a loading buffer, was activated for 4 min at 100 °C. 20 µL for each sample were loaded onto a 12% polyacrylamide gel. The electrophoretic run was performed by placing the gel in the electrophoretic chamber filled with a buffer solution (Buffer Tris Glycine) and setting the amperage and voltage (60 mA in the stacking gel and 120 mA in the running gel at a power of 200 V). Once the run was over, the gels were placed in a Coomassie R-250 solution overnight and then destained by washing in water, ammonium bicarbonate and acetonitrile solution. The gels were scanned using the "GS800" (Biorad, Hercules, CA, USA) densitometer and analyzed with the "QuantityOne" software (Biorad, Hercules, CA, USA) to identify the polypeptide bands.

2.7.3. DNA Detection

To evaluate the nucleic acid occurrence in the three fractions after ultrafiltration, the quantity of DNA molecules has been assessed by the NanoDrop™ One/One^{C} Microvolume UV-Vis Spectrophotometer (Thermo Fisher Scientific Inc., London, UK); absorbances at 260 nm and 280 nm were measured and concentrations were calculated as:

$$\text{DNA concentration } (\text{ng}/\mu\text{L}) = A_{260\text{nm}} \cdot \frac{50\ \text{ng}/\mu\text{L}}{1\text{a.u.}} \tag{10}$$

The quality of the DNA has been evaluated by the direct measurements of the ratios A_{260}/A_{280}.

For the electrophoresis analysis, equal volumes of feed, permeate and retentate from the same UF assay were solubilized in an equal volume of loading buffer under stirring overnight at RT. Samples were separated in a 0.8% agarose gel using 0.5X TBE as the running buffer; bromophenol blue was added to the samples as a tracking dye. The 10 kbp DNA ladder was used (Thermo Fisher Scientific Inc.). After 30 min of running at 70 mA, 120 V, the gel image was digitalized using a UV Transilluminator (Consort bvba, Turnhout, Belgium) equipped with a photocamera.

All analytical measurements were performed in triplicate. Results were expressed as mean value ± SD.

3. Results

3.1. Aqueous Extraction of PC

The variable biomass-solvent ratio strongly influences the PC concentration in the aqueous extract [32]. As illustrated in Figure 2, the maximum yield of PC (0.232 ± 0.0042 mg/mL) in the aqueous extract was obtained by using the largest biomass-solvent ratio (0.02 g/mL). Therefore, UF experiments were performed by using an aqueous extract sample obtained with a biomass-solvent ratio of 0.02 g/mL.

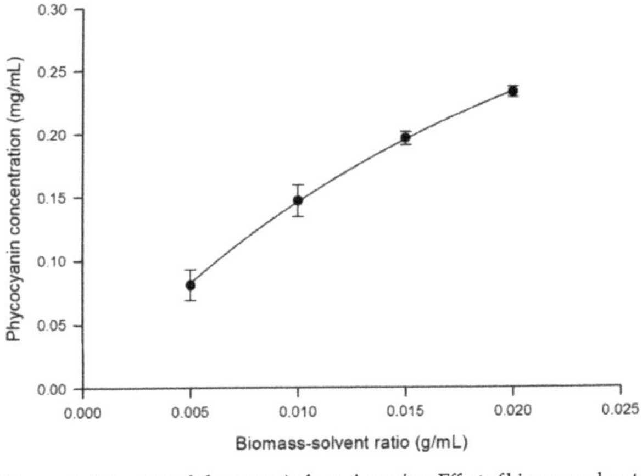

Figure 2. Extraction of phycocyanin from *A. maxima*. Effect of biomass-solvent ratio on phycocyanin concentration.

In Figure 3, the SDS-PAGE pattern of the supernatant from the crude extract at different biomass-solvent ratiosis reported. As it is observed, only a prominent polypeptide band at approx. 19 kDa was resolved in all aqueous extracts. The abundance of this band increased accordingly with the concentration of the extracts. Very weak bands at 50 kDa, 35 kDa and 10 kDa were also detected.The majority of A. maxima proteins, mainly the hydrophobic ones, remained in the biomass pellet, along with all the other water insoluble molecules (see the Supplementary Figure S1).

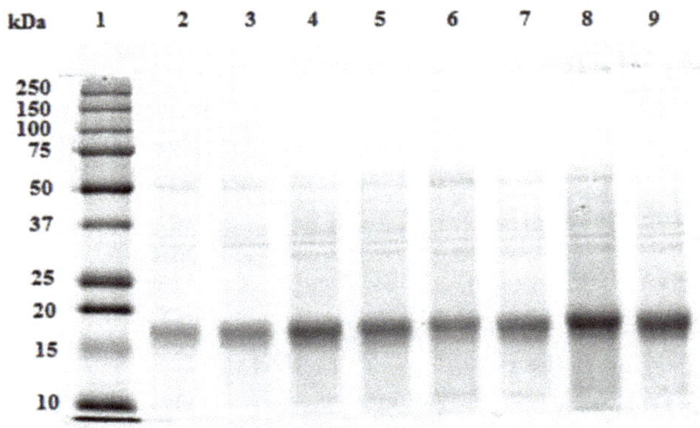

Figure 3. SDS-PAGE of the centrifuged extracts from 0.005 g/mL (lanes 2 and 3), 0.010 g/mL (lanes 4 and 5), 0.015 g/mL (lanes 6 and 7) and 0.020 g/mL (lanes 8 and 9) biomass-solvent ratio. 20 μL of each extract were loaded on the gel. Precision Plus ProteinTM Standards (Biorad, Hercules, CA, USA) were loaded on lane 1.

3.2. *Ultrafiltration/Diafiltration*

The feed solution was ultrafiltered with the selected membrane under operating conditions of pressure and temperature of 2 bar and 24 ± 2 °C, respectively. The initial permeate flux of about 19.2 L/m^{2}h decreased sharply to 26.5% in the first steps of the process to reach a steady-state value of about 11.5 L/m^{2}h at a VCF of 3.1; then, the permeate flux continued to decrease, reaching 9.5 L/m^{2}h when the final VCF of 5 was reached (Figure 4).

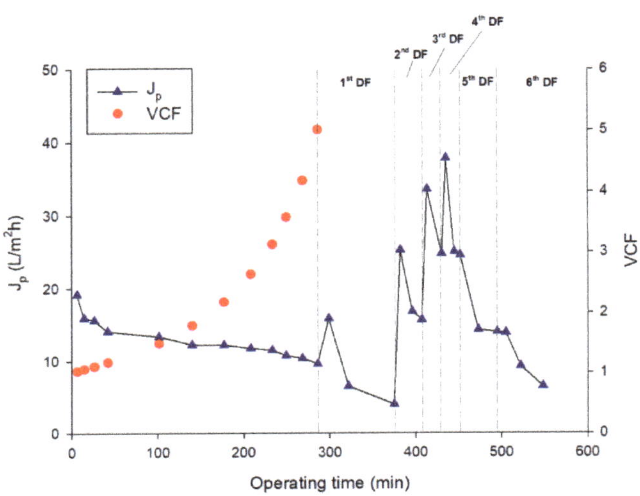

Figure 4. Ultrafiltration of *A. maxima* extract followed by six steps of diafiltration. Time course of permeate flux and volume concentration factor (VCF).

The addition of water in the DF steps produced a dilution of the retentate stream and an increase of the permeate flux at the beginning of each cycle of DF (with the exception of the last two DF steps) (Figure 4), followed by flux decay due to concentration polarization phenomena [22]. Figures 5 and 6 illustrate the samples collected during the UF and DF steps, respectively.

Figure 5. UF samples. (**a**) Aqueous crude extract (supernatant of centrifugation); (**b**) UF permeate; (**c**) UF retentate (at VCF 5).

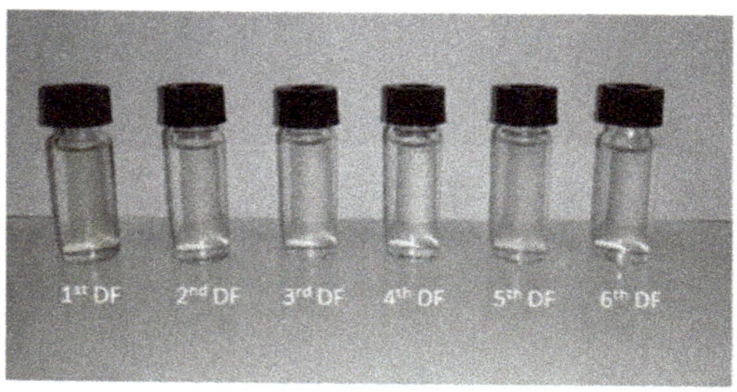

Figure 6. Permeate samples recovered in the course of diafiltration.

Figure 7 shows the pure water flux of the UF membrane before and after the treatment of the crude extract and after cleaning treatments. The initial water permeability of the membrane, of about 140.3 L/m^{2}hbar, dropped to 67.0 L/m^{2}hbar after the treatment of the crude extract. Therefore, the fouling index was estimated to be 52.3%. The cleaning with distilled water and enzymatic detergent at 40 °C allowed an increase in the membrane water permeability at 74.9 L/m^{2}hbar and 114.2 L/m^{2}hbar, respectively. Thus, the cleaning efficiency result was about 81%.

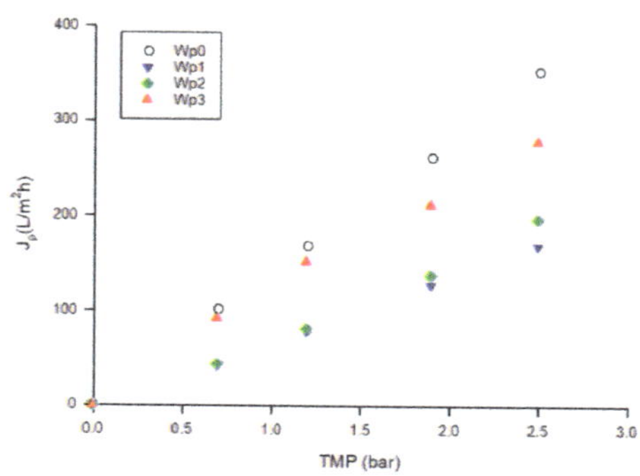

Figure 7. Water permeability of UF membrane before the treatment of aqueous extract and after cleaning procedures (W_{p0}, initial water permeability; W_{p1}, water permeability after treatment of aqueous extract; W_{p2}, water permeability after cleaning with water; W_{p3}, water permeability after enzymatic cleaning).

3.3. Analyses of Phycocyanin

In Figure 8, the electrophoresis of the fractions obtained after the UF treatment of the 0.020 g/mL feed solution is shown. Only one polypeptide band of aprox. 19 kDa was resolved in feed and retentates. This band appeared more concentrated in the retentate withrespect to the starting concentration of the feed fraction. No polypeptide band has been resolved in the permeate fraction.

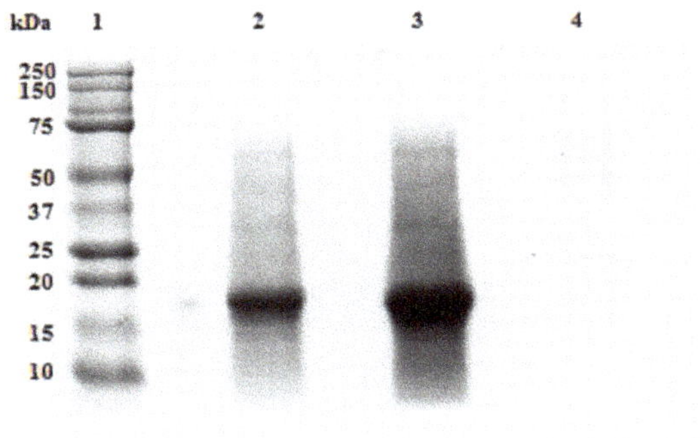

Figure 8. SDS-PAGE of the UF fractions from the aqueous extract of A. maxima at a concentration of 0.02 g/mL.An intense polypeptide band of approximately 19 kDa was resolved in the feed (lane 2) which became more intense in the retentate (lane 3) but was not detectable in the permeated fraction (lane 4). 20 µL of each fraction were loaded on the gel. Precision Plus ProteinTM Standards (Biorad, Hercules, CA, USA) wereloaded on lane 1.

In Figure 9, the absorbance spectra of the UF fractions in the range of 480–750 nm are reported. Higher absorbance was observed in the retentate fraction with respect to feed and permeate samples. The concentration of PC and related purity in the feed in the UF fractions and in the final retentate after diafiltration are reported in Table 1.

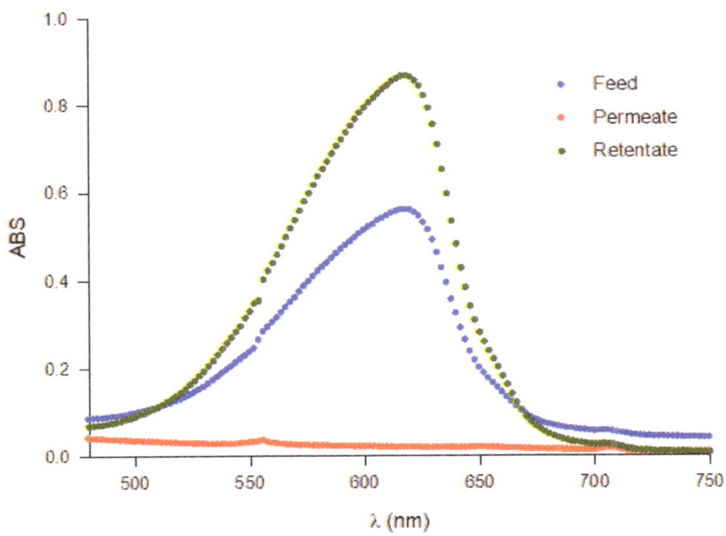

Figure 9. Full scan of the absorbance of UF fractions (feed, retentate and permeate).

Table 1. Concentration and purity of PC obtained by aqueous extraction followed by UF and diafiltration (PF, purification factor).

Samples	PC Concentration (mg/mL)	PC Purity	PF
Feed	0.232 ± 0.0042	0.74 ± 0.030	-
Permeate	0.0089 ± 0.00085	-	-
Retentate	1.124 ± 0.0234	0.93 ± 0.009	4.84
Final retentate	1.174 ± 0.0244	1.16 ± 0.010	5.06

The concentration of PC in the retentate was4.8-fold higher than that of the feed solution and it was in agreement with the VCF of the process. The rejection of the UF membrane towards the PC was 96.2% and according to the mass balance of the process, 96.9% of the PC was recovered in the retentate.

The purity value of the crude extract increased from 0.74 to 0.93 in the UF retentate: accordingly, the UF process allowed a purification factor of 4.84. After six diafiltration cycles, the PC purity of the retentate and the purification factor increased to 1.16 and 5.06, respectively. This behavior can be attributed to the removal of most contaminant particles smaller than the nominal MWCO of the UF membrane (20 kDa) during the first 6 cycles of DF.

3.4. DNA Analyses

The agarose gel electrophoresis of two independent extractions is reported in Figure 10. As can be seen, nucleic acid contamination occurred in the feed solution, suggesting that DNA molecules were co-extracted during solubilization with proteins from *A. maxima* biomass; after ultrafiltration, the DNA molecules were not detected in the retentate, suggesting that the DNA was efficiently removed; on the other hand, DNA was not detected in the permeate fraction, suggesting that during filtration DNA was degraded in very small fragments not detectable through electrophoresis in our conditions. This is corroborated by the evidence of the Abs 260/280 ratio values (Supplementary Table S1).

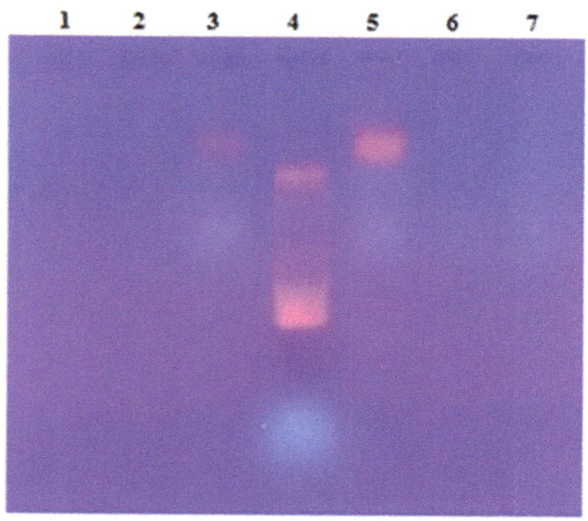

Figure 10. Agarose gel electrophoresis of the UF fractions obtained from the aqueous PC extracts from *A. maxima*. DNA bands were detected in the feeds (lanes 3 and 5) but not in the retentate fractions (lanes 1 and 6) nor in the permeate fractions (lanes 2 and 7). 10bp DNA ladder was used as electrophoresis markers (Lane 4).

The contents of DNA measured in all fractions from the UF and diafiltration process of the feed solution are reported in Table 2.

Table 2. Mass balance of DNA in both UF and DF processes.

Sample	DNA (ng/μL)	Volume	DNA (μg)
Feed	425 ± 8.5	100	42,500.0 ± 637.5
UF Permeate	413 ± 8.2	80	33,040.0 ± 495.6
UF Retentate	473 ± 9.4	20	9460.0 ± 141.9
DF1	181 ± 3.6	15	2715.0 ± 40.7
DF2	107 ± 2.1	15	1605.0 ± 24.1
DF3	52 ± 1.0	15	780.0 ± 11.7
DF4	25 ± 0.5	15	375.0 ± 5.6
DF5	20.5 ± 0.4	15	307.5 ± 4.6
DF6	11 ± 0.2	15	165.0 ± 2.4
DF tot			5947.5 ± 89.2
Permeate			33,040.0 ± 495.6
DFtot + Permeate			38,987.5 ± 584.8

According to the mass balance of the process, about 77% and 14% of DNA wasrecovered in the permeate and diafiltrate samples, respectively. Therefore, the combination of UF and DF allowed the removal of about 91.7% of the DNA from the crude extract, thereby improving the purity of the phycocyanin in the retentate fraction.

4. Discussion

The extraction protocol developed in this work obtained a very high phycocyanin yield from *A. maxima* biomass. The biomass-solvent ratio of 0.02 g/mL makes possible the use of smaller volumes of extractant in the purification steps and it agrees with data reported by Silveira et al. [29] in the optimization of PC extraction from *Spirulina platensis* using factorial design. The authors also reported that the use of a biomass-solvent ratio higher than 0.08 g/mL produced very concentrated suspensions so that the solvent was unable to promote an appropriate interaction with the biomass for efficient extraction.

It is well known that representative phycobiliproteins in *Spirulina* are phycocyanin and allophycocyanin [23,33]. It is interesting to point out that the allophycocyanin concentration measured in the original crude extract of *A. maxima* was about 0.01 mg/mL, 23-fold less than phycocyanin concentration. This unbalanced content of phycobiliproteins could mainly be attributed to our biological source of *A. maxima* cultivated according to the Oilfox® technology [24].

The electrophoretic profile of *A. maxima* PC crude extract has shown that the majority of the water soluble proteins correspond to the molecular weight of approx. 19 kDa; in *Spirulina platensis*, the pure C-phycocyanin was reported to be formed by two subunits corresponding to α and β subunits of 17 and 21 kDa, respectively; as the molecular weight of the native purified C-PC was 115 kDa, the aggregation state of the purified C-PC was accomplished by the assemblage (αβ)3 [34]. Despite the literature on the C-phycocyanin extraction and purification from *A. maxima* [35–37], there is no information, to our knowledge, on the molecular characterization of C-phycocyanin in this species. Under denaturing conditions, our results suggest that the C-phycocyanin in *A. maxima* could consist of monomers with an apparent molecular weight of approximately 19 kDa. Another hypothesis is that, in our conditions, it is not possible to separate the two subunits that therefore appear merged into a single band. Further molecular studies are needed to clarify this finding.

The decrease in permeate flux observed during the UF of the crude extract in selected operating conditions could be attributed to pore blockage and cake formation mechanisms occurring sequentially during the filtration process. A larger drop in the permeate flux followed by long-term decay until reaching a steadystate of about 15 L/m^{2}h was observed by Brião et al. [22] in the UF of a crude PC extract (0.53 mg/L) by using a PES UF membrane with a MWCO of 60 kDa. Higher fluxes (of the order of 50 L/m^{2}h) were obtained by Chaiklahan et al. [38] by using membranes of 50 and 70 kDa; in this case, the crude extract was previously microfiltered in order to remove cellular fragments before UF.

The UF process has proven effective both in the purification of C-phycocyanin and its five-fold concentration in our experimental conditions regardless of the starting biomass-solvent ratio. Furthermore, UF combined with diafiltration solved the contamination by DNA.

The PC purity and the purification factor measured after the UF/DF treatment of the crude extract are in agreement with data reported by Figueira et al. [21] in the UF of a crude extract of *S. platensis* by using a 50 kDa PES membrane in a flat-sheet configuration. The authors reported a PC recovery between 60 and 99% and purification factors between 1.35 and 1.57 in different conditions of pressure (1, 1.5 and 2 kgf/cm^2), pH (6, 6.5 and 7), temperature (4, 14.5 and 25 °C) and diafiltration cycles (0, 2 and 4). Similarly, Chaiklahan et al. [38] reported purity ratios of 1.11, 1.03 and 1.08 after the UF of the microfiltered extract with 50 kDa, 70 kDa and 100 kDa membranes, respectively. The purity ratio of the phycocyanin extract increased approximately 2-fold when compared to the initial purity (0.54).

On the other hand, Brião et al. [22] reported that the MF and UF membranes did not successfully purify the PC extracted from *Spirulina* by freeze-thaw extraction. Indeed, the PC purity of the crude extract decreased from 0.355 to 0.345 in the retentate of the MF process (with a flat-sheet polyvinylidene fluoride membrane of 0.4 μm as pore diameter) and from 0.403 to 0.382 in the retentate of the UF process (with a flat-sheet PES membrane with a MWCO of 60 kDa). These results were attributed to the formation of a thick cake layer which blocked the membrane pores, thus preventing the removal of lower molecular weight contaminants from the crude extract. The extraction with sodium phosphate buffer followed by UF (with a tubular PVDF membrane of 30–80 kDa) and one step of diafiltration allowed theenhancement of the purity of the crude extract up to 0.76, which in any case is lower than the result obtained in our work. In addition, the purification resulted in a 30% loss of antioxidant activity.

Literature data confirm that the combination of UF with diafiltration allow researchers to obtain C-phycocyanin extracts with purity between 0.75 and 1.5, which are suitable for use as food dye [21,38]. Higher purity ratios (up to >4) can be reached through the combination of UF with other technologies (i.e., ammonium sulfate precipitation, expanded-bed and fixed-bed ion exchange chromatography). Extracts suitable as cosmetic dye (purity of 2.1), biomarkers (purity of 3.0) and for therapeutic and biomedicine applications with an analytical grade (purity >4.0) were obtained by Figueira et al. [21] according to different combinations of processes, including UF.

After UF, electrophoresis showed that the band of C-phycocyanin is concentrated in the retentate fraction without changes in the molecular weight. Here, we can speculate that the native form of C-phycocyanin in the feed solution had the least dimeric conformation and therefore was retained by the membrane with a 20 kDa cut-off. Structural modifications might occur during C-phycocyanin ultrafiltration [21]; at the temperature of 25 °C, an increase in C-phycocyanin recovery was reported. This increase can be attributed to higher retention due to a slight increase in the molar mass of C-phycocyanin.

5. Conclusions

A green and sustainable method for the extraction and purification of phycocyanin (PC) from a strain of *Arthrospira maxima* has been developed. Aqueous extracts optimized from different biomass-solvent ratios were purified by ultrafiltration (UF) by using a 20 kDa membrane in a flat-sheet configuration. The UF process allowed the recovery of more than 96% of the PC in the retentate fraction and the removal of about 91.7% of the DNA content from the crude extract. The purity value of the crude extract increased from 0.74 to 0.93 after the UF process. The application of 6 DF cycles enhanced the purification process, thereby obtaining a C-PC extract with a purity degree of 1.16, which is suitable for human food use. The combination of UF with other methodologies (i.e., ammonium sulfate precipitation and ion exchange chromatography) could be a useful approach forenhancing the purity degree of DNA-free ultrafiltered samples, thereby allowing the production of extracts suitable for other uses including therapeutic and biomedicine applications.

Supplementary Materials: The following supporting information can be downloaded at: https://www.mdpi.com/article/10.3390/microorganisms10020308/s1, Figure S1: SDS-PAGE of the pellet after the centrifugation of the extract from 0.005 g/mL (lanes 1), 0.010 g/mL (lanes 2), 0.015 g/mL (lane 4) and 0.020 g/mL (lane 5) biomass-solvent ratio. 20 μL of each extract wasloaded on the gel. Precision Plus ProteinTM Standards (Biorad, Hercules, CA, USA) was loaded on the lane 3; Table S1: Values of DNA absorbance at 260 and 280 nm in the aqueous PC solution (feed), UF and DF fractions and the respective absorbance ratios.

Author Contributions: Conceptualization, S.M. and A.C.; methodology, D.M.N., A.P., D.O., V.O., R.M. and C.C.; formal analysis, A.P., D.M.N. and C.C.; investigation, A.F.; resources, F.A.F.; data curation, S.M., A.P. and A.C.; writing—original draft preparation, D.M.N., S.M. and A.C.; writing—review and editing, S.M. and A.C.; supervision, S.M., A.F. and A.C. All authors have read and agreed to the published version of the manuscript.

Funding: This research was carried out with the financial support of the University of Calabria (ex 60 % research found 2020–2021).

Institutional Review Board Statement: Not applicable.

Informed Consent Statement: Not applicable.

Data Availability Statement: Not applicable.

Conflicts of Interest: The authors declare no conflict of interest.

References

1. Buono, S.; Langellotti, A.L.; Martello, A.; Rinna, F.; Fogliano, V. Functional ingredients from microalgae. *Food Funct.* **2014**, *5*, 669–1685. [CrossRef] [PubMed]
2. Matos, J.; Cardoso, C.; Bandarra, N.M.; Afonso, C. Microalgae as a healthy ingredient for functional food: A review. *Food Funct.* **2017**, *8*, 2672–2685. [CrossRef] [PubMed]
3. Soto-Sierra, L.; Stoykova, P.; Nikolov, Z.L. Extraction and fractionation of microalgae-based protein products. *Algal Res.* **2018**, *36*, 175–192. [CrossRef]
4. Grosshagauer, S.; Kraemer, K.; Somoza, V. The true value of *Spirulina*. *J. Agric. Food Chem.* **2020**, *68*, 4109–4115. [CrossRef]
5. Moraes, C.C.; Kalil, S.J. Strategy for a protein purification design using C-phycocyanin extract. *Bioresour. Technol.* **2009**, *100*, 5312–5317. [CrossRef]
6. Reddy, M.C.; Subliashini, J.; Mahipal, S.V.K.; Bhat, V.B.; Reddy, P.S.; Kiranmai, G.; Madyastha, K.M.; Reddanna, P. C-Phycocyanin, a selective cyclooxygenase-2 inhibitor, induces apoptosis in lipopolysaccharide-stimulated RAW 264.7 macrophages. *Biochem. Biophys. Res. Commun.* **2003**, *304*, 385–392. [CrossRef]
7. Eriksen, N.T. Production of phycocyanin—A pigment with applications in biology, biotechnology, foods and medicine. *Appl. Microbiol. Biotechnol.* **2008**, *80*, 1–14. [CrossRef]
8. Gantar, M.; Simovic, D.; Djilas, S.; Gonzales, W.W.; Miksovska, J. Isolation, characterization and antioxidative activity of C-phycocyanin from *Limnothrix* sp. strain 37-2-1. *J. Biotechnol.* **2012**, *159*, 21–26. [CrossRef]
9. Raposo, M.F.L.; Morais, R.M.S.C.; Morais, A.M.M.B. Health applications of bioactive compounds from marine microalgae. *Life Sci.* **2013**, *93*, 479–486. [CrossRef]
10. Spolaore, P.; Joannis-Cassan, C.; Duran, E.; Isambert, A. Review commercial applications of microalgae. *J. Biosci. Bioeng.* **2006**, *101*, 87–201. [CrossRef]
11. Antelo, F.S.; Costa, J.A.V.; Kalil, S.J. Purification of C-phycocyanin from *Spirulina platensis* in aqueoustwo-phase systemsusing an experimental design. *Braz. Arch. Biol. Technol.* **2015**, *58*, 1–11. [CrossRef]
12. Benedetti, S.; Rinalducci, S.; Benvenuti, F.; Francogli, S.; Pagliarani, S.; Giorgi, L.; Micheloni, M.; D'Amici, G.M.; Zolla, L.; Canestrari, F. Purification and characterization of phycocyanin from the blue-green alga *Aphanizomenonflos-aquae*. *J. Chromatogr. B* **2006**, *833*, 12–18. [CrossRef] [PubMed]
13. Kamble, S.P.; Gaikar, R.B.; Padalia, R.B.; Shinde, K.D. Extraction and purification of C-phycocyanin from dry *Spirulina* powder and evaluating itsantioxidant, anticoagulation and prevention of DNA damage activity. *J. Appl. Pharm. Sci.* **2013**, *3*, 149–153.
14. Kumar, D.; Dhar, D.W.; Pabbi, S.; Kumar, N.; Walia, S. Extraction and purification of C-phycocyanin from *Spirulina platensis* (CCC540). *Indian J. Plant Physiol.* **2014**, *19*, 184–188. [CrossRef] [PubMed]

15. Rito-Palomares, M.; Nuñez, L.; Amador, D. Practical application of aqueous two-phase systems for the development of a prototype process for phycocyanin recovery from *Spirulina maxima*. *J. Chem. Technol. Biotechnol.* **2001**, *76*, 1273–1280. [CrossRef]
16. Sørensen, L.; Hantke, A.; Eriksen, N.T. Purification of the photosynthetic pigment C-phycocyanin from heterotrophic *Galdieriasulphuraria*. *J. Sci. Food Agric.* **2013**, *93*, 2933–2938. [CrossRef] [PubMed]
17. Li, J.; Chase, H.A. Applications of membrane technique for purification of natural products. *Biotechnol. Lett.* **2010**, *32*, 601–608. [CrossRef]
18. Conidi, C.; Drioli, E.; Cassano, A. Membrane-based agro-food production processes for polyphenol separation, purification and concentration. *Curr. Opin. Food Sci.* **2018**, *23*, 149–164. [CrossRef]
19. Herrera, A.; Boussiba, S.; Napoleone, V.; Hohlberg, A. Recovery of c-phycocyanin from the cyanobacterium *Spirulina maxima*. *J. Appl. Phycol.* **1989**, *1*, 325–331. [CrossRef]
20. Jaouen, P.; Lépine, B.; Rossignol, N.; Royer, R.; Quéméneur, F. Clarification and concentration with membrance technology of a phycocyanin solution extracted from *Spirulina platensis*. *Biotechnol. Techol.* **1999**, *13*, 877–881. [CrossRef]
21. Figueira, F.D.; Moraes, C.C.; Kalil, S.J. C-phycocyanin purification: Multiple processes for different applications. *Braz. J. Chem. Eng.* **2018**, *35*, 1117–1128. [CrossRef]
22. Brião, V.B.; Sbeghen, A.L.; Colla, L.M.; Castoldi, V.; Seguenka, B.; Schimidt, G.D.; Costa, J.A.V. Is downstream ultrafiltration enough for production of food-grade phycocyanin from *Arthrospira platensis*? *J. Appl. Phycol.* **2020**, *32*, 1129–1140. [CrossRef]
23. Balti, R.; Zayoud, N.; Hubert, F.; Beaulieu, L.; Massé, A. Fractionation of *Arthrospira platensis* (*Spirulina*) water soluble proteins by membrane diafiltration. *Sep. Purif. Technol.* **2021**, *256*, 117756. [CrossRef]
24. Oil Fox SA. Instalacion para el Cultivo Extensivo e Intensivo de Microalgas en Invernaderos de Estructura Tubular tipo Carrusel. Patent AR070504B1, 19 February 2016.
25. Mittal, R.; Lamdande, A.G.; Sharma, R.; Raghavarao, K.S.M.S. Membrane processing for purification of R-phycoerythrin from marine macro-alga, *Gelidiumpusillum* and process integration. *Sep. Purif. Technol.* **2020**, *252*, 117470. [CrossRef]
26. Bennett, A.; Bogorad, L. Complementary chromatic adaptation in a filamentous blue green alga. *J. Cell Biol.* **1973**, *58*, 419–435. [CrossRef]
27. Patel, A.; Mishra, S.; Pawar, R.; Ghosh, P.K. Purification and characterization of C-Phycocyanin from cyanobacterial species of marine and freshwater habitat. *Protein Expr. Purif.* **2005**, *40*, 248–255. [CrossRef]
28. Abalde, J.; Bentancourt, L.; Torres, E.; Cid, A.; Barwell, C. Purification and characterization of phycocyanin from the marine cyanobacterium *Synechococcus* sp. IO9201. *Plant Sci.* **1998**, *136*, 109–120. [CrossRef]
29. Silveira, S.T.; Burkert, J.F.M.; Costa, J.A.V.; Burkert, C.A.V.; Kalil, S.J. Optimization of phycocyanin extraction from *Spirulina platensis* using factorial design. *Bioresour. Technol.* **2007**, *98*, 1629–1634. [CrossRef] [PubMed]
30. Bradford, M.M. Rapid and sensitive method for the quantitation of microgram quantities of protein utilizing the principle of protein-dye binding. *Anal. Biochem.* **1976**, *72*, 248–254. [CrossRef]
31. Laemli, U.K. Cleavage of structural proteins during the assembly of the head of bacteriophage T4. *Nature* **1970**, *227*, 680–685. [CrossRef]
32. Reis, A.; Mendes, A.; Lobo-Fernandes, H.; Empis, J.A.; Novais, J.M. Production, extraction and purification of phycobiliproteins from *Nostoc* sp. *Bioresour. Technol.* **1998**, *66*, 181–187. [CrossRef]
33. Fernández-Rojas, B.; Hernández-Juárez, J.; Pedraza-Chaverri, J. Nutraceutical properties of phycocyanin. *J. Funct. Foods* **2014**, *11*, 375–392. [CrossRef]
34. Song, W.; Zhao, C.; Wang, S. A large-scale preparation method of high purity C-phycocyanin. *Int. J. Biosci. Biochem. Bioinform.* **2013**, *3*, 293–297.
35. Ruiz-Domínguez, M.C.; Jáuregui, M.; Medina, E.; Jaime, C.; Cerezal, P. Rapid green extractions of C-phycocyanin from *Arthrospira maxima* for functional applications. *Appl. Sci.* **2019**, *9*, 1987. [CrossRef]
36. Antonio-Marcos, E.; Hernández-Vázquez, L.; Olguín, E.J.; Monroy, O.; Navarro-Ocaña, A. C-phycocyanin from *Arthrospira maxima* LJGR1: Production, extraction and protection. *J. Adv. Biotechnol.* **2016**, *5*, 659–666. [CrossRef]
37. Rodríguez-Sánchez, R.; Ortiz-Butrón, R.; Blas-Valdivia, V.; Hernández-García, A.; Cano-Europa, E. Phycobiliproteins or C-phycocyanin of *Arthrospira (Spirulina)* maxima protect against $HgCl_2$-caused oxidative stress and renal damage. *Food Chem.* **2012**, *135*, 2359–2365. [CrossRef]
38. Chaiklahan, R.; Chirasuwan, N.; Loha, V.; Tia, S.; Bunnag, B. Separation and purification of phycocyanin from *Spirulina* sp. using a membrane process. *Bioresour. Technol.* **2011**, *102*, 7159–7164. [CrossRef]

microorganisms

Article

Physiological and Metabolic Response of *Arthrospira maxima* to Organophosphates

Amalia Piro [1,*], Dante Matteo Nisticò [1], Daniela Oliva [1], Francesco Antonio Fagà [2] and Silvia Mazzuca [1]

1 Laboratorio di Biologia e Proteomica Vegetale, Dipartimento di Chimica e Tecnologie Chimiche, Università della Calabria, via P. Bucci 12/C, 87036 Rende, Italy; dante.nistico@unical.it (D.M.N.); daniela.oliva@unical.it (D.O.); silvia.mazzuca@unical.it (S.M.)
2 BIORISI S.r.l.—Oil Fox Europe, Via G. Pinna 78, 88046 Lamezia Terme, Italy; direzione@biorisi.it
* Correspondence: amalia.piro@unical.it

Abstract: The *Spirulina* spp. exhibited an ability to tolerate the organophosphates. This study aimed to explore the effects of the herbicide glyphosate on a selected strain of the cyanobacteria *Arthrospira maxima* cultivated in a company. Experimental cultivations acclimated in aquaria were treated with 0.2 mM glyphosate [N-(phosphonomethyl) glycine]. The culture biomass, the phycocyanin, and the chlorophyll *a* concentrations were evaluated every week during 42 days of treatment. The differentially expressed proteins in the treated cyanobacteria versus the control cultivations were evaluated weekly during 21 days of treatment. Even if the glyphosate treatment negatively affected the biomass and the photosynthetic pigments, it induced resistance in the survival *A. maxima* population. Proteins belonging to the response to osmotic stress and methylation pathways were strongly accumulated in treated cultivation; the response to toxic substances and the negative regulation of transcription seemed to have a role in the resistance. The glyphosate-affected enzyme, chorismate synthase, a key enzyme in the shikimic acid pathway, was accumulated during treatment, suggesting that the surviving strain of *A. maxima* expressed a glyphosate-resistant target enzyme.

Keywords: spirulina; *Arthrospira maxima*; glyphosate; proteomics; resistance

Citation: Piro, A.; Nisticò, D.M.; Oliva, D.; Fagà, F.A.; Mazzuca, S. Physiological and Metabolic Response of *Arthrospira maxima* to Organophosphates. *Microorganisms* **2022**, *10*, 1063. https://doi.org/10.3390/microorganisms10051063

Academic Editors: Patrizia Pagliara and Carmela Caroppo

Received: 9 May 2022
Accepted: 19 May 2022
Published: 21 May 2022

1. Introduction

Literature of the last ten years has focused on the medium and long-term effects of the persistence of organophosphates, revealing their toxicity and the serious impact on ecosystems of all environmental matrices (air, soil, rivers, sea) and consequently on the health of humans [1,2]. By contrast, related literature also reported conflicting results on the role of organophosphates in perturbing or modifying the activity and microbial composition in soil or water [3–5]. Glyphosate is one of the most used organophosphates in industrialized countries; it is a non-selective herbicide with broad-spectrum activity used to eliminate weeds in crops [6]. From the cultivated fields, the glyphosate percolates into the subsoil, thus reaching fresh water and, therefore, the sea, where it causes the global loss of corals and algae [7,8]. Genetically modified crops tolerant to glyphosate were introduced in 1996; however, this has been ineffective in reducing the use of glyphosate and alleviating the severe environmental pollution currently occurring [9]. Therefore, fine-tuning methodologies for the detoxification from these pollutions at a local scale seems to be of great interest for pure and applied research.

Several studies were devoted to the ability of Cyanobacteria to eliminate glyphosate from the wastewater [10]. Various mechanisms for detoxification among species have been reported: tolerant species seem to be able to use glyphosate as a source of inorganic phosphate, other strains showed a glyphosate metabolism different to those using the phosphonate, and some species were found to possess an insensitive form of glyphosate target enzymes. Particularly *Arthrospira fusiformis* and *Spirulina platensis*, two species belonging to the taxonomic group commonly known as spirulina, showed noteworthy resistance up to 10 mM glyphosate. The authors concluded that the two spirulina species were unable to use glyphosate as a source of phosphorus, suggesting that their tolerance might be due to a very low phosphonate uptake into the cells.

Based on this evidence, this study aims to explore the effects of a deliberate treatment with glyphosate on the selected cultivation of *Arthrospira maxima* (synonymy *Spirulina maxima*), a species largely cultivated as animal and human food additives and as a source of c-phycocyanin, a bioactive antioxidant molecule [11,12].

Here we reported the results on the growth and physiological behavior of the cyanobacteria populations treated with glyphosate at a concentration often found in environmental matrices [2]; the molecular adaptations of *A. maxima* during treatment were investigated by means of a free-label semi-quantitative proteomic approach. The analysis of cyanobacterial proteins has traditionally been carried out using approaches based on electrophoresis [13] and has more recently been applied to *Arthrospira platensis* [14,15]. Genome-wide sequencing of the PCC 8005 strain and its annotation have recently been completed and thus provide key resources to facilitate proteomic approaches to this species [16]. Thousands of proteins have been identified in previous works that have expanded the coverage of the *A. platensis* proteome. Taking advantage of this previous knowledge, we used the one-dimensional gel electrophoresis (SDS-PAGE) combined with LC-MS/MS of both cytosolic and membrane protein fractions to identify proteins differentially accumulated under glyphosate treatment. Despite negative effects on biomass and photosynthetic pigments, proteomics highlighted that modulation of various metabolisms, including the shikimic acid pathway, resulted in resistance to the herbicide of a population of *A. maxima* that survived treatment.

2. Materials and Methods

2.1. *Arthrospira maxima* Culture Preparation

Arthrospira maxima strain cultivated according to the Oil Fox® technology and registered trademark Spirulina-fox®, covered by the international patent number AR070504B1, were used [17]. Equal volumes of the original cultivation of *A. maxima* were subcultured in 4 aquaria, each containing 20 L of the Zarrouk culture medium [18]. The culture parameters are as follows: pH 10.0 ± 0.1, temperature 27.5 ± 1 °C, and irradiation of 1350 lumens with a light/dark photoperiod of 12 h/12 h of light and oxygen input equal to 800 L/h for each individual aquarium.

2.2. *A. maxima* Treated with Glyphosate (Gly)

Two independent cultivations were treated with 0.2 mM glyphosate (Bayer RASIKAL PRO). All cultivation parameters were kept constant for 42 days of treatment. Immediately after the treatment (t_0) and every seven days, sampling for biomass analyses, pigment levels, and proteomics was carried out in treated and control cultivations.

2.3. Evaluation of the Biomass of Cultivations

A. maxima biomass concentration in the cultivations was measured as optical density at a wavelength of 680 nm [19] using a digital Vis-spectrophotometer (Jenway 7310, Cole-Parmer, Staffordshire, UK). In detail, 2 mL of cultivation, taken after vigorous stirring, were transferred in a quartz cuvette, and absorbance was read immediately. Five to seven different samples were taken from the cultivations and measured at each time of treatment.

2.4. Analysis of the Concentration of Chlorophyll *a* and Carotenoids

For the concentrations of chlorophyll *a* and carotenoids, 6 mL of culture, taken after vigorous stirring, were transferred into vials and centrifuged at 13,000 rpm for 5 min to RT. A volume of methanol absolute was added, vortexed, and placed in the dark for 30 min. The samples were then centrifuged at 13,000 rpm for 5 min, and the absorbance of the supernatant was measured at wavelengths of 665 and 652 nm for chlorophyll and 480 nm for the carotenoids using the Vis-spectrophotometer (Jenway 7310, Cole-Parmer,

Staffordshire, UK). The concentrations of chlorophyll and carotenoids, expressed as mg/mL, were calculated according to Kumar et al. [20], using the following formulas:

$$\text{Chlorophyll } a\left(\frac{\text{mg}}{\text{L}}\right) = 16.29 \times OD_{665} - 8.54 \times OD_{652}$$

$$\text{Carotenoids}\left(\frac{\text{mg}}{\text{L}}\right) = 4 \times OD_{480}$$

Five to seven independent extractions of chlorophyll and carotenoids were carried out on the same day for each cultivation. These equations are derived from the difference extinction coefficients, as reported in Porra et al. [21]; therefore, all absorbance measurements at the indicated wavelengths must have the absorbance at 750 nm subtracted.

2.5. Extraction of Phycocyanins from A. maxima Cultivation

The repeatability of the extraction of phycocyanin was evaluated with five repetitions of the extraction on the same day. After vigorous stirring, a volume of cultivation was collected and dried in a vacuum centrifuge at 30 °C. The dried biomass, reduced to a very fine powder, was resuspended in a volume of water at a g/mL ratio, vortexed, and centrifuged at 13,000 rpm for 5 min. The absorbance of the supernatant has been read at 615 nm and 652 nm. The phycocyanin concentration (PC) was finally calculated using the formula reported below [22] and expressed in mg/mL and then in mg/g of dry weight:

$$PC\left(\frac{\text{mg}}{\text{mL}}\right) = \frac{OD_{615} - 0.474 \times OD_{652}}{5.34}$$

These equations are derived from the different extinction coefficients, as reported in Bennet et al. [22].

2.6. Extraction of Proteins from A. maxima Cultivation

The protein extraction was carried out following the acetone/trichloroacetic acid (TCA) precipitation method [23] with some modifications. In total, 150 mL of cultivation were divided into 50 mL falcons and centrifuged at 7900 rpm for 15 min at 4 °C to obtain a pellet of about 1 gr of cyanobacteria.

The pellet was washed with 10 mM Na_2-EDTA in water and centrifuged at 13,000 rpm for 5 min at RT; the final pellet was frozen in liquid nitrogen and pulverized in a pre-cooled mortar and pestle. The fine powder obtained was divided into several 2 mL Eppendorf, and 10% (*w*/*v*) TCA and 0.07% (*w*/*v*) dithiothreitol (DTT) were added and incubated at −20 °C overnight. After incubation, samples were centrifuged at 13,000 rpm for 20 min at RT; the pellet was washed several times with pre-cooled acetone, 0.07% (*w*/*v*) DTT was added, incubated at −20 °C for 30 min, and centrifuged at 13,000 rpm for 20 min at RT. The pellet was dried for 20 min at RT in a vacuum centrifuge and subsequently solubilized in the loading buffer [24].

Protein concentration was assayed by Bradford's reagent [25]. A total of 1 µL of the protein extract was added to Bradford's reagent (1:1; *v*/*v* in water) and incubated in the dark for 5 min.

Absorbance at 595 nm was measured using the Vis-spectrophotometer (Jenway 7310, Cole-Parmer, Staffordshire, UK). Protein concentrations were expressed as mg/mL by using a calibration curve (y = 0.0779x; R_2 = 0.9719) of the protein standard at different concentrations. Three biological replicates were analyzed for each experimental set.

2.7. Electrophoretic Separation of A. maxima Proteins by SDS-PAGE

Protein samples were separated by SDS-PAGE [24]. Proteins, solubilized in loading buffer, were activated for 4 min at 100 °C. Between 15 µg and 20 µg were loaded onto 12% polyacrylamide/bisacrylamide gels. The electrophoresis was run in a Tris-glycine buffer at 60 mA in the stacking gel and 120 mA in the running gel at 200 V constant power. After electrophoresis, the gels were placed in a Coomassie R-250 solution overnight and then destained through washing in water/ammonium bicarbonate/acetonitrile solution. Digitalized images of the destained SDS-PAGEs were analyzed by the Quantity One 1-D Analysis Software (Bio-Rad, Berkeley, CA, USA) to measure the band densities at each lane of all biological replicates; the amount of protein at bands of 55, 25, and 10 kDa was measured using the marker reference bands at 75, 50, and 25 kDa that contained 150, 750, and 750 ng of proteins, respectively. Each lane of the same SDS-PAGE was divided into six slices from 200 to 10 kDa and manually excised from the gel. Gel slices were processed by in-gel digestion.

2.8. Sample Preparation for Mass Analysis:Reduction, Alkylation, and in-Gel Digestion of Proteins

The in-gel digestion procedure required three steps: the reduction, the alkylation, and the tryptic digestion of the polypeptides directly on the gel bands.

For the reduction step, polypeptide bands were treated with 100 µL of DTT in 50 mM ammonium bicarbonate for 30 min at 56 °C. The reduced polypeptides were alkylated using the ammonium bicarbonate/acetonitrile/iodoacetamide solution. Then trypsin (Promega, Madison WI, USA) dissolved in 25 mM ammonium bicarbonate was added to the reduced-alkylated-polypeptides and incubated at 37 °C overnight [26]. The tryptic digested peptides were extracted by adding the ammonium/bicarbonate 25 mM/acetonitrile/5% formic acid solution; the liquid phase in each sample was extracted, collected in new tubes, and dried in the vacuum centrifuge. Dried samples were treated with 80% formic acid in H_2O.

2.9. Mass Spectrometry Analysis

Twenty microliters of the tryptic peptides were injected into an inverted phase trap column (BioBasicTM LC18 Analytical Column, 300 Å, 5 µm, 50 µm ID × 1 mm long, Thermo Scientific, Sacramento, CA, USA) and separated by an ultra-chromatographic system (UltiMate 3000 RSLC System, Thermo Scientific, Sacramento, CA, USA), at a constant flow of 100 µL/min and with a gradient of 4% of buffer A (2% ACN and 0.1% formic acid in water) to one at 96% of buffer B (2% water and 0.1% formic acid in ACN) for 60 min. The eluting peptides were on-line sprayed in an LTQ XL mass spectrometer (Thermo Scientific, Sacramento, CA, USA). Full scan mass spectra were collected in the linear ion trap in the mass range of m/z 350 to m/z 1800 Da, and the 10 most intense precursor ions were selected for collision-induced fragmentation. The acquired MS spectra were used for protein identification.

2.10. Bioinformatic Analysis and Proteins Identification of A. maxima

From the MS/MS spectra, protein inference and validation were performed with the Scaffold software 4.8. MS/MS spectra were extracted from raw data by accepting one minimum sequence of three amino acids and fusion scans with the same precursor within one mass window of ± 0.4 m/z over a time interval of ± 30 s. The key parameters of research were Scored Peak Intensity, (SPI) $\geq 50\%$, precursor mass tolerance of ± 10 ppm, and mass tolerance of product ions of ± 20 ppm. The carbamidomethylation of cysteine was fixed as a modification, and trypsin was selected as the enzyme for the digestion, accepting two missing cleavages per peptide.

Through the X!Tandem software, the proteins found in the samples were carefully aligned and identified in all samples analyzed. Automatic thresholds were used for peptide identification in the software Scaffold. Generally, peptide probabilities are evaluated using a Bayesian approach for the estimation of the local FDR (LFDR) up to a value of 1%. The peptide sequences, using Scaffold 4.8 Q + S system software [27], interfaced with both the database of proteins deduced from generalist protein sequences of Cyanobacteria, that of the genus Arthrospira, deposited in the NCBI database (downloaded on 9 May 2020) and in the bank UniProt data (downloaded on 24 June 2021).

2.11. Statistics

Comparison of differences among groups of values for biomass and photosynthetic pigment were analyzed using a *t*-test. All the statistical analyses were performed using XLSTAT (©Addinsoft, Paris, France, released on 2021.3.1.1187) [28]. Significance was defined as $p \leq 0.05$.

For the proteomics results, a comparison of differences among the groups was carried out using the Differentially Expression and Heat map tools (XLSTAT). The Bonferroni test was used to test the assumption of homogeneity of variances. Threshold for significance was $p \leq 0.05$.

The gene enrichment analysis was carried out by BLAST2go software [29] which executes a statistical assessment of differences in functional classes between two groups of sequences based on the Fisher test analysis. By taking a false discovery rate (FDR) significance threshold of 0.05 and a single test *p*-value (Fisher *p*-value), we obtain those functionalities that are strongly significant for proteins in glyphosate-treated samples. The relative frequency of each GO term has been represented for the "biological process" category.

3. Results

3.1. Dynamics of Arthrospira maxima Cultivation and Biochemical Parameters

The biomass and productivity of reference *A. maxima* cultivations were monitored weekly starting from t_0 up to 70 days of cultivation. In Figure 1, the absorbances of the samples from two reference cultivations are reported.

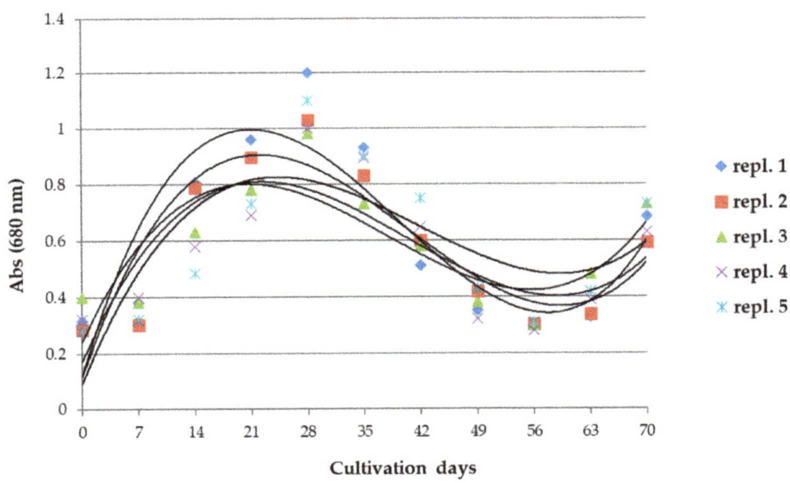

Figure 1. Absorbance values of *A. maxima* samples, taken every 7 days in 2 reference cultivations for a period of 70 days. Growth curves (polynomial of 3rd degree) were calculated for each group of samples. $n = 5$. See Supplementary Table S1A for details.

As can be seen, maximum cell growth occurred at 21–28 days of cultivation, so growth decreased to its minimum after 56–63 days. At 70 days of cultivation, nutrients had to be added to prevent massive cell mortality. Based on these growth dynamics, we set the reference period for treatments at 42 days of cultivation, in which the concentrations of photosynthetic pigments Chl *a*, phycocyanin, and carotenoids were measured.

Concentrations of Chl *a* in the two reference cultivations of *A. maxima* varied between samples at the same sampling time; the polynomial regression lines showed a significant decrease ($p < 0.005$) in concentration after the 35th day of culture, with the exception of a single sample which showed the highest value at 42 days (Figure 2A).

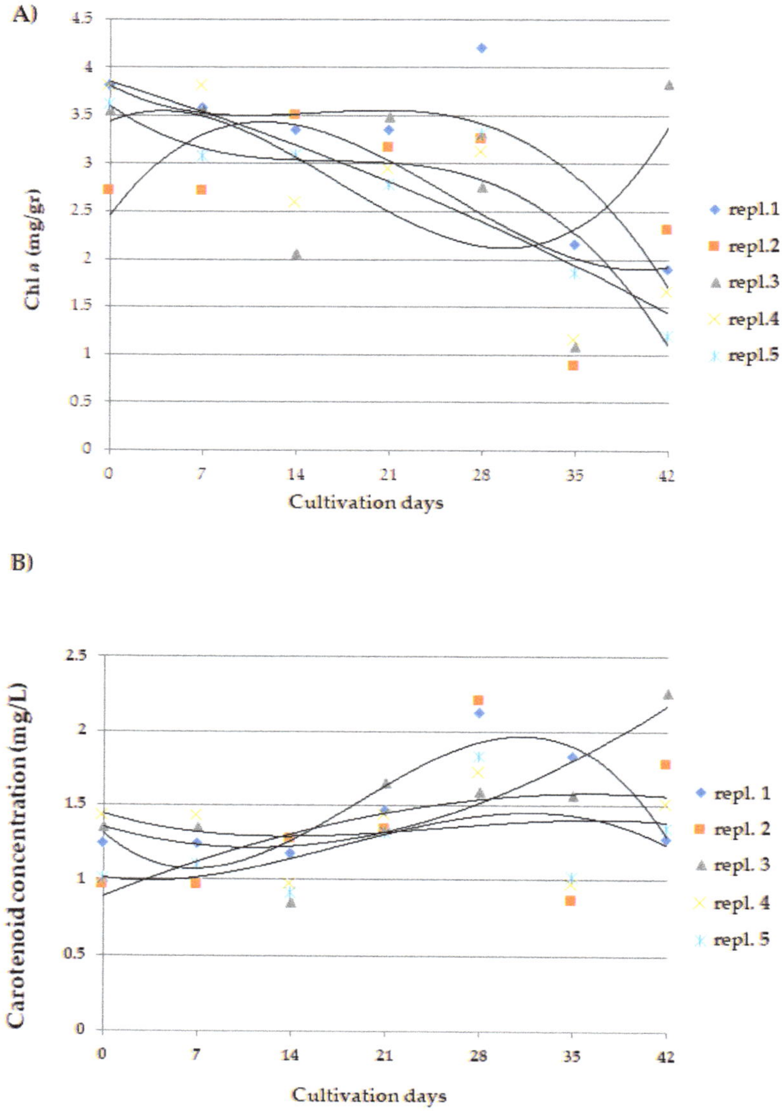

Figure 2. Concentration values of (**A**) Chl *a* and (**B**) total carotenoids in *A. maxima* samples, taken every 7 days in 2 reference cultivations. (**A**) $p < 0.005$ at day 0 and day 35. See Supplementary Table S1B,C for details.

The concentrations of total carotenoids had no statistically significant changes among the treated and reference cultivations (Figure 2B).

In reference cultivations, the concentrations of phycocyanin increased up to 28–35 days (Figure 3); then at 42 days, it decreased significantly in all analyzed samples ($p < 0.005$).

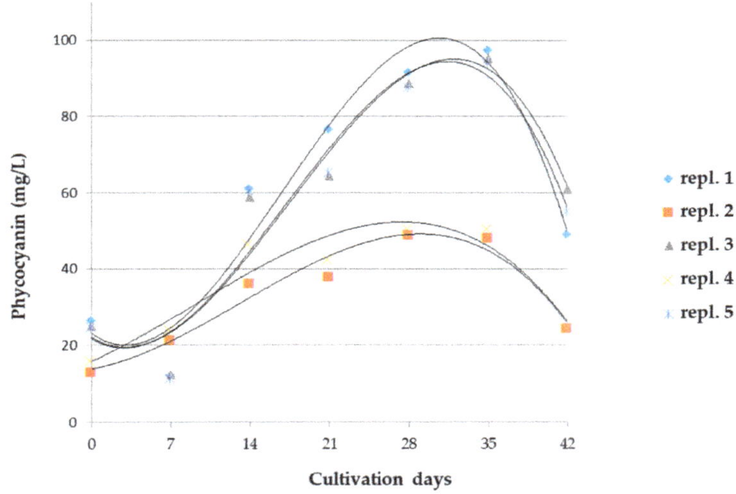

Figure 3. Phycocyanin concentration values in *A. maxima* samples, taken every 7 days in 2 reference cultivations. $n = 5$. Statistical data and parameters are reported in Supplementary Table S1D.

3.2. A. maxima Cultivation Dynamics and Biochemical Parameters during Glyphosate Treatment

Figure 4 shows the biomass values in the cultivations during 42 days of treatments with 0.2 mM glyphosate.

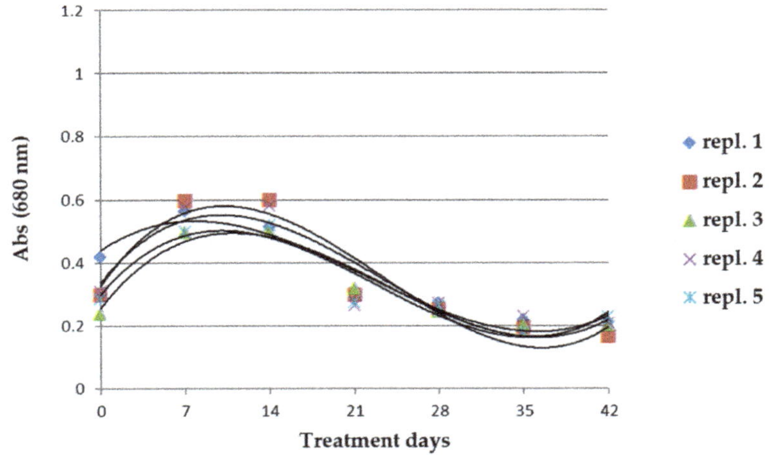

Figure 4. Biomass values evaluated as absorbance of reference samples treated with 0.2 mM glyphosate from cultivations of *A. maxima*. $n = 5$. Details of the data are shown in Supplementary Table S1F.

After the treatment with glyphosate, the biomass underwent a significant reduction when compared to the reference cultivations. The maximum biomass occurred after 14 days of treatment, and then it drastically dropped (Figure 4). At the end of the treatment (42 days), the biomass reduced significantly to 50% of that in the reference cultivations (Figure 1). The residual population of *A. maxima* survived; this population was considered to be resistant to the glyphosate treatment.

Treatment with glyphosate induced an increase in chlorophyll concentration during 14 days of treatment (Figure 5); the increase was statistically significant when compared with the values of the reference cultivations in the same period ($p < 0.05$; Supplementary Table S1A). At the end of the treatment, the Chl *a* resulted three times lower than the reference cultivations.

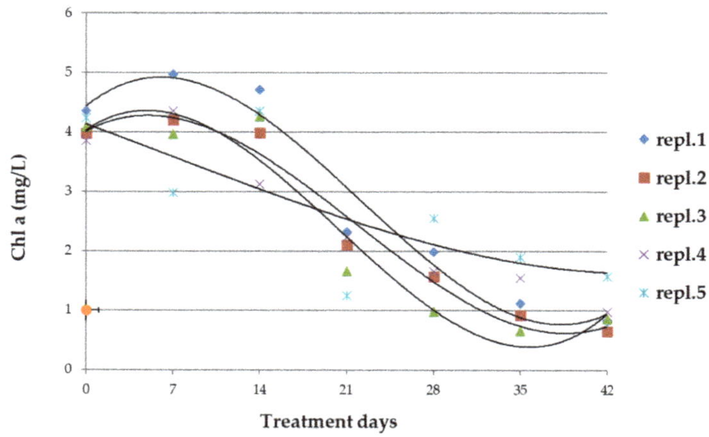

Figure 5. Trend of the concentration of chlorophyll *a* in samples (repl. 1–5) of *A. maxima* cultivations treated with 0.2 mM of glyphosate for 42 days and the mean of the concentration values (±ES) of reference samples. Regression lines (polynomials of 3rd degree) were calculated for each sample group. $n = 5$. The dashed line corresponds to the regression of the values in the reference cultivations. Statistical data and parameters are reported in Supplementary Table S1G.

The concentration of carotenoids was not affected by the treatment; the biological replicates showed variable concentrations at various treatment times, resulting in no significant in comparison with those in the reference cultivations (Figure 6).

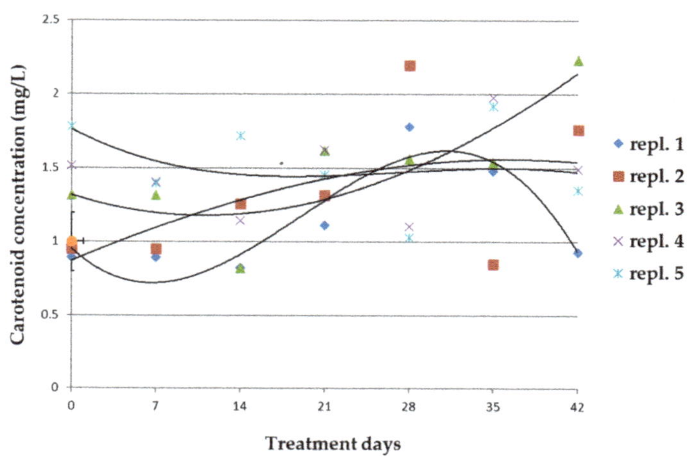

Figure 6. Concentration of carotenoids in biological replicates (n = 5) of *A. maxima* cultivations treated with 0.2 mM glyphosate for 42 days. Mean concentration values (±ES) of the reference cultivations and the regression line (dotted line) have been reported. Regression lines were calculated for each sample group. Statistical data and parameters are reported in Supplementary Table S1H.

Phycocyanin concentrations underwent an exponential reduction in all biological replicates as a result of the treatment (Figure 7). Immediately following the addition of glyphosate to the cultivation (t_0) and during the following 14 days, the concentrations appeared highly variable among replicates. Then values dropped to zero and, at the end of the treatment, the mean phycocyanin concentration was 1.98 ± 1.26 mg/L, a value 20 times lower than that of the reference cultivations (42.8 ± 14.7 mg/L).

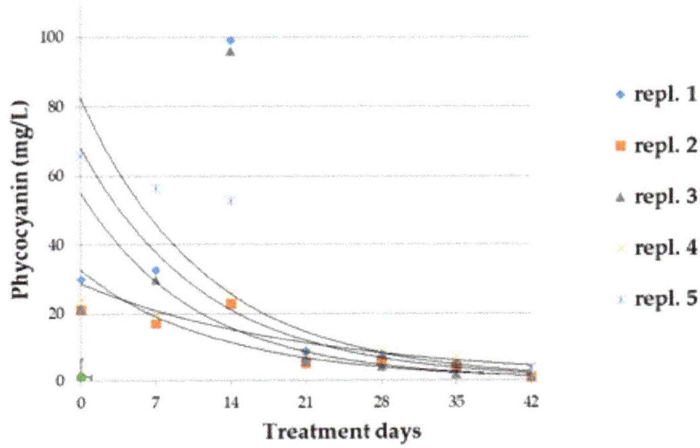

Figure 7. Phycocyanin concentration in biological replicates (n = 5) of *A. maxima* cultivations treated with 0.2 mM glyphosate for 42 days. Mean concentration values (±ES) of phycocyanin and the regression line (dotted line) of the reference samples have been also reported. Regression lines were calculated for each group of samples. Statistical data and parameters are reported in Supplementary Table S1I.

3.3. Proteins Differentially Expressed in Cultivations of Arthrospira maxima following Treatment with Glyphosate

Very high yields of purified proteins (more than 60 mg/g fresh weight) have been obtained in reference samples through the extraction protocol optimized in this work. Treatment with glyphosate caused a reduction in the protein content in all samples; after 21 days of treatment, the mean protein concentration of all biological replicates was 0.10 ± 0.03 mg/g fresh weight.

Samples after 28, 35, and 42 days of treatment yielded tryptic digested peptides, most of which did not receive sequence identification by mass spectrometric analysis (data not shown); therefore, proteomic analysis was performed only in samples up to 21 days of glyphosate treatment.

From mass spectrometry and bioinformatics analyses of the protein extracts of *A. maxima* samples, 660 proteins common to both glyphosate-treated and reference samples were identified (see Supplementary Table S2). Semiquantitative analysis, using the spectral counting method, returned differential expression values with different statistical significances depending on the treatment and sampling times. Figure 8 shows the distribution of proteins as a function of their frequency for each range of p values; as can be seen, 303 proteins were distributed in a range of very significant p-values.

Among these, 143 were differentially expressed with high significance (*** $p < 0.0001$; XLSTAT 2021.3.1.1187—Differential expression feature) in the treated samples compared to the control; another 67 are differentially expressed with $p = 0.0002$. The remainder show differential expression levels with $0.009 < p < 0.001$ (53 proteins) and with $0.001 < p < 0.049$ (40 proteins). A total of 357 proteins had no significant changes in their expression following the treatment.

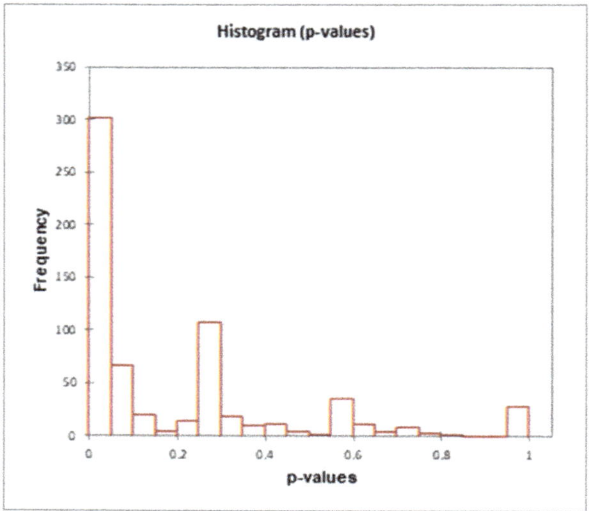

Figure 8. Distribution histograms of differentially expressed proteins in *A. maxima* samples treated with glyphosate compared to control cultivations in the range of *p*-values. XLSTAT 2021.3.1.1187—Differential expression feature. The details of the analysis are shown in Supplementary Tables S2 and S3.

The Volcano plot (Figure 9) shows the statistical significance (*p*-values) with respect to the Fold Change (FC) of the differentially expressed proteins in the treated samples compared to the reference samples.

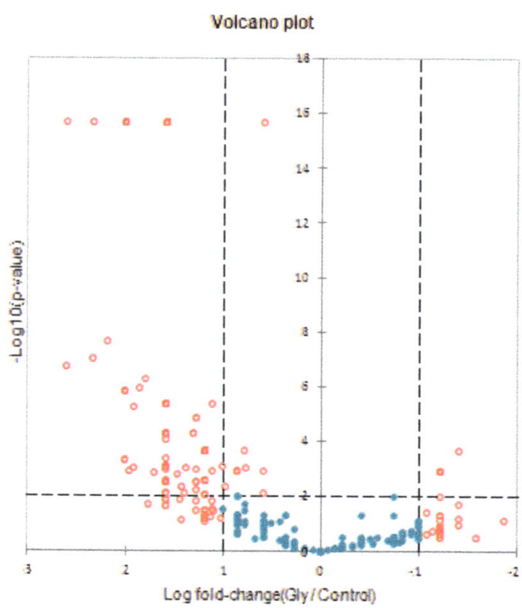

Figure 9. Volcano plot that displays and identifies statistically significant changes in protein expression in *A. maxima* samples treated with 0.2 mM glyphosate compared to the reference samples in terms of change in Log FC (X axis) and *p*-value (Y axis). XLSTAT 2021.3.1.1187- Differential expression feature. The details of the analysis are shown in Supplementary Tables S2 and S3.

The graph allows us to visually identify proteins that had broadly significant FC that were also statistically significant. The most upregulated proteins were distributed on the left side, the most downregulated proteins were distributed on the right, and the most statistically significant proteins were at the top of the graph. As can be deduced, most of the differentially expressed proteins were accumulated with values up to 2.8 LogFC; furthermore, the expression patterns were very significant for five of these proteins (the dots in the upper left of the graph).

The dynamics of the expression levels of the proteins in the various sampling times during the treatment with glyphosate were evaluated and visualized by means of the Heat Map, as shown in Figure 10.

The glyphosate-treated samples grouped together at various sampling times; this suggested that most of the proteins had the same treatment-induced expression pattern. This was not the case with the reference samples because the 7-, 14-, and 21-day samples clustered together as a result of the expression pattern of proteins that occurred in the reference cultivations.

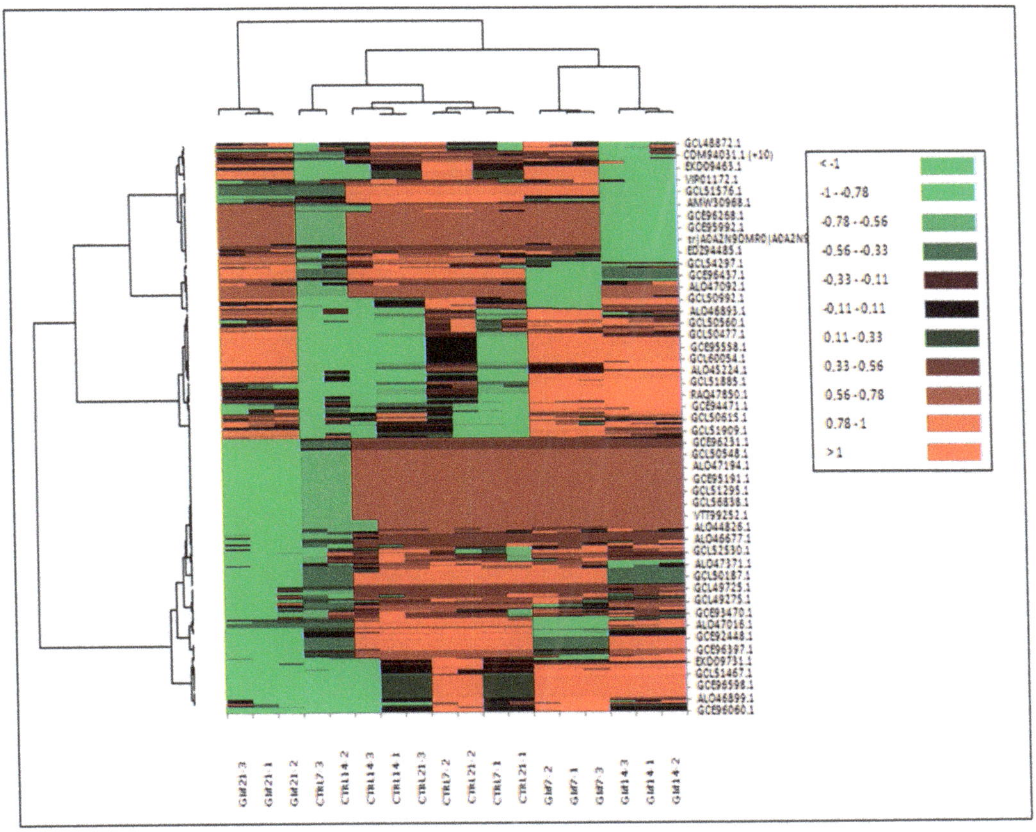

Figure 10. Heat map of proteins differentially expressed in *A. maxima* samples treated with 0.2 mM glyphosate compared to reference samples at various sampling times. Protein expression values were normalized to log2 and cluster analysis was performed using FC levels for proteins with $p < 0.05$. Red indicated a high level of expression; green indicated a low level of expression (XLSTAT 2021.3.1.1187-Heat maps features). The details of the analysis are shown in Supplementary Table S4.

After 7 days of treatment, the majority of proteins did not vary significantly with respect to the reference samples, and their LogFC values ranged between 0.11 and 0.78. At 14 days of treatment, the percentage of proteins with LogFC < −1 increased, and at 21 days, half of the proteins were under-expressed with values of LogFC < −1.

Functional analysis of the DAPs gave significant differences compared to the control only after 21 days of treatment with glyphosate. The Gene Ontology enrichment analysis of the differentially expressed proteins is shown in Figure 11.

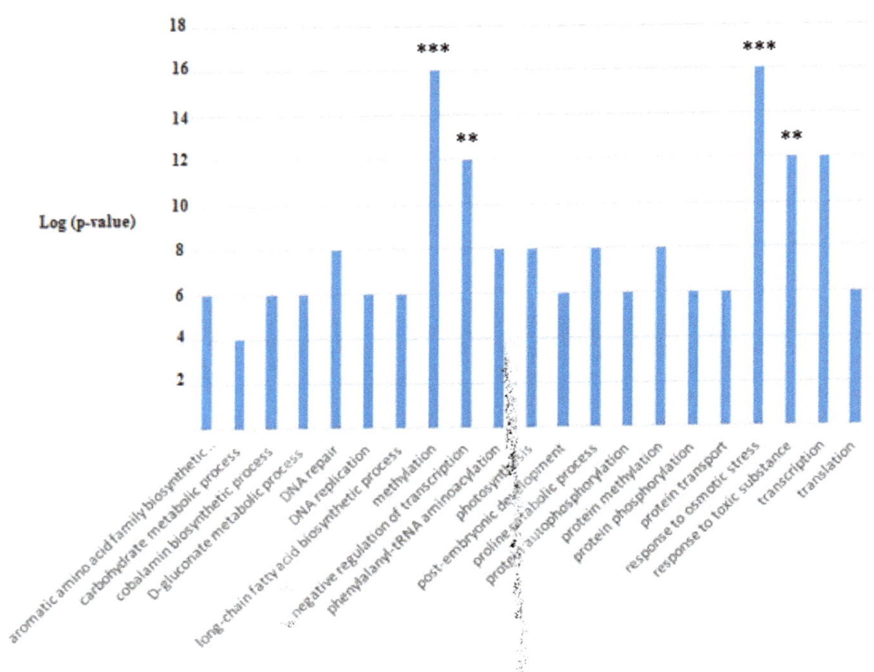

Figure 11. Functional annotation of DAPs in *A. maxima* samples after 21 days of treatment with 0.2 mM glyphosate; the histograms referred to the GO terms to which the enriched proteins, with statistical values of $p < 0.05$, belonged. The data presented are log-transformed p-values (corrected with FDR) of GO terms or KEGG (KG) pathways that were found to be enriched in the group of proteins tested. *** $p < 0.0001$; ** $p < 0.001$). The details of the analysis are reported in Supplementary Table S5.

As can be seen, the functional analysis revealed very significant changes in the metabolism related to Methylation and to Response to Osmotic Stress ($p < 0.0001$); Response to Toxic Substances and Negative Regulation of Transcription were also significantly changed. All these metabolisms were highly accumulated in glyphosate-treated samples; protein belonging to the remaining metabolisms was reduced in abundance with a significance of $p < 0.05$.

In Table 1, the Differentially Accumulated Proteins (DAPs) with the lowest p-values ($p < 0.0001$) are reported; these proteins were found significantly reduced in abundance following the treatment with glyphosate with respect to the reference cultivations (Table 1).

Table 1. Differentially Accumulated Proteins (DAPs) in the Glyphosate treated (Gly) *A. maxima* samples vs. the reference samples after 21 days of treatment. DAPs with the lowest *p*-values and the highest quantitative and qualitative profiles are reported *. The GO- terms, which the DAPs belong to, are also reported.

Protein name and Taxonomy	IDs	*p*-Value	Quantitative Profile ** (Fold Change)		Qualitative Profile	GO Terms
			Reference	Gly		
16S rRNA methyltransferase [*Pseudohongiella spirulinae*]	GCL59050.1	<0.0001	1667 (b)	10 (a)	H.A.	SOS response
HoxH, partial [*Arthrospira platensis* FACHB-440]	EDZ95376.1	<0.0001	1 (b)	6 (a)	H.A.	ADP binding; transferase activity
Penicillin-binding protein 2 [*Arthrospira* sp. PLM2.Bin9]	GCE96397.1	<0.0001	1333 (b)	6 (a)	H.A.	molybdopterin cofactor binding
6-phosphogluconate dehydrogenase [*Arthrospira platensis* NIES-46]	GCE96312.1	<0.0001	1667 (b)	6 (a)	H.A.	glycolytic process
ATP-dependent DNA helicase [*Pseudohongiella spirulinae*]	ALO44826.1	<0.0001	1444 (b)	5 (a)	H.A.	cell division
hypothetical protein NIES46_38410 [*Arthrospira platensis* NIES-46]	ALO47614.1	<0.0001	1 (b)	5 (a)	H.A.	DNA restriction-modification system
Transporter, OMR family [*Pseudohongiella spirulinae*]	GCL52209.1	<0.0001	1 (b)	4 (a)	H.A.	aromatic amino acid family biosynthetic process
Chorismate synthase [*Microcystis aeruginosa* NIES-3804]	GCE92370.1	<0.0001	1 (b)	4 (a)	H.A.	aromatic amino acid family biosynthetic process; chorismate biosynthetic process
hypothetical protein NIES46_48480 [*Arthrospira platensis* NIES-46]	GCL50795.1	<0.0001	1333 (b)	4 (a)	H.A.	ATPase activity; ATP binding; DNA binding
Phosphate acetyltransferase [*Arthrospira platensis* NIES-46]	GCL50871.1	<0.0001	1 (b)	4 (a)	H.A.	cell wall organization; peptidoglycan biosynthetic process; regulation of cell shape
putative DNA helicase [*Microcystis aeruginosa* NIES-3804]	GCL52800.1	<0.0001	1333 (b)	4 (a)	H.A.	DNA metabolism
(myosin heavy-chain) kinase [*Gemmata massiliana*]	GCL49992.1	<0.0001	1 (b)	4 (a)	H.A.	DNA methylation
PBS lyase heat-like repeat protein [*Microcystis aeruginosa* NIES-3804]	GCE96060.1	<0.0001	1667 (b)	4 (a)	H.A.	DNA repair
wd-40 repeat protein: [*Gemmata massiliana*]	GCE92418.1	<0.0001	1333 (b)	4 (a)	H.A.	DNA synthesis involved in DNA repair; double-strand break repair
Acetyltransferase component of pyruvate dehydrogenase complex [*Pseudohongiella spirulinae*]	GCE92614.1	<0.0001	1 (b)	4 (a)	H.A.	glucose metabolic process
Exonuclease SbcC homolog [*Microcystis aeruginosa* NIES-3804]	VTU01960.1	<0.0001	1 (b)	4 (a)	H.A.	integral component of membrane
Putative multifunctional protein glycosyl transferase, tetratricopeptide domains and SAM methyltransferase domains [*Limnospira indica* PCC 8005]	GCE93275.1	<0.0001	1 (b)	4 (a)	H.A.	methylation
PQQ enzyme repeat domain protein [*Gemmataceae bacterium*]	GCE96231.1	<0.0001	1 (b)	4 (a)	H.A.	methylation
Lytic transglycosylase catalytic precursor [*Microcystis aeruginosa* NIES-3804]	GCL46959.1	<0.0001	1333 (b)	4 (a)	H.A.	nitrogen compound metabolic process
Aldolase/epimerase [*Microcystis aeruginosa* NIES-3804]	ALO45161.1	<0.0001	1333 (b)	4 (a)	H.A.	organic acid metabolic process
Transcription-repair coupling factor [*Pseudohongiella spirulinae*]	ALO46521.1	<0.0001	1333 (b)	4 (a)	H.A.	regulation of transcription, DNA-templated
Chromosome segregation protein [*Microcystis aeruginosa* NIES-3804]	GCE95530.1	<0.0001	1333 (b)	4 (a)	H.A.	reproductive process
60 kDa molecular chaperonin 2 [*Microcystis aeruginosa* NIES-3804]	RAQ47071.1	<0.0001	1333 (b)	4 (a)	H.A.	response to stimulus
Transposase [*Microcystis aeruginosa* NIES-3804]	GCE94368.1	<0.0001	1333 (b)	4 (a)	H.A.	signal transduction
hypothetical protein NIES46_27100 [*Arthrospira platensis* NIES-46]	GCL55249.1	<0.0001	1 (b)	4 (a)	H.A.	transmembrane transport
Methionine synthase [*Arthrospira* sp. O9.13F]	ALO46677.1	<0.0001	1 (b)	4 (a)	H.A.	transmembrane transporter activity
hypothetical protein NIES46_44690 [*Arthrospira platensis* NIES-46]	ALO46533.1	<0.0001	1 (b)	4 (a)	H.A.	tricarboxylic acid cycle
hypothetical protein NIES46_24140 [*Arthrospira platensis* NIES-46]	EKD07239.1	<0.0001	1 (b)	4 (a)	H.A.	ubiquitin binding
Two-component sensor histidine kinase [*Microcystis aeruginosa* NIES-3804]	GCL51973.1	<0.0001	1333 (b)	4 (a)	H.A.	viral tail assembly

* the list of all DAPs is reported in Supplementary Table S2; ** Proteins sharing the same letter are not significantly different. a Glyphosate treatment samples, b Reference samples.

In Figure 12, the DAPs that were highly accumulated during 21 days of treatment with glyphosate are reported.

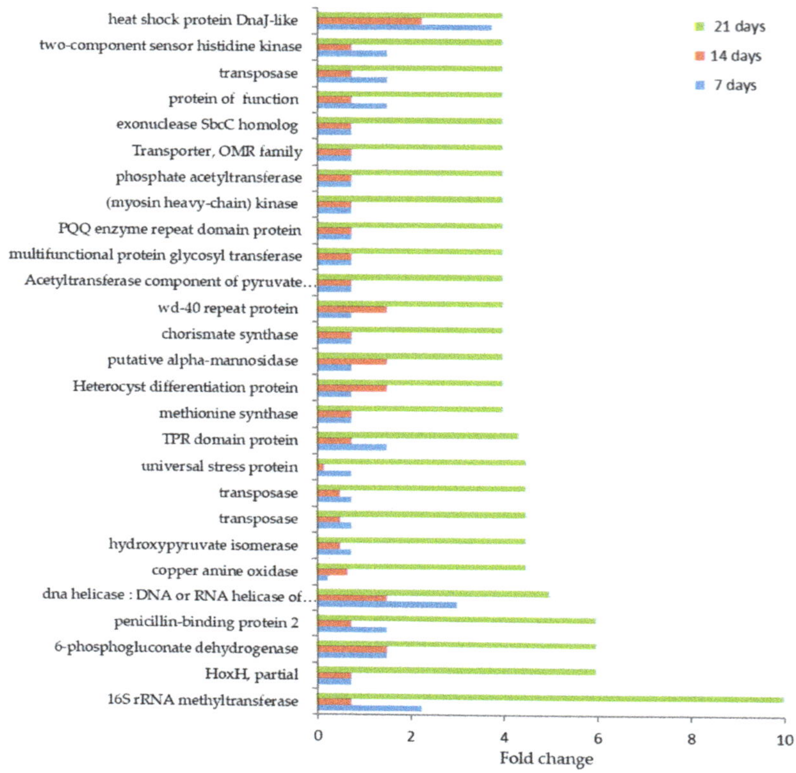

Figure 12. Time course of the highly differentially accumulated proteins during the treatment in treated *A. maxima* cultivation with 0.2 mM glyphosate compared to the reference cultivations. Values are reported as fold changes (Supplementary Table S2).

Among these, the 16S rRNA methyltransferase was found to be 10-fold enriched with respect to the control; the HoxH protein, the 6-phosphogluconate dehydrogenase, and penicillin-binding protein 2 were 6-fold accumulated. The copper amine hydroxypyruvate, the isomerase oxidase, the universal stress protein, and two transposases were 5-fold accumulated in response to the treatment. Finally, the rest of the highly accumulated proteins varied their abundance of 4-fold respect to the reference samples; among these, the chorismate synthase was worthy to note as these proteins are a key enzyme of the shikimic acid pathway, required for the biosynthesis of all three aromatic amino acids (phenylalanine, tyrosine, and tryptophan).

4. Discussion

Previous studies on glyphosate effects on spirulina species reported a strong tolerance of these microorganisms toward the herbicide [10]. Our results did not confirm for *Arthrospira maxima* a high tolerance to a sublethal dose of glyphosate. However, a survival population of *A. maxima* following the glyphosate treatment has been observed under our conditions. When *A. maxima* cultivations were treated with a 0.2 mM glyphosate, after some days, effects on the sensitive cells occurred. Biomass reduction was the first symptom of the toxic effect of the herbicide on these kinds of cells. Physiological adjustments mediated by an initial increase in chlorophyll *a*, phycocyanin, and by the invariance of carotenoid levels appeared to be part of the resistance mechanisms during the first two weeks of treatment. In particular, we can speculate that the increase in chlorophyll content may be linked to an increase in photosynthesis. It is known that abiotic stresses can induce an increase in light phase reactions to cope with the high energy demand imposed by stressors [30]. After that, strong inhibition of phycocyanin occurred, and the survival strain of *A. maxima* after 42 days of treatment still appeared unable to synthesize large amounts of these molecules.

Proteomic results of the survival population suggested that *A. maxima* responded to glyphosate treatment through the modulation of the osmotic regulation since microorganisms upregulated several proteins belonging to this metabolism; the histidine kinase, enzymes belonging to the proline biosynthesis pathway, the glycolate oxidase, and potassium ion transport were accumulated. It is known that the intracellular accumulation of proline is induced by various stresses also in cyanobacteria [31]; in *Synechocystis* sp., a histidine kinase (Hik33) is induced by both osmotic stress and cold stress. As a response to glyphosate stress, several proteins linked to DNA methylation and post-transductional modification of proteins were upregulated; DNA methylation is linked to epigenetic modifications that are well known as being linked to many important biological processes. In the cyanobacterium model, *Synechocystis* sp. PCC 6803, the variation of "methyloma" (which therefore includes both DNA and protein methylation), has been detected, suggesting that nutritional deficiencies, such as nitrogen, can induce epigenetic modifications that are "inherited" across generations [28]. Here we can hypothesize that glyphosate promoted the methylation processes that involved a reorganization of the "methyloma" of *A. maxima* and a reorganization of the transcription; these data were in agreement with the overexpression of proteins involved in the negative regulation of transcription and with the accumulation of proteins involved in the metabolism of transcription. To complete the picture of the stress response induced by glyphosate, there was the accumulation of proteins belonging to the metabolism of the response to toxic substances, such as the Arsenic resistance protein, the ATP-dependent Clp protease ATP-binding subunit ClpC, and the Cation/multidrug efflux pump.

The SOS response was altered by the treatment; in eukaryotes, the SOS is a global response to DNA damage in which the cell cycle is arrested and DNA repair and mutagenesis are induced [32]. The 16S rRNA methyltransferase was the most highly accumulated protein during the glyphosate treatment and, together with the accumulated PBS lyase heat-like repeat protein and the wd-40 repeat protein, was related to SOS response. In prokaryotes, the SOS response is related to antibiotic resistance and genotoxicity testing under various mutagenic conditions [33,34]. During normal growth, the SOS genes are negatively regulated; accordingly, for the general mechanism for mutagens, we suggested that glyphosate activated SOS genes after DNA damage occurred with the accumulation of single-stranded regions (ssDNA) generated at the replication forks, in which DNA polymerase was blocked. This is consistent with the strong reduction of *A. maxima* growth during the treatment. In this view, the glyphosate exhibited a strong mutagen activity against the cyanobacteria.

The primary metabolism of gluconeogenesis was also affected by the treatment as the RuBisCo enzyme level was reduced; although there is no evidence in the literature on this effect in cyanobacteria, in higher plants, glyphosate has, in fact, had negative effects on the assimilation of carbon, on the activity of ribulose bisphosphate carboxylase, and the levels of carbohydrate metabolism [35]. On the other hand, the glycolytic process was positively affected to respond to the cell's high energy demand.

The remarkable tolerance reported in many *spirulina* strains, such as *A. platensis*, a species close to *A. maxima*, was mediated by the ability to degrade glyphosate. High mortality occurred in our conditions suggesting that this was not the case for *A. maxima*. Resistance has been mediated also by the low uptake of organophosphates throughout the cells [10]. In bacteria, it is known that the cell wall acts as a barrier to glyphosate [36]. Interestingly, glyphosate-treated *A. maxima* strongly accumulated the phosphate acetyltransferase, an enzyme involved in the cell wall organization and peptidoglycan biosynthetic process; here we might speculate that reorganization of the cell wall could affect the organophosphate uptake participating in the strategy for resistance.

In other species, tolerance was obtained by the presence of a resistant form of the glyphosate target enzyme EPSP (5-enolpyruvylshikimate-3-phosphate synthase), a key enzyme of the shikimic acid pathway leading to the aromatic amino acid biosynthesis [10]. It is well known that glyphosate blocks this pathway by inhibiting the EPSPS, which catalyzes the reaction of shikimate-3-phosphate (S3P) and phosphoenolpyruvate to form EPSP [37]. The surviving *A. maxima* population largely accumulated enzymes belonging to the aromatic amino acid biosynthesis, such as the transporter, OMR family, and the chorismate synthase. The first enzyme is a well-characterized aromatic amino acid/H+ symport permease, and the second one catalyzes the 1,4-trans elimination of the phosphate group from 5-enolpyruvylshikimate-3-phosphate (EPSP) to form chorismate which can then be used in phenylalanine, tyrosine, or tryptophan biosynthesis.

This evidence strongly suggests that *A. maxima* mediated the resistance by the presence of a glyphosate-resistant form of a key enzyme belonging to the shikimate acid pathway, like other cyanobacteria species, but not currently reported for other *spirulina* species. EPSPS gene sequencing in the original culture of *A. maxima* could elucidate which enzyme form it was, whether during the treatment, mutations at the target site of glyphosate occurred, or whether spontaneous strains that expressed the glyphosate-resistant enzyme were already present.

5. Conclusions

By using glyphosate at concentrations found in environmental matrices (0.2 mM), the *A. maxima* residual population showed degrees of resistance to this pollutant. Although culture growth was significantly reduced, a resistant population survived to pollutants after 42 days of treatment. Molecular mechanisms for resistance were the overexpression of metabolisms linked to the response to toxic substances, homeotic stress, oxidative stress, and the reorganization of the cellular "methyloma". Additionally, the resistance appeared to be mediated by the accumulation of glyphosate-resistant key enzymes belonging to the aromatic amino acid biosynthetic process.

Supplementary Materials: The following supporting information can be downloaded at: https://www.mdpi.com/article/10.3390/microorganisms10051063/s1, Table S1: All biochemical data relating to the graphs inserted in the text; Table S2: Differentially expressed proteins identified in *A. maxima* samples in control cultures (CTRL) and cultures treated with 0.2 mM glyphosate (Gly). The UniProt identification codes, the number of mass spectra assigned to each protein at the different sampling times, and the fold change (FC) for each protein for each sample treated with respect to the control at the different sampling times are reported; the Log2 HR expressed as the logarithm of the HR value for each protein. The matrix of values was sorted by LogFC values (gray column) from the most negative (under-expressed in the treatise) to the most positive (over-expressed in the treatise) with respect to the control. Table S3: Statistical analysis of the "Differential expression" in *A. maxima* samples of cultures treated with 0.2 mM glyphosate compared to control samples. Table S4: Heat map of Differentially Accumulated Proteins (DAP) in *A. maxima* samples from cultures treated with 0.2 mM glyphosate versus expression levels of each protein in control samples. Table S5: Gene ontology (GO): biological process, cellular component, molecular function of proteins extracted from *A. maxima* samples of cultures treated with 0.2 mM glyphosate. Table S6: Sequences of identified peptides assigned to each protein. Statistical parameters for each peptide have been reported.

Author Contributions: Conceptualization, S.M.; methodology, A.P., D.M.N. and D.O.; formal analysis, A.P., D.M.N. and D.O.; investigation, D.M.N.; resources, F.A.F.; data curation, S.M. and A.P.; writing—original draft preparation, A.P. and S.M.; writing—review and editing, S.M. and A.P.; supervision, S.M. All authors have read and agreed to the published version of the manuscript.

Funding: This research received no external funding.

Data Availability Statement: Not applicable.

Conflicts of Interest: The authors declare no conflict of interest.

References

1. Casida, J.E.; Durkin, K.A. Pesticide Chemical Research in Toxicology: Lessons from Nature. *Chem. Res. Toxicol.* **2017**, *30*, 94–104. [CrossRef] [PubMed]
2. ISPRA. *Edizione 2020: Rapporto Nazionale pesticidi nelle acque—Dati 2017–2018. Edizione*; ISPRA: Varese, Italy, 2020.
3. van Bruggen, A.H.C.; Finckh, M.R.; He, M.; Ritsema, C.J.; Harkes, P.; Knuth, D.; Geissen, V. Indirect Effects of the Herbicide Glyphosate on Plant, Animal and Human Health Through its Effects on Microbial Communities. *Front. Environ. Sci.* **2021**, *9*, 464. [CrossRef]
4. Barriuso, J.; Marin, S.; Mellado, R.P. Potential accumulative effect of the herbicide glyphosate on glyphosate-tolerant maize rhizobacterial communities over a three-year cultivation period. *PLoS ONE* **2011**, *6*, e27558. [CrossRef] [PubMed]
5. Motta, E.V.S.; Raymann, K.; Moran, N.A. Glyphosate perturbs the gut microbiota of honey bees. *Proc. Natl. Acad. Sci. USA* **2018**, *115*, 10305–10310. [CrossRef] [PubMed]
6. Lu, T.; Xu, N.; Zhang, Q.; Zhang, Z.; Debognies, A.; Zhou, Z.; Qian, H. Understanding the influence of glyphosate on the structure and function of freshwater microbial community in a microcosm. *Environ. Pollut.* **2020**, *260*, 114012. [CrossRef]
7. Kumar, A.; Singh, J.S. Cyanoremediation: A Green-Clean Tool for Decontamination of Synthetic Pesticides from Agro- and Aquatic Ecosystems. *Agro-Environ. Sustain.* **2017**, *2*, 59–83. [CrossRef]
8. Lee, S.; Kim, J.; Kennedy, I.R.; Park, J.; Kwon, G.; Koh, S.; Kim, J.E. Biotransformation of an organochlorine insecticide, endosulfan, by Anabaena species. *J. Agric. Food Chem.* **2003**, *51*, 1336–1340. [CrossRef]
9. Benbrook, C.M. Trends in glyphosate herbicide use in the United States and globally. *Environ. Sci. Eur.* **2016**, *28*, 3. [CrossRef]
10. Forlani, G.; Pavan, M.; Gramek, M.; Kafarski, P.; Lipok, J. Biochemical Bases for a Widespread Tolerance of Cyanobacteria to the Phosphonate Herbicide Glyphosate. *Plant Cell Physiol.* **2008**, *49*, 443–456. [CrossRef]
11. Nisticò, D.M.; Piro, A.; Oliva, D.; Osso, V.; Mazzuca, S.; Fagà, F.A.; Morelli, R.; Conidi, C.; Figoli, A.; Cassano, A. A Combination of Aqueous Extraction and Ultrafiltration for the Purification of Phycocyanin from *Arthrospira maxima*. *Microorganisms* **2022**, *28*, 308. [CrossRef]
12. Herrera, A.; Boussiba, S.; Napoleone, V.; Hohlberg, A. Recovery of c-phycocyanin from the cyanobacterium *Spirulina maxima*. *J. Appl. Phycol.* **1989**, *1*, 325–331. [CrossRef]
13. Norling, B.; Zak, E.; Andersson, B.; Pakrasi, H. 2D-isolation of pure plasma and thylakoid membranes from the cyanobacterium *Synechocystis sp.* PCC 6803. *FEBS Lett.* **1998**, *436*, 189–192. [CrossRef]
14. Hongsthong, A.; Sirijuntarut, M.; Yutthanasirikul, R.; Senachak, J.; Kurdrid, P.; Cheevadhanarak, S.; Tanticharoen, M. Subcellular proteomic characterization of the high-temperature stress response of the cyanobacterium *Spirulina platensis*. *Proteome Sci.* **2009**, *7*, 33. [CrossRef] [PubMed]
15. Matallana-Surget, S.; Derock, J.; Leroy, B.; Badri, H.; Deschoenmaeker, F.; Wattiez, R. Proteome-Wide Analysis and Diel Proteomic Profiling of the *Cyanobacterium Arthrospira platensis* PCC 8005. *PLoS ONE* **2014**, *9*, e99076. [CrossRef] [PubMed]
16. Janssen, P.J.; Morin, N.; Mergeay, M.; Leroy, B.; Wattiez, R.; Vallaeys, T.; Waleron, K.; Waleron, M.; Wilmotte, A.; Quillardet, P.; et al. Genome sequence of the edible cyanobacterium *Arthrospira* sp. PCC 8005. *J. Bacteriol.* **2010**, *192*, 2465–2466. [CrossRef] [PubMed]
17. United International Bureaux for the Protection of Intellectual Property (BIRPI). *Paris Convention for the Protection of Industrial Property of March 20, 1883, as Revised at Brussels on December 14, 1900, at Washington on June 2, 1911, at The Hague on November 6, 1925, at London on June 2, 1934, at Lisbon on October 31, 1958, and at Stockholm on July 14, 1967*; United International Bureaux for the Protection of Intellectual Property (BIRPI): Geneva, Switzerland, 1883.
18. Zarrouk, C. Contribution à L'étude D'une Cyanophycée. Influence de Divers Facteurs Physiques et Chimiques Sur la Croissance et la Photosynthèse de *Spirulina maxima*. Ph.D. Thesis, Université De Paris, Paris, France, 1966.
19. El-Kassas, H.Y.; Heneash, A.M.M.; Hussein, N.R. Cultivation of *Arthrospira* (*Spirulina*) *platensis* using confectionary wastes for aquaculture feeding. *J. Genet. Eng. Biotechnol.* **2015**, *13*, 145–155. [CrossRef] [PubMed]
20. Kumar, S.; Cheng, J.; Kubar, A.A.; Guo, W.; Song, Y.; Liu, S.; Chen, S.; Tian, J. Orange light spectra filtered through transparent colored polyvinyl chloride sheet enhanced pigment content and growth of *Arthrospira* cells. *Bioresour. Technol.* **2021**, *319*, 124179. [CrossRef]
21. Porra, R.J.; Thompson, W.; Kriedemann, P.E. Determination of accurate extinction coefficients and simultaneous equations for assaying chlorophylls a and b extracted with four different solvents: Verification of the concentration of chlorophyll standards by atomic absorption spectroscopy. *Biochim. Biophys. Acta* **1989**, *975*, 384–394. [CrossRef]
22. Bennett, A.; Bogorad, L. Complementary chromatic adaption in a filamentous bluegreen alga. *J. Cell Biol.* **1973**, *58*, 419–435. [CrossRef]
23. Mostafa, M.S.I.; Piercey-Normorea, M.D.; Rampitsch, C. Proteomic analyses of the cyanobacterium *Arthrospira* (Spirulina) platensis under iron and salinity stress. *Environ. Exp. Bot.* **2018**, *147*, 63–74. [CrossRef]
24. Laemmli, U.K. Cleavage of structural proteins during the assembly of the head of bacteriophage T4. *Nature* **1970**, *227*, 680–685. [CrossRef]
25. Bradford, M.M. Rapid and sensitive method for the quantitation of microgram quantities of protein utilizing the principle of protein-dye binding. *Anal. Biochem.* **1976**, *72*, 248–254. [CrossRef]

26. Wilm, M.; Shevchenko, A.; Houthaeve, T.; Breit, S.; Schweigerer, L.; Fotsis, T.; Mann, M. Fentomole sequencing of proteins from polyacrylamide gels by nanoelectrospray mass spettrometry. *Nature* **1996**, *379*, 466–469. [CrossRef] [PubMed]
27. Searle, B.C. Scaffold: A bioinformatic tool for validating MS/MS-based proteomic studies. *Proteomics* **2010**, *10*, 1265–1269. [CrossRef] [PubMed]
28. Vidal, N.P.; Manful, C.F.; Pham, T.H.; Stewart, P.; Keough, D.; Thomas, R. The use of XLSTAT in conducting principal component analysis (PCA) when evaluating the relationships between sensory and quality attributes in grilled foods. *MethodsX* **2020**, *7*, 100835. [CrossRef]
29. Conesa, A.; Götz, S.; García-Gómez, J.M.; Terol, J.; Talón, M.; Robles, M. Blast2GO: A universal tool for annotation, visualization and analysis in functional genomics research. *Bioinformatics* **2005**, *21*, 3674–3676. [CrossRef]
30. Gutteridge, S.; Cardona, T.; Shao, S.; Nixon, P.J. Enhancing photosynthesis in plants: The light reactions. *Essays Biochem.* **2018**, *62*, 85–94. [CrossRef]
31. Deniz, F.; Saygideger, S.D.; Karaman, S. Response to Copper and Sodium Chloride Excess in *Spirulina* sp. (*Cyanobacteria*). *Bull. Environ. Contam. Toxicol.* **2011**, *87*, 11–15. [CrossRef]
32. Maslowska, K.H.; Makiela-Dzbenska, K.; Fijalkowska, I.J. The SOS system: A complex and tightly regulated response to DNA damage. *Environ. Mol. Mutagenesis* **2019**, *60*, 368–384. [CrossRef]
33. Lee, A.M.; Ross, C.T.; Zeng, B.B.; Singleton, S.F. A molecular target for suppression of the evolution of antibiotic resistance: Inhibition of the Escherichia coli RecA Protein by N6-(1-Naphthyl)-ADP. *J. Med. Chem.* **2005**, *48*, 5408–5411. [CrossRef]
34. Dörr, T.; Vulić, M.; Lewis, K. Ciprofloxacin Causes Persister Formation by Inducing the TisB toxin in Escherichia coli. *PLoS Biol.* **2010**, *8*, e1000317. [CrossRef]
35. Nagib, A.; Dong-Gi, L.; Ki-Won, L.; Iftekhar, A.; Sang-Hoon, L.; Bahk, J.D.; Byung-Hyun, L. Glyphosate-induced oxidative stress in rice leaves revealed by proteomic approach. *Plant Physiol. Biochem.* **2008**, *46*, 1062–1070. [CrossRef]
36. Ravishankar, A.; Pupo, A.; Gallagher, J.E.G. Resistance Mechanisms of Saccharomyces cerevisiae to Commercial Formulations of Glyphosate Involve DNA Damage Repair, the Cell Cycle, and the Cell Wall Structure. *G3 Genes Genomes Genet.* **2020**, *10*, 2043–2056. [CrossRef] [PubMed]
37. Steinrücken, H.C.; Amrhein, N. The herbicide glyphosate is a potent inhibitor of 5-enolpyruvyl-shikimic acid-3-phosphate synthase. *Biochem. Biophys. Res. Commun.* **1980**, *94*, 1207–1212. [CrossRef]

microorganisms

Article

Comparison of the Performance and Microbial Community Structure of Two Outdoor Pilot-Scale Photobioreactors Treating Digestate

Alessia Bani [1,2], Katia Parati [1], Anna Pozzi [1], Cristina Previtali [1], Graziella Bongioni [1], Andrea Pizzera [3], Elena Ficara [3] and Micol Bellucci [3,*]

1 Istituto Sperimentale Lazzaro Spallanzani, Localita' La Quercia, 26027 Rivolta d'Adda (CR), Italy; ab18858@essex.ac.uk (A.B.); katia.parati@istitutospallanzani.it (K.P.); anna.pozzi@istitutospallanzani.it (A.P.); cristina.previtali@istitutospallanzani.it (C.P.); graziella.bongioni@istitutospallanzani.it (G.B.)

2 Gruppo Ricicla labs., Dipartimento di Scienze Agrarie e Ambientali - Produzione, Territorio, Agroenergia (DiSAA), Università degli studi di Milano, Via Celoria 2, 20133 Milano, Italy

3 Dipartimento di Ingegneria Civile e Ambientale (DICA), Politecnico di Milano, P.zza L. da Vinci 32, 20133 Milano, Italy; andrea.pizzera@mail.polimi.it (A.P.) elena.ficara@polimi.it (E.F.)

* Correspondence: micol.bellucci@polimi.it

Received: 24 October 2020; Accepted: 6 November 2020; Published: 8 November 2020

Abstract: This study aimed at examining and comparing the nutrient removal efficiency, biomass productivity and microbial community structure of two outdoor pilot-scale photobioreactors, namely a bubble column and a raceway pond, treating the liquid fraction of an agricultural digestate. Bacterial and eukaryotic communities were characterized using a metabarcoding approach and quantitative PCR. The abundance, composition, diversity, and dynamics of the main microbes were then correlated to the environmental conditions and operational parameters of the reactors. Both photobioreactors were dominated either by *Chlorella* sp. or *Scenedesmus* sp. in function of temperature, irradiance and the nitrogen compounds derived by nitrification. Other species, such as *Chlamydomonas* and *Planktochlorella*, were sporadically present, demonstrating that they have more specific niche requirement. *Pseudomonas* sp. always dominated the bacterial community in both reactors, except in summertime, when a bloom of *Calothrix* occurred in the raceway pond. In autumn, the worsening of the climate conditions decreased the microalgal growth, promoting predation by *Vorticella* sp. The study highlights the factors influencing the structure and dynamics of the microbial consortia and which ecological mechanisms are driving the microbial shifts and the consequent reactor performance. On these bases, control strategies could be defined to optimize the management of the microalgal-based technologies.

Keywords: microalgae-bacteria consortia; predation; raceway pond; bubble column reactors; nitrogen removal; productivity; centrate; molecular tools.

1. Introduction

The water industry is facing a massive transition to transform wastewater treatment plants (WWTPs) into resource recovering factories [1]. In this light, the integration of microalgae-based technologies into WWTPs has gained attention because of their cost-effective bioremediation capability, the relatively low energy requirement, and the possibility to recover valuable biomass [2]. Microalgae-based processes have been successfully applied for the treatment and valorization of various municipal and industrial wastewaters, including secondary and tertiary effluents, piggery wastewater, digestate, and textile wastewater [3–8]. However, several drawbacks, including land requirement, uncompetitive harvesting and valorization costs, and a strict dependency on climatic and operational conditions still impede real implementation of algae-based wastewater treatment, especially in temperate zones [9].

Outdoor open systems (i.e., raceway ponds, bubble columns, flat panels) are considered so far as the most viable method of microalgal cultivation on wastewaters, though the treatment process is less controllable. Such systems are very susceptible to unavoidable environmental changes and contamination from the surrounding environments [10]. These conditions promote the development of complex and dynamic microbial consortia [11,12], which are composed not only of phototrophic organisms, but also of bacteria, protozoa, yeasts and fungi [13]. The community assembly, as well as the antagonistic and synergetic interactions within the organisms, competition for resources, and predation drive the system functions, namely nutrient removal rate and microalgal productivity. Therefore, understanding the biotic and abiotic mechanisms which favor the growth of desired microbial species or consortia might help to manipulate the community in a way to enhance treatment efficiency and productivity of valuable biomass.

Correlations between the composition, diversity and distribution of mixed microalgae-bacteria consortia and biotic, environmental and operational parameters have been the focus of several recent studies [10,14,15]. Light and temperature are reported to control the growth of species belonging to Scenedesmacae or Chlorellacae within a mixed consortia [16], while nutrient depletion (i.e., N, P and CO_2) plays a role in the accumulation of valuable compounds, such as oils and pigments, in the microalgal cells [17]. Grazers could destroy a microalgae cultivation system in a few days, causing important environmental and economic consequences [13,18]. The presence of active nitrifying bacteria (both ammonia-oxidizing bacteria (AOB) and nitrite-oxidizing bacteria (NOB)), which compete with the microalgae for ammonium, phosphorus, and CO_2, might impact the microalgal biomass productivity [19,20]. In addition, nitrifiers oxidize ammonium, which might have an inhibitory effect, to nitrate, which is reported to be less easily assimilable than ammonium by the microalgae [21]. On the contrary, several bacteria are beneficial for the microalgal growth and seem to help in facing environmental perturbations and parasites. Bacteria could provide microalgae with CO_2 through aerobic respiration, macro-micronutrients via nitrogen fixation and mineralization, but also synthetize compounds which stimulate various responses, such as the algal growth (vitamin B12), spore germination, pathogen resistance and cell aggregation [22]. Thanks to these research activities, several ecological mechanisms have been elucidated, with consequent improvements in the functioning and control of microalgae-based technologies. However, such systems are still not operated at their maximum capability and are subjected to unexpected failures.

Further knowledge could be gained by fully characterizing the microbial community within the systems by advanced sequencing approaches. So far, the taxonomic identification of the dominant microalgal and zooplankton species was based on morphology through microscopic visualization [15,16], which is extremely time consuming and dependent on the operator's skills. In addition, the characterization of the bacterial community structure in microalgal ponds is scarce and often performed with culturing-based techniques, which allow covering a limited portion of the biodiversity [23,24]. Amplicon-based metagenomics, also known as metabarcoding, targeting both eukaryotic and bacterial gene markers could provide deep insight into the complex community in the ponds and could help in elucidating the function and interaction between the different taxa. Such approach is still scarcely used in microalgal-based process [14,25–27], though frequently applied to unravel the microbial ecology mechanisms in engineered ecosystems treating waste and wastewater.

Given that, the goal of this study was to investigate how the structure and dynamics of the eukaryotic and bacterial communities in two types of outdoor photobioreactors, namely a raceway pond (RWP) and a column photobioreactor (PBR), treating the liquid fraction of centrifuged digestate, also known as centrate, change with seasonal and operational conditions and their impact on the photobioreactor functioning. A comprehensive characterization of the microbiome in the two reactors was assessed by a metabarcoding approach and compared. The composition, diversity and dynamics were then correlated with the operational parameters, meteorological conditions, and reactor performance (biomass productivity and nutrient removal rates). All data were then gathered to define the main drivers regulating the community assembly, as well as to provide useful information for the design, operation, and control of open real outdoor, open and non-sterile microalgal-based technologies.

2. Materials and Methods

2.1. Pilot Plant System and Operation

The pilot plant for microalgae culturing was located in a piggery farm in the Cremona province (Northern Italy). The system consisted of two bioreactors operated in parallel, a bubble column photobioreactor (PBR) and a raceway pond (RWP), as described in Pizzera et al., 2019 [28]. The bioreactors where placed outdoors, facing the south in order to maximize sun exposure (Figure 1).

(A) (B)

Figure 1. Bubble column photobioreactor (**A**) and a raceway pond (**B**).

Briefly, the RWP consisted of a main polypropylene tank 4 m long, 1 m wide, and 0.3 m deep (working volume = 880 L) with semi-circular edges and a contact cylinder for CO_2 bubbling (named sump, height 0.8 m, diameter 0.44 m, volume 120 L). Mixing was ensured by a paddlewheel (20 rpm) and dissolved oxygen (DO) concentration, pH and temperature of the algal suspension were constantly monitored by two on-line probes connected to a control panel. CO_2 was provided automatically from the bottom of the sump when the pH was above the set-point value of 7.5. The PBR consisted of a Plexiglass cylinder 131 cm high with an inner diameter of 29 cm (working volume of 75 L). Mixing and CO_2 were provided by continuous air-bubbling (2.5 L h^{-1}) through porous stones located at the bottom of the PBR. The pH was not controlled.

The PBR was initially inoculated with 8 L of undiluted centrate, 64 L of tap water and 0.5 L of a suspension of microalgae (2.5 × 10^6 cell mL^{-1} of Chlorellaceae and 13 × 10^6 cell mL^{-1} of Scenedesmaceae) previously grown on centrate from the same farm. After two months of operation, 65 L of algal suspension was transferred to the RWP and diluted in 90 L of centrate and 680 L of tap water to reach the final working volume.

The centrate was derived from the full-scale anaerobic digester in the farm co-digesting piggery wastewater, energy crops and agricultural wastes. The main average characteristics of the centrate were: total Kjeldahl nitrogen (TKN) = 1290 mg L^{-1}; NH_4^+-N = 1250 mg L^{-1}; PO_4^{3-}-P = 54 mg L^{-1}; soluble Chemical Oxygen Demand (COD) = 1520 mg L^{-1}; total suspended solids (TSS) = 470 mg L^{-1}. Both systems were fed on centrate, diluted with tap water (centrate:tap water, 1:5 *v*/*v*; after 143 days the RWP was fed on a centrate:tap water with a dilution of 1:3 *v*/*v*) with a hydraulic retention time (HRT) of 11 ± 4 days. The main characteristics of the diluted centrate feeding the two bioreactors are reported in Table 1.

Table 1. Average physicochemical characteristics of the diluted centrate, which was used to feed the column (PBR) and in the raceway pond (RWP). Significant differences between reactors are also evidenced (*) (Paired -*t* test, *p*-value < 0.05).

	PBR	RWP	*p*-value
COD (mg L^{-1})	298 ± 48.7	376 ± 111	*
NH_4^+-N (mg L^{-1})	260.1 ± 61.9	288.2 ± 86.7	
NO_3^--N (mg L^{-1})	7.9 ± 3.1	11.3 ± 5.1	
PO_4^{3-}-P (mg L^{-1})	14.6 ± 4	15 ± 9.8	

The functioning of the pilot plants had been monitored for 188 days, from May to November, when both reactors were in operation. Samples of the microalgal suspensions and centrate were collected ca. every two weeks from the PBR (for a total of 11 samples) and RWP (for a total of 12 samples) for physicochemical and biological analyses.

2.2. Analytical Methods

Filtered (0.45 μm) samples of the microalgal suspensions (10 mL) were used for the determination of the concentration of ammonium, nitrate, nitrite, orthophosphate phosphorus (PO_4-P) and soluble COD by spectrophotometric test kits (Hach-Lange, DR6000TM UV VIS Spectrophotometer, Hach Lange LT200 Dry thermostat). Total and volatile suspended solids (TSS and VSS) were determined in duplicate according to Standard Methods (APHA, 2005). The pH was measured by a portable pH meter (Hach-Lange HQ40d) in the PBR and with the on-line probe in the RWP. Absorbance at 680 nm (OD_{680}) and turbidity were measured spectrophotometrically (Hach-Lange, DR6000TM UV VIS Spectrophotometer) as described elsewhere [28]. Irradiation and temperature data were obtained from the website of Arpa Lombardia.

2.3. Operational Data Processing

The performance of the reactors was monitored by computing the volumetric production (*pr*, in mg L^{-1} d^{-1}) of each relevant component (C_j with j = VSS, COD, N-forms, P) from the dynamic mass balance, which considers both the PBR and the RWP as completely stirred tank reactors, as follows:

$$pr_{j,ti} = \frac{Cj_{OUT,t_i} - Cj_{OUT,t_{i-1}}}{t_i - t_{i-1}} + \frac{\left(\frac{Cj_{OUT,t_i} + Cj_{OUT,t_{i-1}}}{2}\right)}{HRT} - \frac{\frac{Cj_{IN,t_i} + Cj_{IN,t_{i-1}}}{2}}{HRT} \quad (1)$$

where C_{IN} and C_{OUT} are the concentration of the component in the feeding diluted centrate and in the algal suspension, at two consecutive sampling dates (time t_{i-1} and t_i), and HRT is the average hydraulic retention time in the period, respectively.

Negative results indicate that the component is removed, therefore it is reported as removal rate (*rr*). This is valid for COD and ammoniacal nitrogen. The ammonium oxidation rate (*pr*N-NO_x) was calculated as the sum of the produced oxidized nitrogen forms (*pr*NO_2^--N+ *pr*NO_3^--N).

Finally, removal efficiencies were computed for each relevant contaminant (Cj) as follows:

$$\eta_{j,ti} = \frac{C_{j,INt_i} - C_{j,OUT,t_i}}{C_{j,INt_i}} \times 100 \quad (2)$$

Free ammonia (mg/L) was computed by considering the concentration of the total ammoniacal nitrogen (TAN), temperature and pH, as described elsewhere [29].

2.4. DNA Extraction and Sequencing

For each sampling event, 2 mL of bulk suspensions of each photobioreactor were collected for DNA extraction. The samples were snap frozen in liquid nitrogen at the sampling location and then transported to the laboratory where they were processed. The material was lyophilized in a Christ Alpha 1-4 lsc lyophilizer at −20 °C and 0.520 mbar pressure. DNA was extracted with DNeasy Plant Kit (Qiagen, Milan Italy) according to the user's manual. The resulting DNA was quantified with Qubit Fluorometer (Invitrogen, Monza, Italy) and sent to the IGATech sequencing center (IGA Technology Services s.r.l., Udine, Italy) to perform Illumina sequencing on the 16S rDNA gene for bacteria and the internal transcribed spacer (ITS) of nuclear DNA for eukaryotes. In particular, the V3-V4 region was chosen for the 16S rDNA gene (341F 5′-CCTACGGGNGGCWGCAG-3′; 805R 5′-GACTACHVGGGTATCTAATCC-3′ [30], while the ITS region (ITS1 5′-TCCGTAGGTGAACCTGCGG-3′; ITS4 5′-TCCTCCGCTTATTGATATGC -3′, [31]) was used for the eukaryotes.

2.5. Bacterial Quantification by Real-Time Polymerase Chain Reaction

The total bacteria and ammonia oxidizing bacteria (AOB) were quantified by real time or quantitative PCR (qPCR) using primer set by targeting the 16S rRNA (1055f 5′-ATGGCTGTCGTCAGCT-3′; 1392r 5′-ACGGGCGGTGTGTAC-3′ [32]) and AmoA (AmoA1F 5′-GGGGTTTCTACTGGTGGT-3′; AmoA2R 5′-CCCCTCKGSAAAGCCTTCTTC-3′ [33]) genes, respectively. Each PCR reaction (20 µL final volume) contained 1 × PowerUp SYBR Green Master Mix (Applied Biosystem, Monza, Italy) (10 µL), forward and reverse primers (100 nM each primer, see below), 0.4 mg mL^{-1} bovine serum albumine (BSA, Thermo Fisher, Monza, Italy), distilled water (RNase/DNase free, Thermo Fisher, Monza, Italy) and 2 µL of DNA (ten-fold diluted). The reactions were performed on a 7500 Fast Real-Time PCR system (Applied Biosystem, Monza, Italy) equipped with Applied Biosystems 7500/7500 Fast Real-Time PCR Software (7500/7500 Fast software). PCR reactions were performed as described in Marazzi et al., 2019 [34]. Standard DNA were constructed from purified PCR amplicons from pure cultures of *Nitrosomonas communis* (DSMZ 2843) for bacterial, and *Nitrosomonas eutropha* (DSMZ 101675) for AOB. All standards consisted of ten-fold dilutions ranging from 10^8 to 10^1 copies µL^{-1}, and DNA samples were run in triplicate using the following cycling conditions: 95 °C denaturation for 10 min, followed by 40 cycles of 20 s at 95 °C, 15 s at 58 °C and 30 s at 72 °C (as described in Bani et al., 2019 [35]) and 95 °C for 2 min denaturation followed by 40 cycles of 45 s at 94 °C, 30 s at 56 °C and 60 s at 72 °C (as described by Bellucci et al., 2015, [36]). To check the amplicon quality and potential primer dimer formation, PCR runs were completed with a melting analysis starting from 65 to 95 °C with temperature increments of 0.25 °C and a transition rate of 5 s. The total number of bacteria was estimated from the 16S rRNA gene copy numbers, by assuming that an average of 4.2 rRNA operon exists per cell [37]. The *AmoA* gene copy number was converted in AOB cell number by assuming 2 copies of the gene in each cell [38]. The cell-specific ammonium oxidation rates (CSAOR) of the AOB were calculated by taking into consideration the amount of ammonium oxidized and the number of AOB as previously reported [39,40].

2.6. Bioinformatics

Amplicon sequences were analyzed as follows. Reads were demultiplexed based on Illumina indexing. For 16S sequences, reads were overlapped and the not overlapping discarded. Low quality sequences and sequences shorter than 200 bp were removed. QIIME open-source pipelines [41] were used for the following steps: chimera check, grouping replicate sequences and picking the operational taxonomic units (OTU). The OTU definition was based on the USEARCH algorithm [42] by setting a 97% similarity threshold. Taxonomy assignation was done on the classifier tool of the Ribosomal Database Project (RDP) with the default setting [43]. For the ITS sequences, a different approach was used due to the non-overlapping nature of reads. All sequences were filtered based on the quality and length of the reads (longer than 200 bp). Then, the same steps previously described for the 16S rRNA amplicon were applied, except for the taxonomic classification, which was performed by comparing the sequences with those presents on the National Center for Biotechnology Information (NCBI) database. Sequences were submitted in the NCBI SRA data archive under the code PRJNA634916.

2.7. Statistical Analysis

Statistical analyses were performed using available packages of R (version 3.6.0) [44]. Paired sample *t*-tests were carried out to compare the performance and efficiency between reactors, as well as the total and nitrifying bacterial abundances in the two systems. Matrices based on Pearson correlation were built to find correlations among physicochemical data and operational parameters of the reactors (temperature (Tr), pH, dissolved oxygen; inlet and outlet COD, NH_4^+-N, NO_2^--N, NO_3^--N, PO_4, TSS, VSS, turbidity; N:P ratio in the influent; free ammonia (FA) in the microalgal suspensions; HRT), climatic conditions (averaged daily irradiance and temperature), reactor performance (*rr*COD, *rr*TAN, *pr*NO_x-N, *pr*NO_2^--N, *pr*NO_3^--N, *rr*PO4, *pr*VSS), and microbiological data (microalgal density (OD_{680}), numbers of the total and nitrifying bacteria, species richness of the total bacteria, eukaryotes and microalgae).

The vegan package was used [45] for creating rarefaction curves to test the depth of the sequencing. Species richness of the bacterial (N_{BAC}) and eukaryotic (N_{EK}) communities were estimated by considering each OTU as a singular microbial species. As in Borruso et al., 2018 [46], the number of shared eukaryotic OTUs among PBR and RWP was calculated by considering only the OTUs with a relative abundance higher than 1%. Microalgal species richness was correlated with the biomass productivity, considered as *pr*VSS, and stability, estimated by the standard deviation of three consecutive *pr*VSS values (same sampling day, previous and successive sampling days).

Non-metric multidimensional scaling (NMDS) plots, combined with the function *envfit* of vegan, were used as preliminary assays to elucidate how the bacterial and eukaryotic communities were influenced by the physicochemical and operational parameters, and climatic conditions. The analyses were conducted by using the bacterial and eukaryotic patterns and the operational parameters and climatic conditions detected at the same sampling time.

Clustering analyses based on the Bray–Curtis similarity index, combined with SIMPROF to check for significance (p-value < 0.05), were performed with the package *clustsig* [47] in order to determine significant changes in the main eukaryotic (species level, threshold 1%) and bacterial (genus level, threshold 10%) populations. The heatmaps were built with the same pool of data using the package *ggplot2* [48].

To elucidate which abiotic factors influenced the shape and evolution of the microbial communities, the relative abundance of the dominant eukaryotic (species level, threshold 1%) and bacteria (genus level, threshold 10%) was correlated to the mean of the physicochemical data, operational parameters and climatic conditions, which were detected over the 7 days preceding the sampling time. In this case, the Kendall rank correlation coefficient was used, and matrices were built with the package *corrplot* [49]. Only correlations with a p-value lower than 0.05 are reported.

OTUs with a relative abundance higher than 1% were also used for bipartite networks, which were created using *CoNet* function [50] inside the package *Cytoscape* [51]. Different correlation coefficients were used: Pearson and Spearman, and dissimilarity indices, Bray-Curtis and Kullback-Leibler, permutation and bootstrap score were based on 1000 interactions. All major network indices were calculated using the *igraph* package [52].

3. Results

3.1. Performance of the Bioreactors

Irradiance and temperature values detected during the sampling campaign are reported in Figure 2.

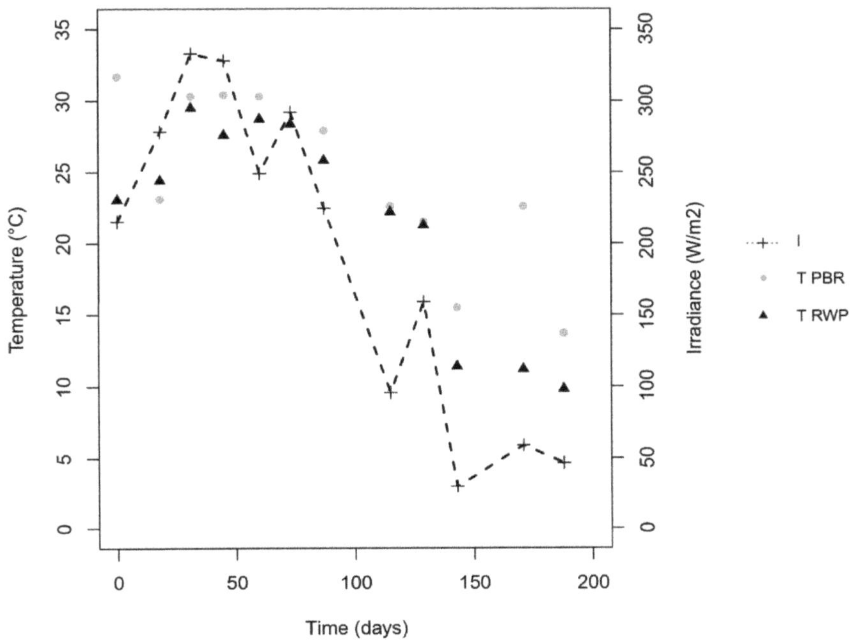

Figure 2. Temperature detected inside the PBR (grey) and RWP (black) and irradiance (dashed line).

The irradiance and temperature figures, which were high at the beginning of the trial (end of May), with peak at the end of June and July (day 31 and 45), and low in September (day 115), are typical of the continental temperate climate of the Padan Plain. The concentrations of ammonium, nitrate, and nitrite measured in the two reactors are shown in Figure 3A,B, while Table 2 summarizes the mean physicochemical parameters measured in the reactors, as well as the nutrient and COD removal and production rates.

Table 2. Average chemical-physical parameters in column (PBR) and in the raceway pond (RWP), as well as the Chemical Oxygen Demand (COD) and nutrient removal rate observed. Significant differences between reactors are also expressed (*) (Paired -*t* test, *p*-value < 0.05).

	PBR	RWP	*p*-value
pH	8.5 ± 0.5	7.2 ± 0.6	*
T_r §	24.5± 6.2	21.9 ± 7.2	*
***rr*COD (mg L^{-1} d^{-1})**	−4 ± 10.7	5.7 ± 6.1	
***rr*TAN (mg L^{-1} d^{-1})**	21.9 ± 7.3	19.5 ± 6.6	
***rp*NO_x-N (mg L^{-1} d^{-1})**	14.4 ± 12.1	18.1 ± 6.9	
***rp*NO_2^--N (mg L^{-1} d^{-1})**	13.5 ± 11.2	1.4 ± 2.2	*
***rp*NO_3^--N (mg L^{-1} d^{-1})**	1 ± 1.3	16.7 ± 7.9	*
Free Ammonia (mg L^{-1})	5.7 ± 8.5	1.8 ± 3.3	
ηTAN (%)	84.1 ± 13.1	79 ± 14.8	
***rr*PO_4^{3-}-P (mg L^{-1} d^{-1})**	0.6 ± 0.9	0.3 ± 0.9	
***rp*VSS (mg L^{-1} d^{-1})**	27 ± 20.3	23 ± 20.3	

§ T_r is the temperature detected inside the reactors.

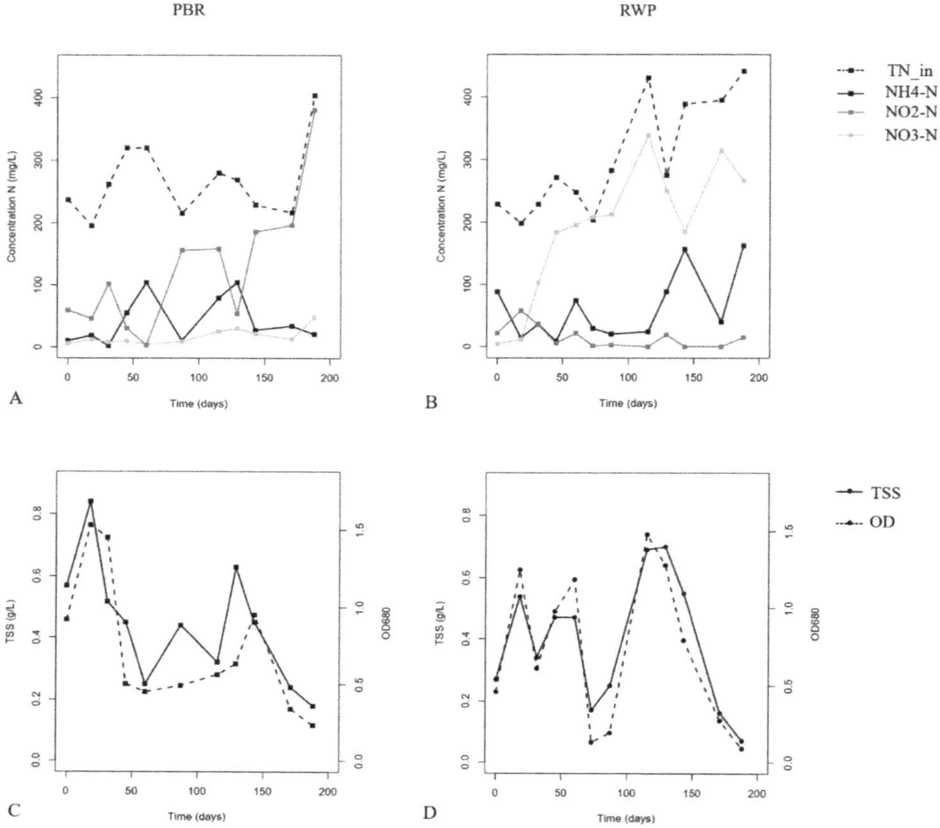

Figure 3. Concentration of the total nitrogen (TN) in the influent (dashed line), ammonium (solid black line), nitrite (solid dark grey line) and nitrate (solid light grey line) measured in PBR (**A**) and RWP (**B**). Microalgal density measured as optical density (OD_{680}) (dashed line) and total suspended solid (TSS) concentration (solid line) detected over time into PBR (**C**) and RWP (**D**).

The average pH in RWP was 7.2 ± 0.6, while the average pH in PBR was 8.5 ± 0.5, because the pH was not controlled by the addition of CO_2 (p-value < 0.05, paired t-test). The ammoniacal nitrogen removal rates and efficiencies were similar in both reactors, averaging 20.6 ± 6.8 mg L^{-1} d^{-1} and 81.5% ± 14% (p-values > 0.1, t-paired test), respectively, indicating that the diverse configurations of the reactors did not impact on the overall NH_4^+-N removal. Most of the ammonium was oxidized ($prNO_x$-N = 16 ± 9 mg L^{-1} d^{-1}) through nitrification. However, the nitrate production rate in RWP was much higher than in PBR ($prNO_3^-$-N = 1 ± 1 mg L^{-1} d^{-1} in the PBR and 16.7 ± 7.9 mg L^{-1} d^{-1} in the RWP), indicating that complete oxidation of ammonium to nitrate could be achieved only in the RWP, while partial nitrification to nitrite occurred in the PBR. No significant differences in the residual concentration of COD (Table 2) between reactors were found and the *rr*COD was negligible.

3.2. Biomass Productivity and Abundance of the Microbial Populations

The microalgal density, which was measured as OD_{680}, varied between 0.09 and 1.53, and the trend over time concurred with the concentrations of the TSS as shown in Figure 3C,D (r = 0.87, p-value < 0.01). The biomass productivity fluctuated, but the average value (25 ± 20 mg VSS L^{-1} d^{-1})

was similar in both configurations (p-value = 0.26, paired t-test) and comparable with the range reported in the literature for microalgae culturing on the liquid fraction of digestate [8].

Figure 4A,B show that the total bacterial concentration ranged between 4.4×10^{10} and 2.8×10^8 cell g VSS^{-1}, significant differences between reactors (p-value > 0.1, paired t-test). However, a significant negative correlation between the bacterial number and microalgal density was observed ($r = -0.72$, p-value = 0.0001). The numbers of AOB, which were always two or three orders of magnitude lower than the total bacteria, were similar in the two reactors (p-value > 0.05, paired t-test), justifying the comparable ammonium oxidation rate observed. Nevertheless, the cell-specific ammonium oxidation rates (CSAOR, logCSAOR ranging between −0.2 and 5 fmol $cell^{-1}$ h^{-1}) were significantly higher than the ones reported for these types of bacteria [53], suggesting that the detected number of AOB was underestimated or other microorganisms contributed to the process.

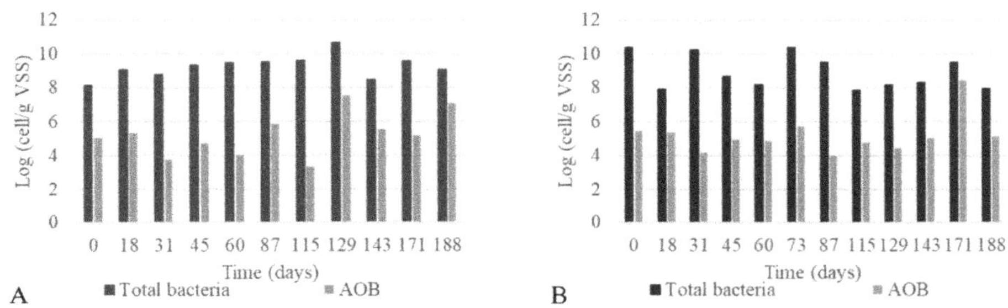

Figure 4. Concentration of the total (dark grey) and ammonia oxidizing (light grey) bacteria detected over time into the PBR (**A**) and the RWP (**B**).

3.3. Composition of the Eukaryotic and Bacterial Communities

At the phylum level, the eukaryotic communities of both reactors were dominated by microorganism belonging to Chlorophyta (58% ± 38% in PBR and 63% ± 33% in RWP), Ciliophora (23% ± 38% in PBR and 26% ± 29% in RWP), and Fungi (5% ±15% in PBR). Chlorophyta are commonly found in outdoor cultivation systems fed on centrate because of their resistance to harsh cultivation conditions [54,55], while members of Ciliophora are algivorous and microalgal grazers, frequently reported in microalgal biomass culturing systems [13,18]. Their high relative abundance, the percentage of which is also negatively correlated with the microalgal one ($r = -0.9$, p-value < 0.001), is an indication of intense predation.

The dominant eukaryotic genera detected by the metabarcoding approach are reported in Figure 5, while in Figure S1 the main microalgae and Ciliophora identified by the optical microscope are shown.

In the PBR, the main microalgal genera were *Chlorella* (23%± 29%) and *Coelastrum* (26% ± 33%) (Figure 5A). *Chlorella* was dominant until day 45, when it was replaced by *Coelastrum*. *Vorticella* contributed to the 6.1±7.7% of the total eukaryotic community most of the time, but it was the sole genus detected in the last two samples. Differently, in the RWP (Figure 5B) the microalgal community was dominated by members belonging to *Chlorella* (31% ± 36%) and *Tetradesmus* (27% ± 36%). *Chlorella* was replaced by *Tetradesmus* between days 45 and 73, when *Chlorella* returned to be the dominant genus. Members closely related to the *Coelastrum* genus were barely detected in the RWP (3% ± 8%), while a severe bloom of *Vorticella* sp. occurred at day 31, when the system faced a drastic reduction in the microalgal density (Figure 3D).

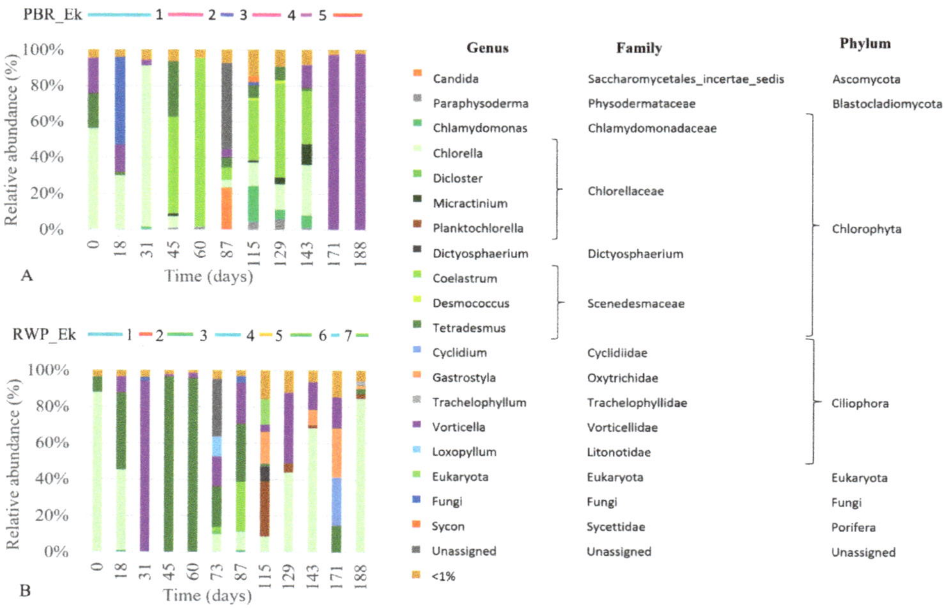

Figure 5. Relative abundance of the main eukaryotic organisms (genus level higher than 1%) detected in the PBR (**A**) and RWP (**B**). Taxonomic ranks (Family and Phylum) are also provided. Above the bar plots, significant population shifts, which were assessed by the combined SIMPROF and clustering analyses, are reported. For line colors refer to the dendrogram in Figure 7A, while the numbers identified the codes in Table 4.

As for the bacterial community (Figure 6A,B), at phylum level, Proteobacteria were dominant in both reactors (59% ± 14% and 49% ± 16% in PBR and RWP respectively), followed by Cyanobacteria (13% ± 11% PBR and 28% ± 15% RWP), and Bacteroidetes (10% ± 5% in PBR and 13% ± 6% in the RWP). Proteobacteria and Bacteroidetes have been already acknowledged to be dominant in high-rate algae ponds [14], as well as in the phycosphere of freshwater microalgal strains [56], while cyanobacteria, being phototrophic, co-exist commonly with microalgae in photobioreactors treating wastewater [10,14]. As for class composition, the PBR was dominated by Gammaproteobacteria (31% ± 9%), followed by Alphaproteobacteria (16% ± 9%) and Nostocophycideae (13% ± 11%), while in the RWP, dominance of Nostocophycideae (26% ± 14%), Gammaproteobacteria (21% ± 9%) and Alphaproteobacteria (18% ± 10%) was revealed. At the genus level, members belonging to *Pseudomonas* were the most abundant in both reactors (20% ± 10% and 14% ± 7%). Their high number is not surprising, as *Pseudomonas* spp. are ubiquitous in freshwater and soil.

Other common genera found in the two photobioreactors were: *Rhodobacter* (4% ± 4% and 4% ± 6%), *Pedobacter* (1% ± 2% and 2% ± 4%), *Opitutus* (2% ± 5% and 0.4% ± 1%). For PBR, the most common genera were instead *Aquimonas* (3% ± 4%), *Denitrobacter* (2% ± 3%), *Oscillochloris* (2% ± 2%), *Segetibacter* (2% ± 2%) and *Stenotrophomonas* (1.5% ± 1%). In RWP, *Calothrix* (12% ±1 8%) became the dominant genera at days 45 and 60. Other specific genera found in the RWP were: *Kaistia* (2% ±3%), *Novosphingobium* (2% ±3%), *Rhodoplanes* (2% ±4%).

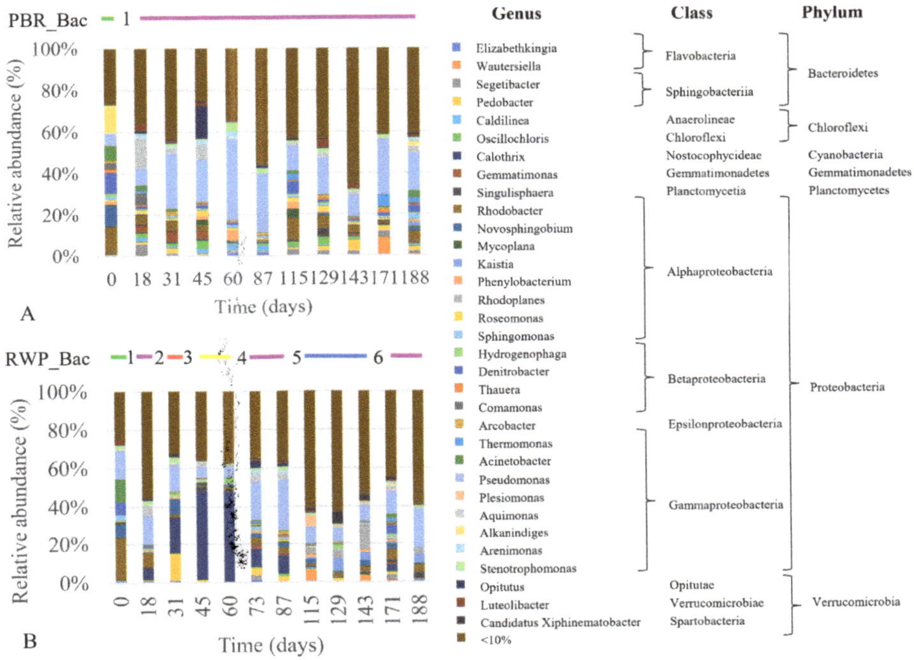

Figure 6. Relative abundance of the main bacterial genera (>10%) detected in the RWP (**A**) and PBR (**B**). Taxonomic ranks (Class and Phylum) are also provided. Above the bar plots, significant shifts of the bacterial communities, assessed by the combined SIMPROF and clustering analyses, are reported. For line colors refer to the dendrogram in Figure 7B, while the numbers identified the codes in Table 4.

3.4. Diversity of the Eukaryotic and Bacterial Communities

The total species richness (N) of the bacteria (N_{BAC} = 3224 ± 1589) was higher than the eukaryotic one (N_{EK} = 574 ± 442) (Figure S2). Yet, a more diverse bacterial community was detected in PBR than in RWP (*p*-value = 0.02; paired-*t* test), while no differences in N_{EK} could be observed between reactors. Both bacterial and eukaryotic species numbers varied over time. The highest values of N_{EK} concurred with the mitigation of the irradiation and temperature conditions after days 87, suggesting that a low to intermediate amount of resources (i.e. irradiance) fosters diversity [57]. A more diverse community in a reactor should increase the productivity and functional stability, especially system resilience to perturbations. However, eukaryotes also include the microalgae predators, which perturbed the ecosystem. Therefore, to better understand the diversity–ecosystem function relationship in the reactors, microalgal and non-microalgal species richness should be differentiated.

The total number of Eukaryotic species with a relative abundance higher than 1% is reported in Table 3. A total of 63 species could be detected, 40 of which were not associated with microalgae, indicating that the reactors were colonized by diverse protozoa, fungi and yeast. Only two species (*Vorticella microstoma* and an undefined fungus) could be found in both reactors, while 11 and 27 species could be detected only in the PBR and RWP, respectively. The higher diversity of non-microalgal eukaryotes found in the RWP than in the PBR indicates that the raceway was more prone to contamination. Besides having a much more open surface than the PBR, sudden variations in the operational parameters, such as the sharp increase of TAN in the influent (between day 87 and 115) and changes in the dilution rate (between day 143 and 171), stressed the microalgal community favoring the growth of grazers and predators.

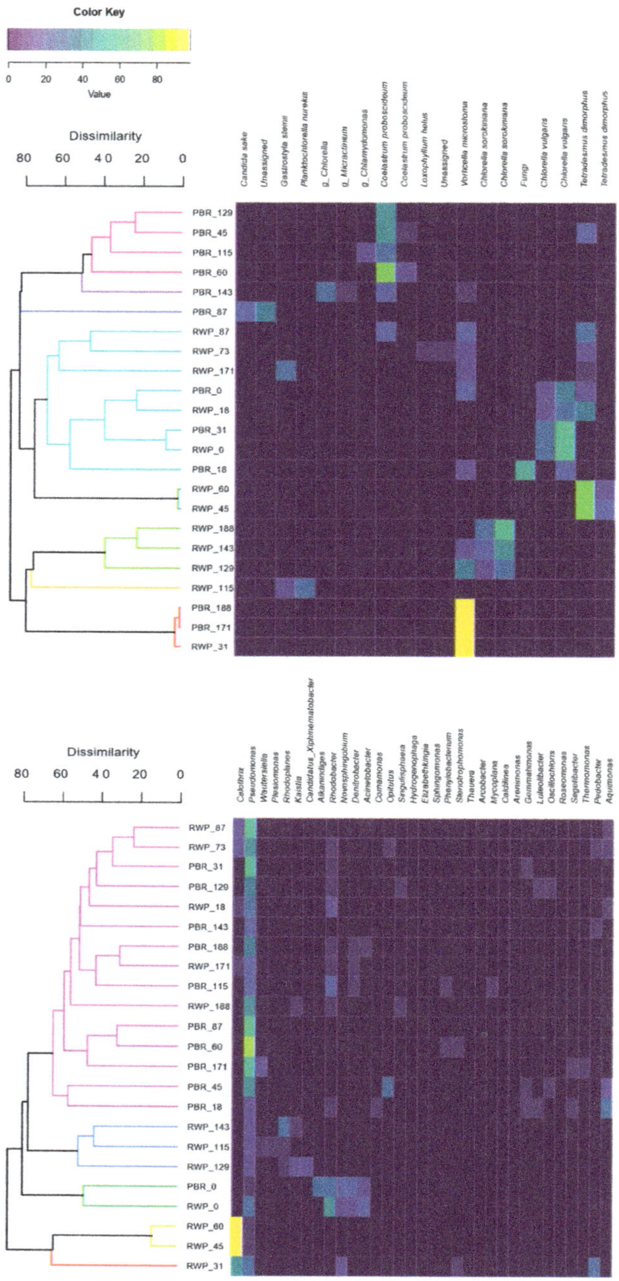

A

B

Figure 7. Heatmap and dendrogram based on Bray-Curtis similarity matrices identifying the main significant eukaryotic (species level) (**A**) and bacterial (genera level) (**B**) population changes detected by SIMPROF analyses. The colored clusters in the dendrogram are representative of statistically similar community patterns.

Table 3. Summary of the total number of Eukaryotic (non-microalgae and microalgae) operational taxonomic units (OTUs) with a percentage above 1%, number of shared OTUs between the column (PBR) and the raceway pond (RWP), and number of OTUs detected in the sole PBR or RWP.

	Total	Shared	Only PBR	Only RWP
Eukaryotes	63	10	18	35
Non-microalgal	40	2	11	27
Microalgae	23	8	7	8

As for microalgae, a total of 23 species could be observed; 34.8% (8 out of 23) of them were shared among reactors. Species belonging to the genera of *Chlamydomonas*, *Chlorella*, *Tetradesmus* and *Coelastrum* were detected in both systems demonstrating their high versatility, while *Desmococcus*, *Micractinium*, *Dicloster*, *Planktonchlorella* and *Dictyosphaerum* were present either in the PBR or RWP, suggesting that they have more definite habitats.

3.5. Associations between the Microbial Community Structure and the Abiotic and Biotic Parameters

Clustering analyses show that the bacterial and eukaryotic patterns had grouped mostly according to the reactor configuration (Figure 7A,B).

This, together with the results of the NMDS analyses (Figure S3), indicates that the microbial community assembly was influenced not only by the meteorological conditions, but also by the geometry of the reactors, their different operational parameters and nitrification pathways. Indeed, the main factors affecting the eukaryotic community composition were irradiance, temperature, and the level of N compounds in the inlet and inside the reactors, whereas the bacterial community composition was mostly influenced by P, irradiance, temperature and level of TAN in the influent.

Nevertheless, several samples collected in the same reactor but at different times were found in statistically different groups, demonstrating significant changes in the communities. The eukaryotic community shifted five and seven times in PBR and RWP, respectively, while one significant change was observed in the bacterial community in PBR and six in the RWP. To establish which factors favored or inhibited the growth of the dominant bacteria, microalgae and predators in each group, giving rise to significant population changes or system failures, the relative abundance of the main microorganisms were evaluated for their correlation with the main operational parameters and meteorological conditions (Figure 8). As regards microalgae, shifts between Chlorellaceae and Scenedesmaceae in the reactors depended on the temperature, irradiance, concentration of solids in the centrate, which in turn affects light penetration, pH and concentration of N forms. High irradiance and T promoted the Scenedesmaceae (i.e., *Tetradesmus* and *Coleastrum*), which are more tolerant to high temperature and have effective quenching mechanisms and high photosynthetic capacity [58,59]. These physiological properties also might explain the significant negative correlation between the abundance of the member of this family with the concentration of solids in the influent. In addition, *Coleastrum* seemed to prefer more alkaline environment than *Tetradesmus* as it grew mainly in the PBR, where the pH was always higher than 8. *Chlorella* spp. exhibited a broad range of niche traits. *Chlorella sorokiniana*, rather than *Chorella vulgaris*, was favored in environments with high TAN, nitrate, and low irradiance, probably because of its more effective photoautotrophic specific growth rates [60]. On the contrary, *Chlamydomonas* sp. and *Micractinium* sp., exhibited higher tolerance toward nitrite. Contamination of *Gastrostyla* occurred in the RWP during the parallel worsening of the meteorological conditions and the increment of the concentration of COD. Such predators coexisted mainly with *Chlorella* sp., confirming that Chlorellaceae are more sensitive to grazing than *Scenedesmus* spp., for which colony formation and the presence of a spine provide a morphological defense against grazers [61].

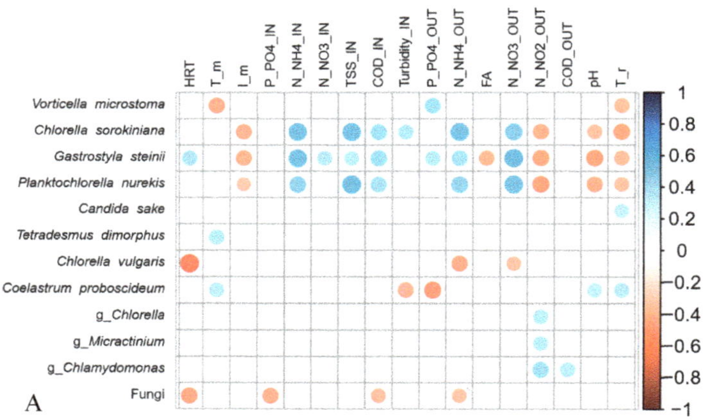

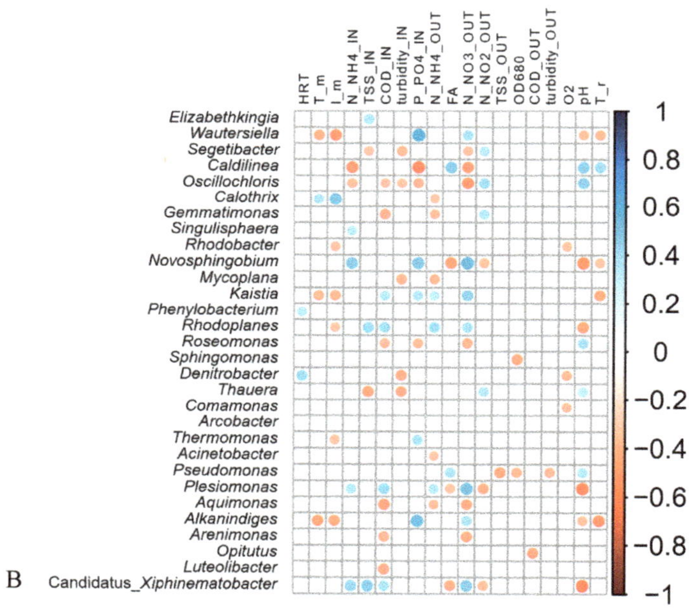

Figure 8. Correlation patterns between the relative abundance of the main eukaryotic (species level, threshold 1%) (**A**) and bacterial (genus level, threshold 10%) (**B**) populations, which were detected in the two reactors, and the mean of the physicochemical and operational parameters measured in the influent (IN) and in the photobioreactors (OUT) the previous seven days based on the Kendall rank correlation coefficient. Only significant positive (blue) and negative (red) correlations are shown (p-value < 0.05) in the matrices. I_m and T_m are the average daily values of the solar irradiance and temperature, respectively; T_r is the temperature detected in the reactors, FA is the free ammonia and O_2 is the dissolved oxygen concentration.

As regards the cyanobacterial genera, *Calothrix* in the RWP was favored by high temperature as it has optimal growth rate at temperatures between 25 and 35 °C [62], as well as in low concentration of TAN being able to N_2-fixation [63]. On the contrary, species belonging to the genera *Oscillochloris* and *Caldilinea* preferred an environment with low turbidity and low concentration of phosphorus.

Several bacterial genera, such as *Kaistia, Plesiomonas, Denitrobacter, Rhodoplanes* and *Wautersiella*, were affected by the concentration of COD in the influent and the level of nitrate in the suspension, being heterotrophs and/or involved in the denitrification process.

All these findings were gathered to delineate the plausible mechanisms (biotic and abiotic) driving the microbial population shifts in the reactors (Table 4). The table highlights how the continuous changes in abiotic conditions and resource availability perturbed the ecosystems, allowing the growth of microalgae and bacteria with more favorable niche traits. Nevertheless, several population changes could not be explained, suggesting a strong role of the stochastic factors in the microbial community assembly [11,12].

3.6. Network Analyses

To untangle the relationship between the bacterial and eukaryotic communities, a bipartite network for each reactor type was constructed (Table 5 and Figure S4). Overall, the RWP had a larger number of interactions than the PBR (see number of interactions, i.e., the number of connected elements, Table 5), though the latter showed a higher complexity of the interactions given the higher network density and heterogeneity. The network heterogeneity is shown by the presence of hubs within the network, while network density is given by the average edges per node, so that a network with many isolated nodes will have a density closer to zero. Communities that are highly connected and have high modularity (i.e., a clustered structured as in the PBR reactor) could be more functionally unstable, since the loss of a node has a higher effect on them [64]. The bloom of predators has a significant impact on the community structure, as predators mutually exclude with almost every other eukaryote as they are microalgae grazers. In general, the PBR experienced less contamination than the RWP over the entire experiment, but once predators entered the reactor they led to a complete disruption of the eukaryotic community. Such fragility can be probably associated to the highly clustered structure of the community [64]. On the contrary, the RWP harbored constantly non-microalgae eukaryotes in the community and the entry of new contaminants was unable to produce drastic damages to the community like in the PBR. In this case, the high diversity and dynamics observed in the RWP (Tables 3 and 4) might help in establishing a complex interactive community (Table 5, interactions number and connected elements) that contrasted the spread of predators over time.

Table 4. Plausible mechanisms of the evolution of the microbial populations in the column (PBR) and the race way pond (RWP). Codes are the same as those reported in Figures 5 and 6, while in brackets the time when the shifts occurred are indicated.

Eukariotes			
Code	**PBR**	**Code**	**RWP**
Inoculum	✓ Coexistence of *C. vulgaris* and *T. dimorphus*	**Inoculum**	✓ Coexistence of *C. vulgaris* and *T. dimorphus*
PBR_Ek_1 (31–45)	✓ *Scenedesmaceae* (*T. dimorphus* and *C. proboscideum*) outcompeted *C. vulgaris*. Increase T and I	**RWP_Ek_1 (18–31)**	✓ Bloom of *Vorticella microstoma* Undefined perturbation
PBR_Ek_2 (60–87)	✓ Increase abundance of yeast and *Vorticella microstoma*. The reactor was covered with a shading net that killed the microalgae	**RWP_Ek_2 (31–45)**	✓ Dominance of *T. dimorphus* Increase T and I
PBR_Ek_3 (87–115)	✓ Most of the populations present before the crashed recovered; increase abundance of *Chlamydomonas* sp. and *Chlorella* sp. High NO_2^--N; decrease I and T	**RWP_Ek_3 (60–73)**	✓ *T. dimorphus* co-existed with *C. vulgaris* and *C. proboscideum*. Presence of predators. Mitigation of I and T conditions
PBR_Ek_4 (129–143)	✓ Increased abundance of *Chlorella* sp. and *Micractinium* sp. and *Vorticella microstoma* High level of NO_2^--N together with the decrease in NH_4^+-N and I	**RWP_Ek_4 (87–115)**	✓ *C. sorokiniana*, *P. nurekis* and their predator *G. steinii*. Drastic increase in NH_4^+-N, NO_3^--N and COD, together with the decrease in I and T
PBR_Ek_5 (143–171)	✓ Dominance of *Vorticella microstoma* Worsening of the climatic conditions reduced the microalgal growth favoring predation	**RWP_Ek_5 (115–129)**	✓ Increase *Chlorella* sp. and *Vorticella microstoma* Decrease influent NH_4^+-N and NO_3^--N
		RWP_Ek_6 (143–171)	✓ Increase number of predators The change of dilution of the centrate disturbed the microalgal community
		RWP_Ek_7 (171–188)	✓ Microalgal community recovered after the system failure due to predation

Table 4. *Cont.*

Bacteria			
Code	**PBR**	**Code**	**RWP**
Inoculum	✓ Bacterial community of the inoculum	**Inoculum**	✓ Bacterial community of the inoculum
PBR_Bac_1 (0_18)	✓ Establishment of a mixed community with *Pseudomonas* spp. as dominant bacteria	**RWP_Bac_1 (0_18)**	✓ Establishment of a mixed community with *Pseudomonas* spp. as dominant bacteria
		RWP_Bac_2 (18_31)	✓ Coexistence of *Pseudomonas* and *Calothrix* spp. Undefined perturbation decreased the number of microalgae; increase in I
		RWP_Bac_3 (31_45)	✓ Dominance of Cyanobacteria.Further increased of I and T
		RWP_Bac_4 (60_73)	✓ Recovery of the original communityMitigation of meteorological conditions
		RWP_Bac_5 (87_115)	✓ Increase abundance of denitrifiers, such as *Kaistia* and *Rhodoplanes* Increase NO_3^--N and COD in the influent, together with the reduction of T
		RWP_Bac_6 (143_188)	✓ Recovery of the previous community

Table 5. Properties of bipartite networks based on the interaction between the most abundant (>1%) bacterial and eukaryotic OTUs.

Network Properties	PBR	RWP
Modularity	0.415	0.508
Connected component	101	117
Average degree	9.373	8.048
Average path length	1.982	2.314
Network centralization	0.2078	0.167
Network heterogeneity	0.7014	0.590
Number nodes	110	125
Network diameter	6	7
Network density	0.086	0.065
Interaction	2480	2588
Co-occurrence	1492	1488
Mutual exclusion	988	1100

4. Discussion

4.1. Environmental Parameters, Biomass Productivity and Nutrients Removal in the Photobioreactors

The biomass productivity and overall nutrient removal were comparable in the two photobioreactors, though influenced by the environmental conditions. Both systems reached a total ammoniacal nitrogen and PO_4^{3-}-P removal rates of 21 and 0.45 mg L^{-1} d^{-1}, respectively, which are in line with similar systems treating digestate located in the same geographical area. Low

COD removal rate could be observed in the PBR and RWP, demonstrating that centrate was mainly composed of recalcitrant and not easily biodegradable compounds. This suggests that this type of wastewater is suitable for preventing the contamination of heterotrophic microbes (i.e., bacteria, fungi and protozoa) in outdoor microalgae cultivation. The microalgal growth and productivity (average 25 mg L^{-1} d^{-1}) followed the variation of the temperature and light, which were typical of the area, and was impacted by erroneous operational strategies. The sudden drop in the microalgal biomass between day 60 and 73 in the RWP could be attributed to the coverage of the system with a shading net, which was placed to prevent photoinhibition. Meanwhile, the worsening of the meteorological conditions, such as low temperature and light at the end of the trial, led to a reduction in the microalgal growth. Overall, these finding suggest that, in suboptimal climate conditions, the performance of outdoor photobioreactors could be improved by the implementation of equipment to mitigate the irradiance and temperature variations, such as the use of external light and heating systems. Although the operational costs would increase, the system are expected to produce more biomass to valorize.

On the other hand, the biological nitrogen removal pathways in the systems were linked to the different configurations and operations of the reactors. Complete nitrification could be achieved only in the RWP, while partial nitrification occurred in the PBR. High ammonium oxidation is quite common in microalgae-based technologies treating municipal and agro-zootechnical digestate [34,54,55], especially if phosphorus is readily available [65]. The elevated P concentration in the centrate favored the activity of the AOB [66], which are the main responsible for ammonium oxidation. This hypothesis is supported also by the positive correlation between the ammonium oxidation rate and the concentration of PO_4^{3-}-P (PO_4^{3-}-P $_{out}$ = 0.86 *pr*N-NO_x + 6.8, R^2 = 0.3623, *p*-value = 0.0038). Nevertheless, the number of detected AOB was lower than expected if considering the ammonium oxidation rate. It is plausible that a fraction of AOB, which prefer growing in colonies, often attached on cellular aggregate and/or within biofilm developing on the reactor walls or on the paddlewheel in the case of the raceway pond [54], were not sampled. Another plausible reason is that other microorganisms contributed to the process. For instance, *Pseudomonas veronii*, which was the dominant bacterial species, might have a role, as it possesses the *AmoA* gene encoding for the oxygen monooxygenase [67–69], and its function as heterotrophic nitrifier was already observed in natural and engineered ecosystems. However, more studies are required to confirm this hypothesis.

Nitrite oxidation occurred in the RWP but it was seriously impaired in the PBR. Most likely, the higher pH of the column, shortage of CO_2 and the slightly higher level of free ammonia than in the RWP prevented the NOB growth [28]. However, if complete nitrification is desired, the implementation of pH control systems and longer HRT might help in overcoming these issues.

4.2. Microbial Community Structure

Species belonging to Scenedesmaceae and Chlorellaceae were the dominant microalgae observed in both systems, confirming their suitability for the bioremediation of digestate [8]. The non-axenic and open conditions of the bioreactors allowed for the entry of protozoa and bacteria, which actively interacted with the microalgae. As commonly observed in microalgal cultivation systems, predation, synergistic, antagonist and competitive mechanisms could be detected [13,70]. In the RWP, algivorous and microalgal grazers, belonging to *Ciliophora*, constantly co-existed with the microalgae, while in the PBR, predators were mainly detected when the microalgal growth was impaired. Nitrifiers and cyanobacteria, which are known to compete with the microalgae for CO_2 and ammoniacal nitrogen, colonized both reactors to different extent. *Pseudomonas* sp., which are known to be plant growth-promoting bacteria [71], were probably beneficial for the microalgae by producing siderophores for iron absorption, solubilize phosphate and synthetize phytohormones, such as indole-3-acetic acid (IAA) [72,73]. On the other hand, antagonist interactions between bacteria and microalgae occurred, as demonstrated by the negative correlation between the number of bacteria and microalgal counts. This evidence contrasts with the frequently stated symbiotic relationship between microalgae and bacteria in microalgae-based technology treating wastewaters [70,74]. However, most of the synergetic interaction concerns the exchange of O_2 and CO_2 during the photosynthetic activity and the aerobic degradation of the organic matter by the heterotrophic bacteria. In our systems, most of the COD in the centrate was not easily biodegradable, as shown by the negligible COD removal rate. Therefore, the growth of the heterotrophs was fostered only when organic compounds were available by microalgal cells' decay.

The non-parallel evolution of the microbial populations in the two reactors indicates that the configuration and operational parameters had a key role in regulating the composition and dynamics of the microalgal and bacterial communities. Without a doubt, high irradiance and temperature were responsible for the shifts between Chlorellaceae and Scenedesmaceae in both reactors due to the more advantageous physiological traits of the latter in such conditions [58,59], but the diverse pH and concentrations of N compounds in each reactor promoted the selection of specific populations. In the RWP, the high level of nitrate, which was derived by complete nitrification, and the concentration of ammonium in the influent seems to favor *C. sorokiniana, P. nurekis* and their predator *G. steinii*, as well as denitrifiers. On the contrary, in the PBR, nitrite accumulation caused by the incomplete nitrification, promoted the growth of *Chlamydomonas* spp. and *Micractinium* sp., while the alkaline pH supported the establishment of members of *Chloroflexi*. Although not all the population shifts could be explained by the variations of the abiotic conditions because of the influence of stochastic factors [12], the association found between the microbes displaying specific functional traits and the environmental conditions established in each reactor might help to predict which populations are favored under various environmental and operational scenarios.

The communities in the RWP were more dynamic than those in the PBR, as demonstrated by the higher number of significant population shifts observed in both microalgal and bacterial communities. Nevertheless, the PBR was more subject to drastic events, while the RWP seemed to be more resilient to perturbations. In the RWP, the negative effect of the shading net on the microalgal composition could be barely detected, while a drastic increase in non-microalgal populations could be observed in the PBR. Yet, the microalgal community in the RWP recovered when the climate conditions got worse, but not in the PBR. The continuous perturbations of the ecosystems due to changes in the abiotic conditions and resource availability favored the growth of microalgae and bacteria with more favorable niche traits and migration of new microbes in both reactors. Nevertheless, the RWP constantly harbored non-algal eukaryotes suggesting more trophic interactions; the coexistence of predators and microalgae was more evident in the RWP than in the PBR, as well as predator diversity was higher in RWP than PBR (only one in PBR and three in RWP). It seems that the more complex community established in the RWP helped in contrasting the spread of the predators over time by maintaining the predator/prey dynamics under control, even if the raceway system has a wider open area and a design more prone to external contamination.

The establishment of a diversified microalgal community derived by the variable abiotic conditions seemed to play a role in maintaining the productivity of the reactors in agreement with other aquatic ecosystems [75,76] and microalgal cultivation systems [25,77]. The trends of microalgal density showed two peaks in both reactors. In the PBR, the highest and first peak corresponded to day 18 and 31, when the community was dominated by *Chlorella*. The extremely high temperature and irradiance of the following days led to a switch to the Scenedesmaceae, but a drastic decrement of the biomass. Biomass concentration and diversity increased again only with the mitigation of the climatic factors. Similarly, in the RWP, the initial biomass productivity was sustained by the Chlorellaceae and Scenedesmaceae; nevertheless, the highest concentration of biomass could be detected when the climatic and operational conditions were more variable. These findings endorse the use of microalgal polycultures for a more stable production of biomass in outdoor cultivation systems, especially if located in regions with a temperate climate where the environmental conditions are subject to seasonal variations

Supplementary Materials: The following are available online at http://www.mdpi.com/2076-2607/8/11/1754/s1, Figure S1: Images of the main microalgae, protozoa and cyanobacteria observed in the PBR and RWP over time; Figure S2: Number of the total eukariotic (N_{EK}) (A) and bacterial species (N_{BAC}) (B) detected in the column (PBR) and raceway pond (RWP) over time; Figure S3: Non-metric multidimensional scaling analyses of the eukaryotic (A and B) and bacterial OTU patterns (C and D) in the PBR (blue) and in the RWP (black). Numbers indicate the sampling date, while arrows represent directions of forcing of each environmental parameter measured in the influent (IN) and reactors (OUT). Torb is the turbidity. T_{PBR} and T_{RWP} is the temperature in the bubble column and in the raceway pond, respectively. I is the irradiance; Figure S4: Bipartite network based on the interaction between the most abundant (>1%) eukaryotic and bacterial OTUs in bubble columns and raceways pond (A and B).

Author Contributions: Conceptualization, K.P., E.F. and M.B.; Data curation, A.B., K.P., A.P. (Anna Pozzi), G.B., A.P. (Andrea Pizzera), E.F. and M.B.; Formal analysis, A.B., C.P., A.P. (Andrea Pizzera) and M.B.; Funding acquisition, K.P. and E.F.; Investigation, A.B., A.P. (Andrea Pizzera) and M.B.; Methodology, A.B., A.P. (Anna Pozzi), C.P. and M.B.; Project administration, K.P. and E.F.; Resources, K.P., G.B. and E.F.; Software, A.B.; Supervision, K.P., E.F. and M.B.; Validation, A.B., A.P. (Anna Pozzi), C.P. and M.B.; Visualization, A.B. and M.B.; Writing—original draft, A.B. and M.B.; Writing—review and editing, E.F. All authors have read and agreed to the published version of the manuscript.

Funding: This work was supported by Fondazione Cariplo within the project "Polo delle Microalghe". A.B. was supported by SABANA Horizon 2020 grant (University of Milan, Grant No. 727874).

Acknowledgments: The authors would like to thank Davide Scaglione, Lorenzo Proietti, Luciano Foglio and Federico Castillo Cascino for their logistic help and the images taken through the microscope. The authors are also grateful to Luigimaria Borruso for fruitful discussion.

Conflicts of Interest: The authors declare no conflict of interest.

References

1. Puyol, D.; Batstone, D.J.; Hülsen, T.; Astals, S.; Peces, M.; Krömer, J.O. Resource Recovery from Wastewater by Biological Technologies: Opportunities, Challenges, and Prospects. *Front. Microbiol.* **2017**, *7*, 1–23. [CrossRef] [PubMed]
2. Subashchandrabose, S.R.; Ramakrishnan, B.; Megharaj, M.; Venkateswarlu, K.; Naidu, R. Consortia of cyanobacteria/microalgae and bacteria: Biotechnological potential. *Biotechnol. Adv.* **2011**, *29*, 896–907. [CrossRef] [PubMed]
3. Bellucci, M.; Marazzi, F.; Naddeo, L.S.; Piergiacomo, F.; Beneduce, L.; Ficara, E.; Mezzanotte, V. Disinfection and nutrient removal in laboratory-scale photobioreactors for wastewater tertiary treatment. *J. Chem. Technol. Biotechnol.* **2019**. [CrossRef]
4. Marazzi, F.; Bellucci, M.; Fantasia, T.; Ficara, E.; Mezzanotte, V. Interactions between Microalgae and Bacteria in the Treatment of Wastewater from Milk Whey Processing. *Water* **2020**, *12*, 297. [CrossRef]
5. Huy, M.; Kumar, G.; Kim, H.W.; Kim, S.H. Photoautotrophic cultivation of mixed microalgae consortia using various organic waste streams towards remediation and resource recovery. *Bioresour. Technol.* **2018**, *247*, 576–581. [CrossRef] [PubMed]
6. Lim, S.L.; Chu, W.L.; Phang, S.M. Use of Chlorella vulgaris for bioremediation of textile wastewater. *Bioresour. Technol.* **2010**, *101*, 7314–7322. [CrossRef]
7. Trentin, G.; Bertucco, A.; Sforza, E. Mixotrophy in Synechocystis sp. for the treatment of wastewater with high nutrient content: Effect of CO_2 and light. *Bioprocess Biosyst. Eng.* **2019**, *42*, 1661–1669. [CrossRef]
8. Xia, A.; Murphy, J.D. Microalgal cultivation in treating liquid digestate from biogas systems. *Trends Biotechnol.* **2016**, *34*, 264–275. [CrossRef]
9. Lavrinovičs, A.; Juhna, T. Review on Challenges and Limitations for Algae-Based Wastewater Treatment. *Constr. Sci.* **2018**, *20*, 17–25. [CrossRef]
10. Fulbright, S.P.; Dean, M.K.; Wardle, G.; Lammers, P.J.; Chisholm, S. Molecular diagnostics for monitoring contaminants in algal cultivation. *Algal Res.* **2014**, *4*, 41–51. [CrossRef]
11. Ofiţeru, I.D.; Lunn, M.; Curtis, T.P.; Wells, G.F.; Criddle, C.S.; Francis, C.A.; Sloan, W.T. Combined niche and neutral effects in a microbial wastewater treatment community. *Proc. Natl. Acad. Sci. USA* **2010**, *107*, 15345–15350. [CrossRef] [PubMed]
12. Van Der Gast, C.J.; Ager, D.; Lilley, A.K. Temporal scaling of bacterial taxa is influenced by both stochastic and deterministic ecological factors. *Environ. Microbiol.* **2008**, *10*, 1411–1418. [CrossRef] [PubMed]
13. Day, J.G.; Gong, Y.; Hu, Q. Microzooplanktonic grazers—A potentially devastating threat to the commercial success of microalgal mass culture. *Algal Res.* **2017**, *27*, 356–365. [CrossRef]

14. Mark Ibekwe, A.; Murinda, S.E.; Murry, M.A.; Schwartz, G.; Lundquist, T. Microbial community structures in high rate algae ponds for bioconversion of agricultural wastes from livestock industry for feed production. *Sci. Total Environ.* **2017**, *580*, 1185–1196. [CrossRef]
15. Sutherland, D.L.; Turnbull, M.H.; Craggs, R.J. Environmental drivers that influence microalgal species in fullscale wastewater treatment high rate algal ponds. *Water Res.* **2017**, *124*, 504–512. [CrossRef]
16. Cho, D.-H.; Choi, J.-W.; Kang, Z.; Kim, B.-H.; Oh, H.-M.; Kim, H.; Ramanan, R. Microalgal diversity fosters stable biomass productivity in open ponds treating wastewater. *Sci. Rep.* **2017**, *7*, 1979. [CrossRef]
17. Rodolfi, L.; Chini Zittelli, G.; Bassi, N.; Padovani, G.; Biondi, N.; Bonini, G.; Tredici, M.R. Microalgae for oil: Strain selection, induction of lipid synthesis and outdoor mass cultivation in a low-cost photobioreactor. *Biotechnol. Bioeng.* **2009**, *102*, 100–112. [CrossRef] [PubMed]
18. Wang, L.; Yuan, D.; Li, Y.; Ma, M.; Hu, Q.; Gong, Y. Contaminating microzooplankton in outdoor microalgal mass culture systems: An ecological viewpoint. *Algal Res.* **2016**, *20*, 258–266. [CrossRef]
19. González-Camejo, J.; Aparicio, S.; Ruano, M.V.; Borrás, L.; Barat, R.; Ferrer, J. Effect of ambient temperature variations on an indigenous microalgae-nitrifying bacteria culture dominated by Chlorella. *Bioresour. Technol.* **2019**, *290*, 121788. [CrossRef]
20. González-Camejo, J.; Montero, P.; Aparicio, S.; Ruano, M.V.; Borrás, L.; Seco, A.; Barat, R. Nitrite inhibition of microalgae induced by the competition between microalgae and nitrifying bacteria. *Water Res.* **2020**, *172*, 115499. [CrossRef]
21. Xin, L.; Hong-ying, H.; Ke, G.; Jia, Y. Growth and nutrient removal properties of a freshwater microalga *Scenedesmus* sp. LX1 under different kinds of nitrogen sources. *Ecol. Eng.* **2010**, *36*, 379–381. [CrossRef]
22. Lian, J.; Wijffels, R.H.; Smidt, H.; Sipkema, D. The effect of the algal microbiome on industrial production of microalgae. *Microb. Biotechnol.* **2018**, *11*, 806–818. [CrossRef] [PubMed]
23. Krustok, I.; Odlare, M.; Shabiimam, M.A.; Truu, J.; Truu, M.; Ligi, T.; Nehrenheim, E. Characterization of algal and microbial community growth in a wastewater treating batch photo-bioreactor inoculated with lake water. *Algal Res.* **2015**, *11*, 421–427. [CrossRef]
24. Lakaniemi, A.-M.; Hulatt, C.J.; Wakeman, K.D.; Thomas, D.N.; Puhakka, J. A Eukaryotic and prokaryotic microbial communities during microalgal biomass production. *Bioresour. Technol.* **2012**, *124*, 387–393. [CrossRef] [PubMed]
25. Beyter, D.; Tang, P.; Becker, S.; Hoang, T.; Bilgin, D.; Lim, W.; Peterson, T.C.; Mayfield, S.; Haerizadeh, F.; Shurin, J.B.; et al. Diversity, Productivity, and Stability of an Industrial Microbial Ecosystem. *Appl. Environ. Microbiol.* **2016**, *82*, 2494–2505. [CrossRef]
26. Eland, L.E.; Davenport, R.J.; dos Santos, A.B.; Mota Filho, C.R. Molecular evaluation of microalgal communities in full-scale waste stabilisation ponds. *Environ. Technol.* **2019**, *40*, 1969–1976. [CrossRef]
27. Ferro, L.; Hu, Y.O.O.; Gentili, F.G.; Andersson, A.F.; Funk, C. DNA metabarcoding reveals microbial community dynamics in a microalgae-based municipal wastewater treatment open photobioreactor. *Algal Res.* **2020**, *51*, 102043. [CrossRef]
28. Pizzera, A.; Scaglione, D.; Bellucci, M.; Marazzi, F.; Mezzanotte, V.; Parati, K.; Ficara, E. Digestate treatment with algae-bacteria consortia: A field pilot-scale experimentation in a sub-optimal climate area. *Bioresour. Technol.* **2019**, *274*, 232–243. [CrossRef] [PubMed]
29. Anthonisen, A.; Loehr, R.; Prakasam, T.; Srinath, E. Inhibition of Nitrification by Ammonia and Nitrous Acid. *J. Water Pollut. Control Fed.* **1976**, *48*, 835–852.
30. Herlemann, D.P.R.; Labrenz, M.; Jürgens, K.; Bertilsson, S.; Waniek, J.J.; Andersson, A.F. Transitions in bacterial communities along the 2000 km salinity gradient of the Baltic Sea. *ISME J.* **2011**, *5*, 1571. [CrossRef]
31. White, T.; Bruns, T.; Lee, S.; Taylor, J. Amplification and direct sequencing of fungal ribosomal RNA genes for phylogenetics. In *PCR Protocols: A Guide to Methods and Applications*; Innis, M., Gelfand, D., Shinsky, J., White, T., Eds.; Academic Press: San Diego, CA, USA, 1990; pp. 315–322.
32. Ferris, M.J.; Muyzer, G.; Ward, D.M. Denaturing gradient gel electrophoresis profiles of 16S rRNA-defined populations inhabiting a hot spring microbial mat community. *Appl. Environ. Microbiol.* **1996**, *62*, 340–346. [CrossRef]
33. Rotthauwe, J.H.; Witzel, K.P.; Liesack, W. The ammonia monooxygenase structural gene amoA as a functional marker: Molecular fine-scale analysis of natural ammonia-oxidizing populations. *Appl. Environ. Microbiol.* **1997**, *63*, 4704–4712. [CrossRef] [PubMed]

34. Marazzi, F.; Bellucci, M.; Fornaroli, R.; Bani, A.; Ficara, E.; Mezzanotte, V. Lab-scale testing of operation parameters for algae based treatment of piggery wastewater. *J. Chem. Technol. Biotechnol.* **2020**, *95*, 967–974. [CrossRef]
35. Bani, A.; Borruso, L.; Matthews Nicholass, K.J.; Bardelli, T.; Polo, A.; Pioli, S.; Gómez-Brandón, M.; Insam, H.; Dumbrell, A.J.; Brusetti, L. Site-Specific Microbial Decomposer Communities Do Not Imply Faster Decomposition: Results from a Litter Transplantation Experiment. *Microorganisms* **2019**, *7*, 349. [CrossRef]
36. Bellucci, M.; Ofiţeru, I.D.; Beneduce, L.; Graham, D.W.; Head, I.M.; Curtis, T.P. A preliminary and qualitative study of resource ratio theory to nitrifying lab-scale bioreactors. *Microb. Biotechnol.* **2015**, *8*, 590–603. [CrossRef]
37. Klappenbach, J.A. rrndb: The Ribosomal RNA Operon Copy Number Database. *Nucleic Acids Res.* **2001**, *29*, 181–184. [CrossRef]
38. McTavish, H.; Fuchs, J.A.; Hooper, A.B. Sequence of the gene coding for ammonia monooxygenase in Nitrosomonas europaea. *J. Bacteriol.* **1993**, *175*, 2436–2444. [CrossRef]
39. Bellucci, M.; Ofiteru, I.D.; Graham, D.W.; Head, I.M.; Curtis, T.P. Low-dissolved-oxygen nitrifying systems exploit ammonia-oxidizing bacteria with unusually high yields. *Appl. Environ. Microbiol.* **2011**, *77*, 7787–7796. [CrossRef]
40. Coskuner, G.; Ballinger, S.J.; Davenport, R.J.; Pickering, R.L.; Solera, R.; Head, I.M.; Curtis, T.P. Agreement between theory and measurement in quantification of ammonia-oxidizing bacteria. *Appl. Environ. Microbiol.* **2005**, *71*, 6325–6334. [CrossRef]
41. Caporaso, J.G.; Kuczynski, J.; Stombaugh, J.; Bittinger, K.; Bushman, F.D.; Costello, E.K.; Fierer, N.; Peña, A.G.; Goodrich, J.K.; Gordon, J.I.; et al. QIIME allows analysis of high-throughput community sequencing data. *Nat. Methods* **2010**, *7*, 335–336. [CrossRef]
42. Edgar, R.C. Search and clustering orders of magnitude faster than BLAST. *Bioinformatics* **2010**, *26*, 2460–2461. [CrossRef]
43. Wang, Q.; Garrity, G.M.; Tiedje, J.M.; Cole, J.R. Naïve Bayesian Classifier for Rapid Assignment of rRNA Sequences into the New Bacterial Taxonomy. *Appl. Environ. Microbiol.* **2007**, *73*, 5261–5267. [CrossRef]
44. R Core Team; R Development Core Team. R: A Language and Environment for Statistical Computing [Internet]; 2019. Available online: https://www.r-project.org/ (accessed on 20 October 2020).
45. Oksanen, J.; Blanchet, F.G.; Friendly, M.; Kindt, R.; Legendre, P.; McGlinn, D.; Minchin, P.R.; O'Hara, R.B.; Simpson, G.L.; Solymos, P.; et al. Vegan: Community Ecology Package. R Packag Version 2.4-1. 2016. Available online: https://cran.r-project.org/web/packa (accessed on 20 October 2020).
46. Borruso, L.; Sannino, C.; Selbmann, L.; Battistel, D.; Zucconi, L.; Azzaro, M.; Turchetti, B.; Buzzini, P.; Guglielmin, M. A thin ice layer segregates two distinct fungal communities in Antarctic brines from Tarn Flat (Northern Victoria Land). *Sci. Rep.* **2018**, *8*, 6582. [CrossRef]
47. Whitaker, D.; Christman, M. Package 'Clustsig'. R Top Doc. 2015; pp. 1–7. Available online: http://www.douglaswhitaker.com (accessed on 20 October 2020).
48. Wickham, H. *ggplot2: Elegant Graphics for Data Analysis*; Springer: New York, NY, USA, 2016; ISBN 978-3-319-24277-4.
49. Wei, T.; Simko, V. R Package "Corrplot": Visualization of a Correlation Matrix. 2017. Available online: https://github.com/taiyun/corrplot (accessed on 20 October 2020).
50. Faust, K.; Raes, J. CoNet app: Inference of biological association networks using Cytoscape. *F1000Research* **2016**, *5*, 1519. [CrossRef]
51. Shannon, P.; Markiel, A.; Ozier, O.; Baliga, N.S.; Wang, J.T.; Ramage, D.; Amin, N.; Schwikowski, B.; Ideker, T. Cytoscape: A Software Environment for Integrated Models of Biomolecular Interaction Networks. *Genome Res.* **2003**, *13*, 2498–2504. [CrossRef]
52. Csardi, G.; Nepusz, T. The igraph software package for complex network research. *Inter J. Complex Syst.* **2006**, *1695*, 1–9.
53. Daims, H.; Ramsing, N.B.; Schleifer, K.H.; Wagner, M. Cultivation-independent, semiautomatic determination of absolute bacterial cell numbers in environmental samples by fluorescence in situ hybridization. *Appl. Environ. Microbiol.* **2001**, *67*, 5810–5818. [CrossRef] [PubMed]
54. Mantovani, M.; Marazzi, F.; Fornaroli, R.; Bellucci, M.; Ficara, E.; Mezzanotte, V. Outdoor pilot-scale raceway as a microalgae-bacteria sidestream treatment in a WWTP. *Sci. Total Environ.* **2020**, *710*, 135583. [CrossRef] [PubMed]

55. Marazzi, F.; Bellucci, M.; Rossi, S.; Fornaroli, R.; Ficara, E.; Mezzanotte, V. Outdoor pilot trial integrating a sidestream microalgae process for the treatment of centrate under non optimal climate conditions. *Algal Res.* **2019**, *39*, 101430. [CrossRef]
56. Krohn-molt, I.; Alawi, M.; Förstner, K.U.; Wiegandt, A.; Burkhardt, L.; Indenbirken, D.; Thieß, M.; Grundhoff, A.; Kehr, J.; Tholey, A.; et al. Insights into Microalga and Bacteria Interactions of Selected Phycosphere Biofilms Using and Proteomic Approaches. *Front. Microbiol.* **2017**, *8*, 1–14. [CrossRef]
57. Tilman, D. Resource Competition between Plankton Algae: An Experimental and Theoretical Approach. *Ecology* **1977**, *58*, 338–348. [CrossRef]
58. Hodaifa, G.; Martínez, M.E.; Sánchez, S. Influence of temperature on growth of Scenedesmus obliquus in diluted olive mill wastewater as culture medium. *Eng. Life Sci.* **2010**, *10*, 257–264. [CrossRef]
59. Masojídek, J.; Torzillo, G.; Koblížek, M.M.; Kopecký, J.; Bernardini, P.; Sacchi, A.; Komenda, J. Photoadaptation of two members of the Chlorophyta (Scenedesmus and Chlorella) in laboratory and outdoor cultures: Changes in chlorophyll fluorescence quenching and the xanthophyll cycle. *Planta* **1999**, *209*, 126–135. [CrossRef]
60. Rosenberg, J.N.; Kobayashi, N.; Barnes, A.; Noel, E.A.; Betenbaugh, M.J.; Oyler, G.A. Comparative analyses of three Chlorella species in response to light and sugar reveal distinctive lipid accumulation patterns in the Microalga C. sorokiniana. *PLoS ONE* **2014**, *9*, e92460. [CrossRef]
61. Mayeli, S.M.; Nandini, S.; Sarma, S.S.S. The efficacy of Scenedesmus morphology as a defense mechanism against grazing by selected species of rotifers and cladocerans. *Aquat. Ecol.* **2005**, *38*, 515–524. [CrossRef]
62. Litchman, E.; Pinto, P.D.T.; Klausmeier, C.A.; Thomas, M.K.; Yoshiyama, K. Linking traits to species diversity and community structure in phytoplankton. *Hydrobiologia* **2010**, 15–28. [CrossRef]
63. Issa, A.; Mohamed, H.A.; Ohyama, T. Nitrogen fixing cyanobacteria: Future prospect. In *Advances in Biology and Ecology of Nitrogen Fixation*; Ohyama, T., Ed.; InTech: Leeds, UK, 2014; pp. 23–48.
64. Yang, G.; Peng, M.; Tian, X.; Dong, S. Molecular ecological network analysis reveals the effects of probiotics and florfenicol on intestinal microbiota homeostasis: An example of sea cucumber. *Sci. Rep.* **2017**, *7*, 4778. [CrossRef]
65. Marcilhac, C.; Sialve, B.; Pourcher, A.M.; Ziebal, C.; Bernet, N.; Béline, F. Control of nitrogen behaviour by phosphate concentration during microalgal-bacterial cultivation using digestate. *Bioresour. Technol.* **2015**, *175*, 224–230. [CrossRef]
66. Bellucci, M.; Marazzi, F.; Ficara, E.; Mezzanotte, V. Effect of N:P ratio on microalgae/nitrifying bacteria community in agro-digestate treatment. *Environ. Clim. Technol.* **2020**, *24*, 136–148. [CrossRef]
67. Montes, C.; Altimira, F.; Canchignia, H.; Castro, Á.; Sánchez, E.; Miccono, M.; Tapia, E.; Sequeida, Á.; Valdés, J.; Tapia, P.; et al. A draft genome sequence of Pseudomonas veronii R4: A grapevine (*Vitis vinifera* L.) root-associated strain with high biocontrol potential. *Stand. Genom. Sci.* **2016**, *11*, 76. [CrossRef]
68. Liu, H.; Ding, Y.; Zhang, Q.; Liu, X.; Xu, J.; Li, Y.; Di, H. Heterotrophic nitrification and denitrification are the main sources of nitrous oxide in two paddy soils. *Plant Soil* **2019**, *445*, 39–53. [CrossRef]
69. Trung Tran, T.; Bott, N.J.; Dai Lam, N.; Trung Nguyen, N.; Hoang Thi Dang, O.; Hoang Le, D.; Tung Le, L.; Hoang Chu, H. The Role of Pseudomonas in Heterotrophic Nitrification: A Case Study on Shrimp Ponds (*Litopenaeus vannamei*) in Soc Trang Province. *Microorganisms* **2019**, *7*, 155. [CrossRef]
70. Ramanan, R.; Kim, B.-H.H.; Cho, D.-H.H.; Oh, H.-M.M.; Kim, H.-S.S. Algae-bacteria interactions: Evolution, ecology and emerging applications. *Biotechnol. Adv.* **2016**, *34*, 14–29. [CrossRef]
71. Glick, B.R. Plant Growth-Promoting Bacteria: Mechanisms and Applications. *Scientifica* **2012**, *2012*, 963401. [CrossRef]
72. Amavizca, E.; Bashan, Y.; Ryu, C.M.; Farag, M.A.; Bebout, B.M.; De-Bashan, L.E. Enhanced performance of the microalga Chlorella sorokiniana remotely induced by the plant growth-promoting bacteria Azospirillum brasilense and Bacillus pumilus. *Sci. Rep.* **2017**, *7*, 1–11. [CrossRef]
73. Saranraj, P. Biocontrol potentiality of plant growth promoting bacteria (PGPR)—Pseudomonas fluorescens and Bacillus subtilis: A review. *Afr. J. Agric. Res.* **2014**, *9*, 1265–1277.
74. Seymour, J.R.; Amin, S.A.; Raina, J.-B.; Stocker, R. Zooming in on the phycosphere: The ecological interface for phytoplankton–bacteria relationships. *Nat. Microbiol.* **2017**, *2*, 17065. [CrossRef]
75. Cardinale, B.J. Biodiversity improves water quality through niche partitioning. *Nature* **2011**, *472*, 86–89. [CrossRef]
76. Godwin, C.M.; Lashaway, A.R.; Hietala, D.C.; Savage, P.E.; Cardinale, B.J. Biodiversity improves the ecological design of sustainable biofuel systems. *GCB Bioenergy* **2018**, *10*, 752–765. [CrossRef]

77. Stockenreiter, M.; Graber, A.K.; Haupt, F.; Stibor, H. The effect of species diversity on lipid production by micro-algal communities. *J. Appl. Phycol.* **2012**, *24*, 45–54. [CrossRef]

microorganisms

MDPI

Review

Algae as Feedstuff for Ruminants: A Focus on Single-Cell Species, Opportunistic Use of Algal By-Products and On-Site Production

Diogo Fleury Azevedo Costa [1,*], Joaquín Miguel Castro-Montoya [2], Karen Harper [3], Leigh Trevaskis [1], Emma L. Jackson [4] and Simon Quigley [1]

1 School of Health, Medical and Applied Science, Institute for Future Farming Systems, CQUniversity, Rockhampton, QLD 4701, Australia
2 Faculty of Agricultural Sciences, University of El Salvador, San Salvador 01101, El Salvador
3 School of Agriculture and Food Sciences, The University of Queensland, Gatton, QLD 4343, Australia
4 Coastal Marine Ecosystems Research Centre, School of Health, Medical and Applied Science, CQUniversity, Gladstone, QLD 4680, Australia
* Correspondence: d.costa@cqu.edu.au; Tel.: +61-409445454

Abstract: There is a wide range of algae species originating from a variety of freshwater and saltwater habitats. These organisms form nutritional organic products via photosynthesis from simple inorganic substances such as carbon dioxide. Ruminants can utilize the non-protein nitrogen (N) and the cell walls in algae, along with other constituents such as minerals and vitamins. Over recent decades, awareness around climate change has generated new interest into the potential of algae to suppress enteric methane emissions when consumed by ruminants and their potential to sequester atmospheric carbon dioxide. Despite the clear potential benefits, large-scale algae-livestock feedstuff value chains have not been established due to the high cost of production, processing and transport logistics, shelf-life and stability of bioactive compounds and inconsistent responses by animals under controlled experiments. It is unlikely that algal species will become viable ingredients in extensive grazing systems unless the cost of production and practical systems for the processing, transport and feeding are developed. The algae for use in ruminant nutrition may not necessarily require the same rigorous control during the production and processing as would for human consumption and they could be grown in remote areas or in marine environments, minimizing competition with cropping, whilst still generating high value biomass and capturing important amounts of atmospheric carbon. This review will focus on single-cell algal species and the opportunistic use of algal by-products and on-site production.

Keywords: biofilters; cyanobacteria; microalgae; single-cell protein; rumen function

Citation: Costa, D.F.A.; Castro-Montoya, J.M.; Harper, K.; Trevaskis, L.; Jackson, E.L.; Quigley, S. Algae as Feedstuff for Ruminants: A Focus on Single-Cell Species, Opportunistic Use of Algal By-Products and On-Site Production. *Microorganisms* **2022**, *10*, 2313. https://doi.org/10.3390/microorganisms10122313

Academic Editor: Assaf Sukenik

Received: 2 November 2022
Accepted: 21 November 2022
Published: 22 November 2022

1. Introduction

Algae and cyanobacteria are found in both marine and freshwater environments and are classified as either eukaryotic organisms or prokaryotic cyanobacteria (blue-green algae). Many species of algae have been in the diet of non-Western civilizations for centuries [1] and have been a relatively common feed source in livestock and aquaculture for many years [2]. However, one of the first experiments with single-cell microalgae species as a ruminant feed was conducted less than twenty years ago [3] with larger feeding trials only happening in the past decade with sheep [4] and cattle [5]. Since then, there has been limited adoption in their use in nutrition of ruminant animals mainly due to the high costs involved with algal production and harvesting. Despite this, recent awareness around climate change has renewed interest in the use of macroalgae, e.g., the red seaweed *Asparagopsis* sp., as mitigators of enteric methane emissions from ruminants [6]. To be market viable, algae products require value adding through either carbon reduction funds or marketing linked to the sustainability of beef produced under low emission schemes. Parallel to this, other algae species have been included in ruminant diets to promote good health-related outcomes to consumers of meat [4] and milk products [7]. Again, price is the main challenge for the effective adoption of algae as an orthodox and off the shelf feedstuff. The alternative could come in the form of algae by-products sourced from biotechnological industries. Industries of importance include fuel production [8] and when algae are used as biofilters to remove nutrients or contaminants [9,10] or to ameliorate carbon emission from various factories [11]. The concept of biofilters is of particular interest in extensive grazing systems of northern Australia since it has the potential to be linked to the development of silvopastoral plans for rehabilitating mine sites with the algae produced in wastewater being used as feed for livestock or organic fertilizer during the implantation of trees or to increase pastures productivity.

In summary, algae can be used in ruminant diets in their raw form if produced in excess, or as by-products from other industries. The differences in nutrient composition cause variable effects within the rumen and animal response. Reviewing the available literature on algae species and products used in ruminant nutrition will provide a greater understanding of the effects on ruminant production and physiology. As algae and algal byproducts become more available at a global scale, this review hopes to recommend new opportunities for efficient in ruminant production systems.

2. Species of Algae and Their Nutritional Composition

Algae chemical composition varies from strain to strain and from batch to batch, particularly in protein and lipid content and in the composition of their fatty acids [12,13]. A particular interest has risen around the use of the single-cell microalgae and cyanobacteria as some have the potential to produce more lipids per area than traditional plants used for biofuel production. Another recent interest is focused on bioactive compounds present in red seaweed, such as bromoform, that directly affects methanogenesis. The latter topic will be discussed in greater depth later in Section 6. Nevertheless, the higher cost of cultivating and harvesting the biomass of algal species remains the most critical barrier to market deployment of large-scale commercial production [8]. These microorganisms occur widely in a variety of natural and man-made environments [14] and despite a large number of species identified, i.e., 15,000, only about 15 were in use in industry at the time [15]. Although not currently prevalent, the increased use of microalgae as biofuel has led to the use of the resulting by-product, i.e., post-lipid extraction algae residue (PEAR), as a ruminant feed [16]. The ruminant's ability to upcycle by-products into high-quality human-edible protein has been highlighted [17] and PEAR retains about 25% of the protein fraction of the original biomass [18]. The latter creates opportunities for inclusion of PEAR in ruminant diets as a protein source, such as with other by-products such as distiller's grain and soybean meal. A recent study suggested that PEAR could be included up to 60% of dietary dry matter in beef cattle rations [16]. However, the issue with any by-product is that the final nutrient composition is subject to the extraction process, which is highly variable.

Table 1 presents the nutritional composition of algae species that have been used in ruminant in vivo trials in their raw form only. Some examples include the marine red seaweed *Asparagopsis taxiformis*, the brown algae *Sargassum* spp. and *Macrocystis pyrifera* and the single-cell microalgae *Schizochytrium* spp., *Crypthecodinium cohnii*, *Dunaliella salina* and the freshwater *Arthrospira platensis.* and *Chlorella* spp.

The mineral composition of algae can be of high relevance depending on a number of factors. The ash content in Table 1 indicates the amount of minerals present in these algal species. However, the mineral profile is expected to greatly change amongst them. A recent paper by Neville et al. [19] demonstrated the benefits of replacing limestone for the calcareous marine algae *Lithothamnion calcareum* as a source of Ca in the diet of dairy cows during the transition period. Mineral-related challenges may be more pronounced in those high producing animals, in which potential differences in bioavailability between mineral sources may play a stronger role. In the latter work, the algae species was added in the diet as a feed additive containing 95% ash composed of 30% Ca.

Table 1. Composition of macro and micro-algae species used for ruminants in their raw form.

	Protein	Lipids	Ash	NDF	Ruminant Species	References
	(g/kg Dry Matter)					
Macroalgae						
Asparagopsis sp.	183	3	504	272	cattle	[6,20]
Chaetomorpha linum	103–182	14–20	120–319	319	sheep	[21,22]
Macrocystis pyrifera	128	22	386	199	goats	[23]
Sargassum sp.	86	6	277	141	cattle	[24]
Ulva lactuca	95–211	5–17	175–181	216–415	goats, sheep	[22,25]
Microalgae						
Arthrospira platensis	460–744	20–150	47–257	35–87	cattle	[5,7,12,26,27]
Chlorella pyrenoidosa	548–600	20–143	64–202	4	cattle	[5,12,28]
Chlorella vulgaris	586	123	51	15	cattle	[7]
Crypthecodinium cohnii	194	575	69	50	sheep, cattle	[29–31]
Dunaliella salina	62–570	60–281	90–787	0	cattle	[5,12,28]
Nannochloropsis gaditana	385	192	158	219	cattle	[7]
Nannochloropsis oculata	289–292	197–292	81–89	69.5	cattle, goats	[32,33]
Prototheca moriformis	38–76	81–109	5–70	114	cattle, sheep	[34,35]
Schizocythrium sp.	130–208	38–577	74–139	263–369	cattle, sheep	[3,4,36,37]

It is important to emphasise that neither the search for new species nor evaluations of the chemical composition of algae currently in use has reached an end. The biomass of naturally occurring single-cell microalgae species found in northern Australia were evaluated and it was concluded that some protein-rich strains could be used for animal feed [14]. Despite considering the latter information relevant, in this review the authors attempted to focus the discussion on the limited data available from in vivo experiments, relying less on data from in vitro trials or experiments using non-ruminant animal species.

3. Effects of Algae on Dry Matter Intake and Apparent Total Tract Digestibility

Ten manuscripts, involving 12 feeding trials reporting dry matter intake were reviewed (Table 2). Five evaluated lactating dairy cattle [7,34,36–38] in which the inclusion of *Arthrospira*, *Chlorella*, a mixture of *Chlorella* and *Nannochloropsis* (1:1) [7] and *Prototheca* [35] at levels between 1.72 and 2.50 g/kg BW, had no effect on dry matter intake, or digestibility of dry matter, organic matter, crude protein and neutral detergent fibre. Similarly, both Till et al. [37] and Till et al. [38] observed no effects on dry matter intake with the inclusion of *Schizochytrium limacinum* at levels of between 50 and 150 g/d. Interestingly, a decrease in the concentrate intake was observed when microalgae were included in the diet, but this reduction was compensated by an increase in silage intake [7]. In the latter, the microalgae was included in the concentrate, which could lead to speculation that there was a palatability issue. In this regard, it was reported a clear fishy odour associated with *Nannochloropsis* whereas the smell of *Arthrospira* and *Chlorella* appeared to be more neutral [7]. As stated by the authors, the smell resembling fish might not be common for all algae products, but it is important to keep in mind that the aroma can change with the oxidation of unsaturated fatty acids during processing and storage [39]. Algal processing and storage conditions are variable and therefore odour may differ for each experiment or region.

Table 2. Summary of in vivo studies testing the effects of algae on dry matter intake (kg/d) and apparent total tract digestibility (g/kg) in ruminants.

Study	Animals [1]	Basal Diet [2]	Algae Species	Doses			Diet Composition [3]			DMI (kg/d) [4]	Digestibility (g/kg) [3,4]			
				g/kg DM	kg/d	g/kg BW	CP	EE	NDF		DM	OM	CP	NDF
[7]	Finnish Ayrshire lactating cows (112 + 21. 6 DIM; 36.2 + 3.77 kg/d MY; 652 + 79.5 kg BW)	Grass silage and concentrate. (Forage = 50%)	Control	0	0	0	154	81.1	410	21.5	651	659	617	474
			Arthrospira platensis	50.9	1.12	1.72	153	87.3	424	22.0	641	650	602	504
			Chlorella vulgaris	64.6	1.35	2.07	154	92.4	409	20.9	650	661	609	491
			C. vulgaris + *Nannochloropsis gaditana* [5]	75.5	1.63	2.50	150	88.0	421	21.6	651	661	606	516
										ns	ns	ns	ns	ns
[36]	Holstein lactating cows (163 + 9.2 DIM; 20.0 + 3.11 kg/d MY; 571 + 48.1 kg BW)	Alfalfa hay and concentrate (Forage = 75%)	Control	0	0	0	199	30.9	370	22.1				
			Schizochytrium DHA-Gold	5.58	0.125	0.22	199	33.6	370	22.4				
			Schizochytrium DHA-Gold	11.7	0.25	0.44	200	37.2	366	21.3				
			Schizochytrium DHA-Gold	18.3	0.375	0.66	201	28.9	363	20.5				
										Linear decrease				
[34]	Holstein lactating cows (57.7 + 49.4 DIM; 25.3 + 5.3 kg/d MY; 590 + 71 kg BW)	TMR corn silage-based (Forage = 50%)	Control	0	0	0	166	37.6	333	22	737	760	737	668
			Prototheca moriformis	52.3	1.18	2.00	163	39.5	345	22.6	736	758	723	666
										ns	ns	ns	ns	ns
[37]	Holstein lactating cows (77 + 17 DIM; 44 + 1.9 kg/d MY; 654 + 42.4 kg BW)	TMR corn silage-based (Forage = 55%)	Control	0	0	0	166		452	23.7				
			Schizochytrium limacinum	2.15	0.05	0.076	170		455	23.3				
			Schizochytrium limacinum	4.33	0.1	0.153	165		452	23.1				
			Schizochytrium limacinum	6.44	0.15	0.229	164		460	23.3				
										ns				
[38]	Holstein lactating cows (22 + 0.5 kg/d MY)	TMR corn silage-based (Forage = 55%)	Control	0	0	0	163		419	22.1				
			Schizochytrium limacinum	4.55	0.1	0.153	161		419	22				
										ns				
[5]	*Bos indicus* steers (187 + 7.5 kg BW)	Speargrass (24 g CP/kg DM, 695 g NDF/kg DM) (Forage > 66%)	Control	0	0	0	24	20	695	2.35a	418ab			
			Arthrospira platensis	188.7	0.748	4	168	38.8	564	3.96c	455ab			
			Chlorella pyrenoidosa	258.2	0.879	4.7	186	52.5	497	3.40b	479b			
			Dunaliella salina	52.2	0.131	0.7	35.6	24.6	650	2.51a	412a			
			Cottonseed meal	279.1	1.12	6	172	26.5	537	4.02c	476b			
[5]	*Bos indicus* steers (236 kg BW)	Speargrass (33 g CP/kg DM, 689 g NDF/kg DM) (Forage > 66%)	Control	0	0	0	33	20	689	2.35a	418ab			
			Arthrospira platensis	188.7	0.133	0.71				Quadratic increase	Linear in-crease			
			Arthrospira platensis	258.2	0.264	1.41								
			Arthrospira platensis	52.2	0.529	2.83								
			Arthrospira platensis	279.1	0.79	4.23								
[27]	Brahman-Shorthorn cross steers (250.1 + 10.86 kg BW)	Mitchell grass (38.1 g CP/kg DM; 746 g NDF/kg DM) (Forage > 98%)	Control		0	0				Quadratic increase		Quadratic increase		
			Arthrospira platensis		0.125	0.5								
			Arthrospira platensis		0.35	1.4								
			Arthrospira platensis		0.625	2.5								
			Arthrospira platensis		1.525	6.1								
[40]	Steers (292 + 22.4 kg BW)	Wet corn gluten feed + Bromegrass hay (Forage = 15%)	Control	0	0	0	177	21	467	7.19				
			Algae meal	150	1.14	0.44	164	27	450	7.57				
			Algae meal	300	2.53	0.99	150	36	433	8.42				
			Algae meal	450	3.98	1.56	136	43	416	8.85				
										Linear increase				
[35]	Whiteface cross wethers (23.0 + 0.54 kg BW)	Grass hay and concentrate (Forage = 8%)	Control	0	0	0	120	35.9	484	1.31	727	736	602	655
			Prototheca moriformis	100	0.114	4.96	122	41.7	442	1.14	721	729	589	613
			Prototheca moriformis	200	0.254	11.0	121	37.3	389	1.27	703	710	580	536
			Prototheca moriformis	300	0.36	15.7	120	40.8	323	1.2	684	691	572	390
										ns	L	L	L	L
[35]	Whiteface cross wethers (33.7 + 0.55 kg BW)	Grass hay and concentrate (Forage = 10%)	Control	0	0	0	110	28.9	252	1.04	751	764	685	375
			Prototheca moriformis	150	0.173	5.12	113	32.9	297	1.15	733	745	670	429
			Prototheca moriformis	300	0.387	11.5	110	39.1	330	1.29	698	707	618	447
			Prototheca moriformis	450	0.536	15.9	112	43.8	351	1.19	680	689	591	449
			Prototheca moriformis	600	0.696	20.7	112	47.6	402	1.16	675	680	593	507
										ns	L	L	L	L

Lowercase letters indicate significant differences between treatments ($p < 0.05$); [1] DIM = days in milk; MY = milk yield; BW = body weight. [2] TMR = total mixed ratio; CP = crude protein; DM = dry matter; NDF = neutral detergent fiber. [3] CP = crude protein; EE = ether extract; NDF = neutral detergent fiber; DM = dry matter; OM = organic matter. [4] The effects of the treatments within a study are portrayed as ns = not significant effect of the algae meal; L = linear effect of algae meal inclusion; superscripts for [5]; "Linear or quadratic increase/decrease" = when a regression equation was reported in the study without specification of treatment means. [5] *Chlorella vulgaris* and *Nannochloropsis gaditana* in a 1:1 ratio.

The only study with dairy cows reporting an effect of microalgae on dry matter intake used *Schizochytrium* fed at 0.22, 0.44 and 0.66 g/kg BW to Holstein cattle and showed a linear decrease in intake [36]. These authors did not discuss the decreased intake, however *Schizochytrium*, the microalgae present in their supplement DHA Gold, is a source rich in fat, which at high levels reduces intake. However, the ether extract of all diets in the study did not vary significantly and remained in the normal ranges for dairy cows.

It is also important to note that the purpose of the addition of microalgae differed between studies: Microalgae species had been included as source of polyunsaturated fatty acids to modulate rumen fermentation [36–38]; added as a source of protein to replace soybean meal [7]; and *Prototheca* (crude protein = 77.9 g/kg dry matter) was added as energy replacing corn [34]. The sample size of microalgae studies on intake and digestibility in dairy cattle is small, and the ration ingredients were diverse between studies. It is therefore difficult to ascertain the effect of microalgae supplementation on intake and digestibility in dairy cows and more studies involving dose responses of different algae and ingredient interactions are required. Nevertheless, providing dietary composition of rations for dairy cows are not novel, microalgae can be included in the diet of dairy cows at inclusion levels of 2.15 to 75.5 g/kg dry matter, without deleterious effects on intake.

Another five studies involving seven trials have been conducted with either growing cattle or sheep. Costa et al. [5] fed *Arthrospira platensis*, *Chlorella pyrenoidosa* and *Dunaliella salina* (4, 4.7 and 4 g/kg BW, equivalent to 35, 48 and 56 g/kg dry matter) as well as cottonseed meal as a positive control. In this study *Arthrospira* and *Chlorella* supplementation increased the dry matter intake of the basal diet which consisted of a low-quality tropical grass Speargrass (*Heteropogon contortus*) hay. Only *Arthrospira* supplementation increased dry matter intake to the extent of cottonseed meal supplementation. *Dunaliella* supplementation did not increase intake of the basal diet. In this study there were no differences in dry matter digestibility between *Arthrospira*, *Chlorella* and cottonseed meal supplement treatments, while *Dunaliella* showed a similar dry matter digestibility to the control diet. A follow up response trial using either *Arthrospira* or cottonseed meal as supplements [5], found an equivalent quadratic increase in intake with each supplement in steers fed a basal diet of poor-quality hay. An identical linear increase in dry matter digestibility was observed when either *Arthrospira* or the conventional cottonseed meal was supplemented.

The second trial in Costa et al. [5] confirmed previous research of Panjaitan et al. [27] where *Arthrospira* was deposited directly in the rumen of fistulated steers fed a low-quality hay. In this experiment there were quadratic increases in dry matter intake and dry matter digestibility which was associated with an increase in the supply of rumen degradable nitrogen that enhanced microbial activity. It was suggested that the main mode of action of *Arthrospira* was the increase in the protein to energy ratio that led to a higher passage rate in the rumen [27]. This higher passage rate promotes a higher dry matter intake, increases microbial protein flow to the intestine, and likely reduces the maintenance energy requirements of the rumen microbes [41]. When supplemented to steers fed a low-quality basal forage, *Arthrospira* increased dry matter intake and digestibility in a similar fashion to that of other conventional protein meals [5].

Another study with growing steers, fed a commercial microalga between 0.44 and 1.56 g/kg BW in a feedlot-type diet (forage to concentrate ratio of 15 to 85) [40]. Dry matter intake linearly increased with increasing microalgae inclusion, but no changes were observed for average daily gain, leading to a decreased feed conversion rate when algae meal was fed. The authors hypothesized that supplement improved palatability, but an increased passage rate may have led to a reduced nutrient utilization and, therefore, a lower feed conversion efficiency. This is supported by an increase in dry matter disappearance rate of microalgae when incubated in situ in the rumen of steers fed increasing algae levels (from 0 to 45%). Another factor that could have contributed to the decreased average daily gain was that the crude protein concentration significantly differed within the study of Van Emon et al. [40] (ranging from 177 to 136 g/kg DM), which may have impaired rumen fermentation and nutrients digestibility, leading to a reduced supply of nutrients for absorption in the duodenum, hence, a lower supply of building blocks for growth. Digestibility, however, was not reported in the latter study.

A study with feedlot diets for sheep reported no effects of *Prototheca moriformis* dietary inclusion on dry matter intake, however a linear reduction of dry matter, organic matter and crude protein digestibility were apparent [35]. Interestingly, both experiments in the latter study found contrasting effects regarding neutral detergent fibre digestibility. In the first experiment, where microalgae replaced soybean hulls, neutral detergent fibre digestibility linearly decreased with increasing algae inclusion, but in the second experiment, where microalgae replaced corn, neutral detergent fibre digestibility increased. As proposed by the authors, these differences relate to the nature of the fibre present in algae, which could be more of soluble nature, but apparently with a digestibility ranking somewhere between that of soybean hulls and corn. Further studies are required to better understand the nature and degradability characteristics of fibre from microalgae.

In general, due to the limited number of studies evaluating the effects of microalgae on dry matter intake and digestibility, as well as the differences in the experimental conditions, e.g., diets, animals, and the microalgae being tested, it is difficult to define the extent of the effects of microalgae supplementation. However, some microalgae can be safely included in diets of dairy cattle, growing steers or sheep as a source of protein or energy without any deleterious effects on intake and digestibility, as long as the basal diet composition remains similar, and for levels of inclusion to be between 2.15 and 75.5 g/kg dry matter. Interestingly, there is evidence that microalgae interact with diet ingredients and affects neutral detergent fibre digestibility. For example, when substituting microalgae for corn, neutral detergent fibre digestibility improves when the concentrate proportion is above 90%. It is known that high starch diets tend to decrease fibre degradation and therefore adding microalgae could potentially increase fibre utilization in feedlot diets

4. Effects of Algae on Rumen Parameters

4.1. Volatile Fatty Acids

The volatile fatty acids acetic, butyric and propionic are the main short chain fatty acids produced in ruminants through enteric fermentation. These acids are of paramount importance for providing most of the ruminant's energy supply. However, there is not much information regarding the effects of algae supplementation on their production nor on other short chain fatty acids and the links to microbial protein synthesis in vivo. A negative relationship of *Arthrospira* supplementation level with butyrate proportion, and a positive relationship with proportions of other branched-chain fatty acids has been reported in the work of Panjaitan et al. [27]. Additionally, the latter authors found a quadratic relationship between *Arthrospira* intake and propionate proportions. Subsequently, Costa et al. [5] observed a positive relationship of total volatile fatty acids using the same microalgae. Panjaitan et al. [27] found a positive linear relation of *Arthrospira* with molar proportions of propionate and branched-chain fatty acid, and a negative relationship between *Arthrospira* inclusion and acetate proportion. Interestingly, a quadratic relationship between *Arthrospira* and branched-chain fatty acid reduced in proportion with increasing *Arthrospira* inclusion in the work of Costa et al. [5]. In addition, in the latter work, no differences in total short chain fatty acid concentration, or individual short chain fatty acid proportions between *Arthrospira* and cottonseed meal when supplemented as sources of protein. It was also found by these authors that *Chlorella* and *Dunaliella* both increased the acetate proportions and decreased the branched-chain fatty acid proportions compared to cottonseed meal inclusion. Conversely, Moate et al. [36] found no effects of microalgae inclusion in the diet of dairy cows on total volatile fatty acids concentration or individual short chain fatty acid proportions, except for a linear increase in butyrate with increasing algae inclusion. Importantly, Costa et al. [5] and Panjaitan et al. [27] supplemented microalgae to a basal diet of poor-quality grass, thus changing significantly the composition of the diet consumed, whereas the rations tested in the study of Moate et al. [36] remained similar in composition.

Microbial protein synthesis and rumen ammonia-N both increase in a quadratic fashion with increasing *Arthrospira* inclusion in the diet in Panjaitan et al. [27]. However, no differences were found between *Arthrospira*, *Chlorella* and cottonseed meal in ammonia-N concentration in the rumen nor in microbial protein synthesis between these true protein sources in Costa et al. [5]. Despite this, in the feeding, it was observed a quadratic response of ammonia-N and branched-chain fatty acid proportion to increasing inclusion of *Arthrospira* in the diet in the latter study. Microalgae inclusion in the diet had no effect on the pH of rumen fluid in cattle fed forage based diets [5,27,36].

It is a fact that more research is required to address the effect of microalgae on rumen function. The current information available in the literature does not highlight any obvious negative effect on rumen function in cattle or sheep however, there is not sufficient information to conclude on the possible effects of these algae on the rumen fermentation and the supply of microbial protein post rumen in various feeding conditions and in different ruminant species production systems.

4.2. Microbial Synthesis in the Rumen

Suitable conditions, such as absence of oxygen, relatively constant pH, appropriate nutrients, and the absence of growth-preventing inhibitors, facilitate microbial growth in the rumen environment [42]. The main nutrients required for the growth of microbes are fermentable carbohydrates, as source of energy, and N. The energy required for their growth comes from structural or non-structural carbohydrates, depending on diet type. Protein can be also used as energy, but it is usually the most expensive ingredient of the diet [43]. Other nutrients are also required by rumen microbes, such as minerals, e.g., sulfur, phosphorus and magnesium [44] and vitamins [45]. Insufficient amounts of nutrients result in a lower efficiency of microbial protein synthesis in the rumen. Microalgae, besides being a source of both N and energy, are a potential source of these other nutrients for microbes. In addition, Costa et al. [5] indicated that both *Arthrospira* and *Chlorella* were high in phosphorus, often a limiting nutrient in grazing systems across the globe. The most important nutrient supplied by these single-cell microorganisms is likely to be N released on the degradation of algae protein. The extent of degradation of the microalgae within the rumen is not fully known but the lysis and fermentation of microalgae within the rumen may be presumed to follow the normal fermentative process outlined below. A higher efficiency of microbial protein synthesis in steers fed a low protein basal diet supplemented with *Arthrospira* compared to the equivalent rumen degradable N intake supplied by a non-protein N source, i.e., urea, was attributed to the package of nutrients supplied by the microalgae [27]. Despite a higher branched-chain amino acid content in *Arthrospira*, which theoretically could benefit the growth of some bacterial species in the rumen, this microalga achieved the same efficiency of microbial protein synthesis values observed for cottonseed meal, a by-product traditionally used as protein supplement for ruminant animals [5].

Microbial protein can provide all the amino acids (AA) required by the host animal, i.e., ruminant animal, but some microbes are able to use pre formed AAs. A spray-dried *Arthrospira* from two different sources had 333 and 361 g/kg of their protein composed of essential AAs [46]. A more recent study reported total AA concentration ranging from 855 to 930 g/kg CP for *Arthrospira*, *Chlorella* and *Nanochloropsis*, including 416 to 464 g/kg CP of essential AA [7]. Therefore, microalgae have the potential also to provide dietary AA for absorption by the animal but the nutritional value will remain contingent on the proportion of this protein that remains undegraded in the rumen (RUP). In this regard, a recent study reported the 48 h in vitro RUP of four species of microalgae: *Arthrospira* (n = 2), *Chlorella* (n = 7), *Nannochloropsis* (n = 7), and *Phaeodactylum* (n = 2) [47]. The RUP ranged between 40 and 61% of RUP of the total protein in non-cell disrupted microalgae, with the highest rumen undegradable protein being found for *Nannochloropsis* and the lowest for *Arthrospira* [47]. These results are in agreement with Costa et al. [5], who found that algal protein has a higher resistance to degradation in the rumen compared to soybean. Importantly, the in vitro intestinal digestibility of the rumen undegradable protein ranged between 270 and 430 g/kg RUP [47], significantly lower than the intestinal digestibility for soybean meal (between 700 and 880 g/kg RUP) and rapeseed meal (between 500 and 820 g/kg RUP) [48]. If the results of these in vitro studies are replicated in vivo, the supply of dietary AA from microalgae would be of a lesser quantity compared to other protein supplements. However, in vitro study results do not necessarily reflect the in vivo rumen degradability of protein and therefore, the subsequent intestinal utilization of the resulting RUP becomes a critical area that deserves further attention.

Moreover, the combination of protein and the resulting branched-chain amino acids and branched-chain fatty acids, minerals, and vitamins within microalgae contribute as a source of nutrients to microbes and potentially directly to the host. In the work of Panjaitan et al. [27] *Arthrospira platensis* was fed to animals offered a basal diet of low-quality hay, i.e., 3.8% protein, ad libitum. This resulted in a higher efficiency of microbial protein synthesis than animals fed urea, a non-protein N source, with the same basal diet. This indicates that specific nutrients, not just N, influenced rumen microbe activity allowing them to reach microbial protein synthesis up to 550 g/d and efficiency of microbial protein synthesis of 179 g microbial protein/kg digestible organic matter, similar to those parameters reported for cattle grazing temperate grasses [49]. The exclusive use of a non-protein N source results in microbial protein synthesis of approximately 130 g/kg of digestible organic matter [50]. The nutritional attributes of this microalgae with respect to the efficiency of microbial protein synthesis and aspects of its protein degradation in the rumen, need to be further evaluated to better understand the mechanisms by which these microalgae stimulate microbial activity and increase microbial efficiency.

5. Fatty Acid Composition

Lipids sources in the form of fats and oils often depress fibre digestion when present at high concentrations in the diet. This must be accounted for when considering the use of algae since they can be rich sources of lipids, with the most diverse fatty acids profiles. Numerous health benefits can be attributed to specific long chain polyunsaturated fatty acids when present in human diets [51–53]. Algae are good sources of polyunsaturated fatty acids such as linoleic and linolenic, both essential for life of all mammals, but they vary in content and composition. When fed as a supplement to animals, algae may alter the fatty acids composition of meat [4,54]. Costa et al. [12] examined the FA profile in the rumen fluid of cattle fed three microalgae species and found that *Chlorella pyrenoidosa* increased polyunsaturated fatty acids concentration in the rumen fluid of fistulated steers, which if transferred to meat, could have health related benefits to consumers. When feeding unsaturated fatty acids to ruminants, the fatty acids profile encountered in the meat will be different to the one present in the diet because of the biohydrogenation process in the rumen [55]. Although some polyunsaturated fatty acids, such as C20:5 n-3, EPA and C22:6 n-3, DHA can escape rumen biohydrogenation [54]. Microalgae that are rich in longer chain polyunsaturated fatty acids with 20 and 22 Cs, include *Schizochytrium* [56,57], and *Crypthecodinium cohnii* [29,58,59]. For humans these long chain polyunsaturated fatty acids are physiologically important, helping retinal and cortical development during early life [52,53]. Pickard et al. [29] fed a marine alga rich in DHA (C 22:6n-3) to pregnant ewes in the final weeks of gestation, i.e., 10 to 6 weeks prior to birth, and reported that lambs from these ewes stood significantly sooner after birth demonstrating an improved vigour compared to the control treatment.

Long chain polyunsaturated fatty acids are derived from linoleic (C18:2n-6) and alpha-linolenic (C18:3n-3) acids by enzymatic desaturation and chain elongation and cannot be synthesized in the body [53]. The polyunsaturated fatty acids, linoleic (C18:2n-6) and alpha-linolenic (C 18:3n-3) are present in the lipids of microalgae, e.g., *Arthrospira*. However, if these sources reach the rumen they are exposed to transformation by microbial enzymes leading to formation of other free fatty acids, such as C18:3n-3 converted to stearic acid, C18:0, or C18:2n-6, often incompletely biohydrogenated, converting into C18:0 and monounsaturated isomers [54].

The longer chain length of C20:5n-3 and C22:6n-3 present in fish oil and some other algae, such as the marine microalgae *Dunaliella* and *Schizochytrium* potentially is the reason for their low rumen biohydrogenation [60]. However, the main fatty acids of *Arthrospira* are the already saturated palmitic acid 16:0 and the polyunsaturated fatty acids C18:1, C18:2 and C18:3 that would be completely hydrogenated to stearic and monoenoic acids in the rumen [61], most likely why Costa et al. [12] did not observe polyunsaturated fatty acids increments with inclusion of this microalgae.

Dairy and beef are the major sources of conjugated linoleic acid for humans and there are various isomers resulting from rumen biohydrogenation, e.g., *cis*9, *trans*11, or *trans*10, *cis*12) [55]. Sehat et al. [62] found rumenic acid, i.e., *cis*9, *trans*11, to be the predominant isomer (78 to 84%) present in cheese products; however other isomers of conjugated linoleic acid were also identified in small percentages. A portion of these conjugated linoleic acids escape from the rumen and affect lipid metabolism in the mammary gland [55], subcutaneous and intramuscular fat [63]. More importantly, the *trans*10 *cis*12 isomer of conjugated linoleic acid can markedly inhibit fat synthesis in all three tissue types [55,63]. The extent of formation of these inhibitory isomers within the rumen of ruminants supplemented with algae is not well known. More importantly, none of microalgae led to the formation of conjugated linoleic acid isomers known to inhibit fat synthesis [12].

Another issue is that dietary fatty acids of the n-3 family could potentially delay parturition in sheep [64]. Although, Staples et al. [65] suggested an increase in fertility due to effects of linoleic acid and other longer chain fatty acids on the pituitary, ovaries and uterus, rather than from a higher energy status. Nonetheless, algal supplementation improved lamb vigour at birth as a result of the long chain polyunsaturated fatty acids in the algal species *Crypthecodinium* [29]. It is important to highlight that the fatty acids composition of microalgae varies with species and the types of fatty acids have variable effects in which the long chain polyunsaturated fatty acids and n-3 forms are generally associated with positive effects on ruminant production, e.g., newborn vigour, and fatty acids profile of the fat deposited. In addition, no major effects are expected on rumen function from the basal diet in grazing systems [66] nor from the addition of small quantities of lipids to the diet [67,68]. Although, for more substantial inclusions, i.e., above 6% dry matter, fibre degradation can be negatively affected. It is anticipated that further in vivo studies are required to draw more realistic conclusions about algae as a feedstuff for ruminants on this regard.

6. Algae as Enteric Methane Mitigators

Another important parameter associated with the feeding of algae to ruminant animals is the influence on enteric methane emissions. Methane synthesis in the rumen is directly related to the presence of methanogens but also linked to the efficiency of energy utilization which is interrelated with the methanogenic pathway. Methane emissions from livestock account for a considerable proportion of greenhouse gas emissions in the agricultural sector and it is therefore a relevant parameter to study when investigating the potential use of a feedstuff for ruminants. Recently, McCauley et al. [69] reviewed the use of algal-derived feed additives and their influence on enteric methanogenesis in ruminant animals. One of the genera highlighted by the latter authors includes the macroalgae *Asparagopsis*. These red seaweeds contain bioactive compounds, with special emphasis on bromoform that directly affects methanogenesis. Bromoform is a halogenated analogue of methane that inhibits enzymatic activity of methyltransferase by reacting with the reduced vitamin B_{12} cofactor [70]. The latter enzyme is required in methane formation, directly affecting enteric methane emissions. In contrast, the single-cell microalgae species with potential as methane mitigators, such as *Schizochytrium* are high in polyunsaturated fatty acids which work as a H sink, competing with methane formation pathways [3]. Despite being very promising in suppressing methane emissions, production of these algae currently inhibits price competitiveness with other supplements, unless there is value added to beef or milk produced under low emission schemes. Furthermore, as highlighted in a recent meta-analysis evaluating the use of *Asparagopsis*, there was marked heterogeneity in the results of methane reductions [71]. The latter authors found differences in responses which were evident for the red seaweed at the species level. These authors concluded that while there were practical applications to reduce methane emissions, more in vivo experiments are required to strengthen the evidence and to evaluate potential risks for the use of the different seaweed. Both *Asparagopsis taxiformis* and *Asparagopsis armata* have shown to be effective in reducing methane but they contrast in efficacy most likely because of the concentration of bromoform in those species. For example, the concentration of bromoform in *A. armata* was 1.32 mg/g in the work of Roque et al. [6] compared to 6.55 mg/g in *A. taxiformis* in Kinley et al. [20]. Roque et al. [6] observed a reduction of

67.2% at an inclusion rate of 18.3 g/kg dry matter in lactating dairy cows whilst Kinley et al. [20] reported reductions of enteric CH_4 production of up to 98% at a much lower inclusion, i.e., 3.26 g/kg dry matter, in beef cattle fed a high grain diet. Recently, Glasson et al. [72] discussed some of the benefits and risks involved in the feeding of *Asparagopsis* for the reduction of methane production from ruminants, including the effects it might have on atmospheric chemistry. Compounds recognized as ozone-depleting substances are listed in annexes of the Montreal Protocol and whilst compounds such as methyl bromide and bromochloromethane are listed there, bromoform is not listed as an ozone depleting compound as such. Bromoform itself is classified as a very short-lived substance and therefore has a relatively low ozone depletion potential overall. Bromoform is released slowly during the natural life cycle of algae and during their senescence and decay. It is during this natural transport of volatilized bromoform through the ozone that it will react with radicals and hence undergoes photolysis, which results in the production of water-soluble reactive product gases and inorganic bromine. It is these latter products that contribute to the decomposition of the ozone layer and that are the ozone depleting substances, rather than the compound bromoform itself, and hence why not bromoform, but these other reactive compounds. In the recent work of Glasson et al. [72], the authors concluded that large scale production and use of this red algae as ruminant feed would not negatively impact animal health, food quality, nor cause ozone depletion.

From the few in vivo studies with cattle focusing on microalgae species, only one peer reviewed article and a report presented results on methane emissions [36,73]. Moate et al. [36] reported no effect on methane in g/d, but a linear increase in g/kg dry matter intake with increasing levels of *Schizochytrium* in DHA-Gold; whilst Klieve et al. [73] indicated reductions on methane emissions on a liveweight and dry matter intake basis by around 22% and 19.4%, respectively. Although, these authors did not clarify which algae species composed the commercial product Algamac. In summary, the interest of studying methane production when feeding microalgae has developed from in vitro studies [74–76] which reported reductions in methane concentration in the gas produced from those fermentations. However, in vitro studies utilize only a known amount of feedstuff and therefore do not take into account feed intake, passage rate and average daily gain. To our knowledge, in contrast to recent work done with the marine macroalgae *Asparagopsis*, evidence supporting meaningful enteric methane reductions in vivo using microalgae species is lacking.

7. On-Farm or On-Site Production of Algae Species

Algae production through mariculture remains a challenge for some species and may limit their distribution to farms located near coastal areas. For example, complex life histories associated with the red algae *Asparagopsis* were restricting the large-scale commercial farming, although significant research funding for this genus has led to recent research breakthroughs with life stage transitioning triggers. Whereas most macroalgae are readily harvested using straightforward and less expensive mechanical methods, it is far more energy intensive to harvest single-cell species that are less than 5 μm in diameter at a concentration of often no greater than 0.5 g/L [77]. The single-cell species are normally so dispersed in nature that they can only be seen when optimal growing conditions generate the blooming of a very dense population of cells. Though formation of blooms is unpredictable under natural conditions [78], controlled cultivations systems (e.g., photobioreactors; open raceway ponds) optimize light, nutrient, temperature conditions to achieve a continual state of bloom [79]. This approach maximizes biomass yield by harvesting microalgae during a continuous exponential growth phase. More control over production variables such as temperature, or the addition of nutrients into the aquaculture system allows greater uniformity of composition of the algae being produced. Some of the most important commercially produced microalgae are the freshwater *Arthrospira* and *Chlorella* [80]. They can all grow on open systems with relatively no contamination by other microorganisms [81]. This is a very important consideration, since harmful algal blooms could potentially compromise the health of animals [82] and affect the microbial community in the water supply [83]. The production of microalgae on-farm could be an important source of protein for cattle especially during the drier months of the year when energy and protein are limiting. *Arthrospira* has been harvested from its natural habitats, such as Lake Texcoco in Mexico and Lake Chad in Central Africa for centuries [46], hence why it was selected for early studies on the potential use of microalgae as a protein source for cattle [5,26,27]. In the work of Costa et al. [5] the freshwater microalga *Arthrospira* was successfully fed to cattle with positive outcomes in liveweight gain, but the authors highlighted the need to devise harvesting methods suitable for those extensive and remote locations that do not require constant maintenance nor high labour input. Beef cattle production systems in northern Australia are predominantly represented by animals raised in extensive grazing systems [84]. Geographical remoteness, large property sizes, and other characteristics of the production systems impose several management challenges on producers [85]. Access to protein-rich supplements and distributing these to livestock grazing tropical grasses with a low protein content presents a challenge within these systems. With this context in mind, microalgae such as *Arthrospira*, *Chlorella*, and *Nannochloropsis* species appear to be the most suitable species for cultivation as protein supplements in situ on remote cattle properties.

A more opportunistic use of microalgae produced in the semi-arid regions near extensive grazing systems for production of red meat could be linked to recovery and improvement in water quality of pit and recycled mine waters. The main objective of the use of waste water from mines is to avoid having to rely on the scarcely available fresh water within a semi-arid context. The risk, however, is that microalgae adsorb contaminants that could end up in the food chain if fed to ruminants. Therefore, careful attention must be given before use as animal feed since microalgae have the potential to accumulate toxic metals [86]. As previously indicated, these organisms can work as biofilters removing both nutrients and contaminants [9]. Despite this, these organisms could not only be used as feedstuffs but also as organic fertilizers [87] with considerations to avoid accumulation of contaminants into the areas where they are applied. For these options to become reality, a few key elements must be in place which include the production and harvest methods, delivery mechanisms and adequate technical support for producers adopting the technology. In Figure 1, two hypothetical options illustrate the ideas for the use of microalgae as by-products from mine waste water recovery and from on-farm production, where in both the delivered to animals would occur via water systems. It is important to note that the microalgae being produced could either occur on raceway ponds or covered systems, the latter aiming to prevent evaporation.

The main advantage of on-farm production and feeding systems devised on-site is that less-intensive post-harvest technologies would be required and the issues of shelf-life and stability of bioactive compounds would be minimized, considerably decreasing requirement for freeze-drying or new alternative methods of processing such as oil immersions to deliver bioactive compounds [88]. The final algae mix is then offered to animals in existing infrastructure such as water troughs.

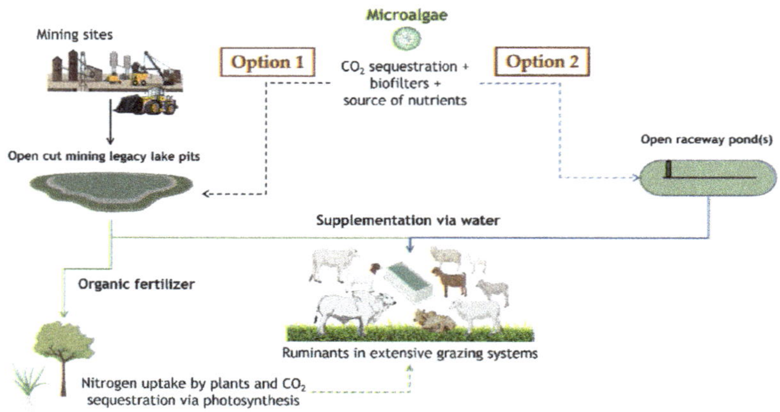

Figure 1. Hypothetical opportunistic on-farm production and water delivery feeding systems.

8. Challenges with Algae

Macro- and microalgae produce different biologically active compounds. A number of those compounds can result in anti-nutritive and occasionally even more profound detrimental effects. There are specific algae toxins, such as domoic acid [89,90] that could result in serious health problems. Other components, such as high iodine concentration was found in some macroalgae [91] that could be toxic if consumed in excess by livestock. Iodine should not be in amounts greater than 8 mg I/kg DM of total diet [92]. These compounds may be beneficial in some circumstances, but if present in excess they may be detrimental to the rumen microorganisms and/or the animal. Recent literature around the use of macroalgae *Asparagopsis* show great promise because of their effect in enteric methane emissions but they highlight the need to evaluate and understand the consequences of bromoform or other halogens transferring into meat or milk. Muizelaar et al. [93] did not detect the accumulation of bromoform in animal tissue but the compound was excreted in urine and milk of dairy cows. The latter is a concern and needs attention soon prior to large scale commercialisation already taking place.

Freshwater toxic algae have been reported in over 45 countries [94] including Australia, e.g., *Cylindrospermopsis raciborskii* [95]. Through skin contact and ingestion, these toxins can result in many forms of human illness [96]. Therefore, one of the important issues when devising a way of feeding algae on-farm is to closely monitor any potential health issues when feeding animals. Generally, the fresh algae *Arthrospira* has low contamination by other algae and protozoa in open-air cultures [97].

The high nucleic acid content of rapidly growing microorganisms such as microalgae represents a limitation for its use in humans and other monogastrics, due to possible hyperuricemia (presence of high levels of uric acid in the blood) [98]. Ruminants overcome this, as the microbes in their rumen can utilize these nucleic acids. Nucleic acids are totally degraded in the rumen by extracellular bacterial nucleases and then captured and metabolized by ruminal microbes [99].

9. Final Considerations

The demand for food and feed is increasing and will so for the foreseeable future. The need to sustain animal production by making a more efficient use of available resources is a major challenge that animal scientists face to prevent the further degradation of land resources. The production of microalgae in close proximity to grazing ruminants and its utilization as feed represents an important opportunity to produce high quality biomass without utilizing arable land. These microalgae exist in a range of environments, and selection and production of naturally occurring, regionally adapted species may be a more robust approach than the production of introduced species. There is a diversity of microalgae species that have been utilized in ruminant diets, and depending on their nature, they could be added to the diets as sources of protein, energy (in the form of fat or in the form of carbohydrates) or other nutrients (e.g., minerals, essential fatty acids) that can enhance the nutritional and health status of the animals and consumers consuming their meat or milk products.

Despite the small number of studies available, the evidence suggests potential for microalgae to be successfully included in the diets of ruminants. When incorporated in the diet without significant changes in the ration's nutrients composition, microalgae can be added without substantial changes in intake or digestibility. When supplied as a source of dietary protein, microalgae had similar effects on rumen function and animal liveweight gain to conventional protein sources and better responses than non-protein nitrogen supplements. Microalgae supplementation may reduce the digestibility of nutrients in the diet, particularly at high levels of inclusion, an aspect worthy of further attention. Moreover, some evidence suggests a low intestinal digestibility of rumen undegradable protein with microalgae meal, which is a potentially detrimental feature of these supplements if the protein is lost in faeces. Importantly, not all species appear to cause identical responses on the animal, and their effects are largely related to their nutrient composition.

Author Contributions: D.F.A.C.: visualization, writing—original draft preparation, review and editing; J.M.C.-M.: visualization, writing—original draft preparation, review and editing; K.H.: visualization, writing, review and editing; L.T.: review and editing; E.L.J.: review and editing; S.Q.: visualization, review and editing. All authors have read and agreed to the published version of the manuscript.

Funding: This research received no external funding.

Institutional Review Board Statement: Not applicable.

Informed Consent Statement: Not applicable.

Data Availability Statement: This is a review of data already published and cited accordingly.

Acknowledgments: The researchers acknowledge Dennis Poppi and Stuart McLennan for valuable insight in the topic over the years.

Conflicts of Interest: The authors declare no conflict of interest.

References

1. Gantar, M.; Svircev, Z. Microalgae and Cyanobacteria: Food for thought. *J. Phycol.* **2008**, *44*, 260–268. [CrossRef] [PubMed]
2. Rajauria, G. Chapter 15—Seaweeds: A sustainable feed source for livestock and aquaculture. In *Seaweed Sustainability*; Tiwari, B.K., Troy, D.J., Eds.; Academic Press: San Diego, CA, USA, 2015; pp. 389–420.
3. Boeckaert, C.; Mestdagh, J.; Clayton, D.; Fievez, V. Micro-algae as potente rumen methane inhibitors and modifiers of rumen lipolysis and biohydrogenation of linoleic acid. *Commun. Agric. Appl. Biol. Sci.* **2004**, *69*, 127–130.
4. Meale, S.J.; Chaves, A.V.; He, M.L.; McAllister, T.A. Dose-response of supplementing marine algae (*Schizochytrium* sp.) on production performance, fatty acid profiles and wool parameters of growing lambs. *J. Anim. Sci.* **2014**, *92*, 2202–2213. [CrossRef] [PubMed]
5. Costa, D.F.A.; Quigley, S.P.; Isherwood, P.; McLennan, S.R.; Poppi, D.P. Supplementation of cattle fed tropical grasses with microalgae increases microbial protein production and average daily gain. *J. Anim. Sci.* **2016**, *94*, 2047–2058. [CrossRef] [PubMed]
6. Roque, B.M.; Salwen, J.K.; Kinley, R.; Kebreab, E. Inclusion of *Asparagopsis armata* in lactating dairy cows' diet reduces enteric methane emission by over 50 percent. *J. Clean. Prod.* **2019**, *234*, 132–138. [CrossRef]
7. Lamminen, M.; Halmemies-Beauchet-Filleau, A.; Kokkonen, T.; Jaakkola, S.; Vanhatalo, A. Different microalgae species as a substitutive protein feed for soya bean meal in grass silage based dairy cow diets. *Anim. Feed Sci. Technol.* **2019**, *247*, 112–126. [CrossRef]
8. Laurens, L.; Glasser, M.; McMillan, J. A perspective on renewable bioenergy from photosynthetic algae as feedstock for biofuels and bioproducts. *Algal Res.* **2017**, *24*, 261–264. [CrossRef]
9. Bani, A.; Parati, K.; Pozzi, A.; Previtali, C.; Bongioni, G.; Pizzera, A.; Ficara, E.; Bellucci, M. Comparison of the Performance and Microbial Community Structure of Two Outdoor Pilot-Scale Photobioreactors Treating Digestate. *Microorganisms* **2020**, *8*, 1754. [CrossRef]

10. He, P.; Xu, S.; Zhang, H.; Wen, S.; Dai, Y.; Lin, S.; Yarish, C. Bioremediation efficiency in the removal of dissolved inorganic nutrients by the red seaweed, *Porphyra yezoensis*, cultivated in the open sea. *Water Res.* **2008**, *42*, 1281–1289. [CrossRef]
11. Suresh, A.; Benor, S. Chapter 17—Microalgae-based biomass production for control of air pollutants. In *From Biofiltration to Promising Options in Gaseous Fluxes Biotreatment*; Soreanu, G., Dumont, É., Eds.; Elsevier: Amsterdam, The Netherlands, 2020; pp. 345–372.
12. Costa, D.F.A.; Quigley, S.P.; Isherwood, P.; McLennan, S.R.; Sun, X.Q.; Gibbs, S.J.; Poppi, D.P. *Chlorella pyrenoidosa* supplementation increased the concentration of unsaturated fatty acids in the rumen fluid of cattle fed a low-quality tropical forage. *Rev. Bras. Zootec.* **2020**, *49*, e20200042. [CrossRef]
13. Kholif, A.E.; Olafadehan, O.A. Microalgae in Ruminant Nutrition: A Review of the Chemical Composition and Nutritive Value. *Ann. Anim. Sci.* **2021**, *21*, 789–806. [CrossRef]
14. Duong, V.T.; Ahmed, F.; Thomas-Hall, S.R.; Quigley, S.; Nowak, E.; Schenk, P.M. High Protein- and High Lipid-Producing Microalgae from Northern Australia as Potential Feedstock for Animal Feed and Biodiesel. *Front. Bioeng. Biotechnol.* **2015**, *3*, 53. [CrossRef] [PubMed]
15. Raja, R.; Hemaiswarya, S.; Kumar, N.; Sridhar, S.; Rengasamy, R. A Perspective on the Biotechnological Potential of Microalgae. *Crit. Rev. Microbiol.* **2008**, *34*, 77–88. [CrossRef]
16. Drewery, M.L.; Sawyer, J.E.; Wickersham, T.A. Post-extraction algal residue as a protein supplement for beef steers consuming forage: Palatability and nutrient utilization. *Anim. Feed Sci. Technol.* **2021**, *273*, 114796. [CrossRef]
17. Baber, J.R.; Sawyer, J.E.; Wickersham, T.A. Estimation of human-edible protein conversion efficiency, net protein contribution, and enteric methane production from beef production in the United States. *Transl. Anim. Sci.* **2018**, *2*, 439–450. [CrossRef] [PubMed]
18. Bryant, H.L.; Gogichaishvili, I.; Anderson, D.; Richardson, J.W.; Sawyer, J.; Wickersham, T.; Drewery, M.L. The value of post-extracted algae residue. *Algal Res.* **2012**, *1*, 185–193. [CrossRef]
19. Neville, E.W.; Fahey, A.G.; Meade, K.G.; Mulligan, F.J. Effects of calcareous marine algae on milk production, feed intake, energy balance, mineral status, and inflammatory markers in transition dairy cows. *J. Dairy Sci.* **2022**, *105*, 6616–6627. [CrossRef]
20. Kinley, R.D.; Martinez-Fernandez, G.; Matthews, M.K.; de Nys, R.; Magnusson, M.; Tomkins, N.W. Mitigating the carbon footprint and improving productivity of ruminant livestock agriculture using a red seaweed. *J. Clean. Prod.* **2020**, *259*, 120836. [CrossRef]
21. Ktita, S.R.; Chermiti, A.; Mahouachi, M. The use of seaweeds (*Ruppia maritima* and *Chaetomorpha linum*) for lamb fattening during drought periods. *Small Rumin. Res.* **2010**, *91*, 116–119. [CrossRef]
22. Ktita, S.R.; Chermiti, A.; Bodas, R.; France, J.; López, S. Aquatic plants and macroalgae as potential feed ingredients in ruminant diets. *J. Appl. Phycol.* **2017**, *29*, 449–458. [CrossRef]
23. Castro, N.M.; Valdez, M.C.; Alvares, A.; Ramírez, R.N.Á.; Rodríguez, I.S.; Contreras, H.H.; García, L.S. The kelp *Macrocystis pyrifera* as nutritional supplement for goats. *Rev. Cient.* **2009**, *19*, 63–70.
24. Choi, Y.Y.; Lee, S.J.; Kim, H.S.; Eom, J.S.; Kim, D.H.; Lee, S.S. The potential nutritive value of *Sargassum fulvellum* as a feed ingredient for ruminants. *Algal Res.* **2020**, *45*, 101761. [CrossRef]
25. Ventura, M.R.; Castañón, J.I.R. The nutritive value of seaweed (*Ulva lactuca*) for goats. *Small Rumin. Res.* **1998**, *29*, 325–327. [CrossRef]
26. Panjaitan, T.; Quigley, S.P.; McLennan, S.R.; Poppi, D.P. Effect of the concentration of Spirulina (*Spirulina platensis*) algae in the drinking water on water intake by cattle and the proportion of algae bypassing the rumen. *Anim. Prod. Sci.* **2010**, *50*, 405–409. [CrossRef]
27. Panjaitan, T.; Quigley, S.P.; McLennan, S.R.; Swain, A.J.; Poppi, D.P. Spirulina (*Spirulina platensis*) algae supplementation increases microbial protein production and feed intake and decreases retention time of digesta in the rumen of cattle. *Anim. Prod. Sci.* **2015**, *55*, 535–543. [CrossRef]
28. Becker, E.W. Micro-algae as a source of protein. *Biotechnol. Adv.* **2007**, *25*, 207–210. [CrossRef]
29. Pickard, R.M.; Beard, A.P.; Seal, C.J.; Edwards, S.A. Neonatal lamb vigour is improved by feeding docosahexaenoic acid in the form of algal biomass during late gestation. *Animal* **2008**, *2*, 1186–1192. [CrossRef]
30. Fievez, V.; Boeckaert, C.; Vlaeminck, B.; Mestdagh, J.; Demeyer, D. In vitro examination of DHA-edible micro-algae: 2. Effect on rumen methane production and apparent degradability of hay. *Anim. Feed Sci. Technol.* **2007**, *136*, 80–95. [CrossRef]
31. Valente, L.M.P.; Cabrita, A.R.J.; Maia, M.R.G.; Valente, I.M.; Engrola, S.; Fonseca, A.J.M.; Ribeiro, D.M.; Lordelo, M.; Martins, C.F.; Cunha, L.F.e.; et al. Microalgae as feed ingredients for livestock production and aquaculture. In *Microalgae—Cultivation, Recovery of Compounds and Applications*; Galanakis, C.M., Ed.; Academic Press: Cambridge, MA, USA, 2021; pp. 239–312.
32. Archibeque, S.L.; Ettinger, A.; Willson, B.D. *Nannochloropsis oculata* as a source for animal feed. *Acta Agron. Hung.* **2009**, *57*, 245–248. [CrossRef]
33. Kholif, A.E.; Gouda, G.A.; Hamdon, H.A. Performance and milk composition of nubian goats as affected by increasing level of *Nannochloropsis oculata* Microalgae. *Animals* **2020**, *10*, 2453. [CrossRef]
34. Da Silva, G.G.; de Jesus, E.F.; Takiya, C.S.; del Valle, T.A.; da Silva, T.H.; Vendramini, T.H.A.; Yu, E.J.; Rennó, F.P. Short communication: Partial replacement of ground corn with algae meal in a dairy cow diet: Milk yield and composition, nutrient digestibility, and metabolic profile. *J. Dairy Sci.* **2016**, *99*, 8880–8884. [CrossRef] [PubMed]
35. Stokes, R.S.; van Emon, M.L.; Loy, D.D.; Hansen, S.L. Assessment of algae meal as a ruminant feedstuff: Nutrient digestibility in sheep as a model species. *J. Anim. Sci.* **2015**, *93*, 5386–5394. [CrossRef] [PubMed]

36. Moate, P.J.; Williams, S.R.; Hannah, M.C.; Eckard, R.J.; Auldist, M.J.; Ribaux, B.E.; Jacobs, J.L.; Wales, W.J. Effects of feeding algal meal high in docosahexaenoic acid on feed intake, milk production, and methane emissions in dairy cows. *J. Dairy Sci.* **2013**, *96*, 3177–3188. [CrossRef] [PubMed]
37. Till, B.E.; Huntington, J.A.; Posri, W.; Early, R.; Taylor-Pickard, J.; Sinclair, L.A. Influence of rate of inclusion of microalgae on the sensory characteristics and fatty acid composition of cheese and performance of dairy cows. *J. Dairy Sci.* **2019**, *102*, 10934–10946. [CrossRef] [PubMed]
38. Till, B.E.; Huntington, J.A.; Kliem, K.E.; Taylor-Pickard, J.; Sinclair, L.A. Long term dietary supplementation with microalgae increases plasma docosahexaenoic acid in milk and plasma but does not affect plasma 13,14-dihydro-15-keto PGF2α concentration in dairy cows. *J. Dairy Res.* **2020**, *87*, 14–22. [CrossRef]
39. Sérot, T.; Regost, C.; Arzel, J. Identification of odour-active compounds in muscle of brown trout (*Salmo trutta*) as affected by dietary lipid sources. *J. Sci. Food Agric.* **2002**, *82*, 636–643. [CrossRef]
40. Van Emon, M.L.; Loy, D.D.; Hansen, S.L. Determining the preference, in vitro digestibility, in situ disappearance, and grower period performance of steers fed a novel algae meal derived from heterotrophic microalgae. *J. Anim. Sci.* **2015**, *93*, 3121–3129. [CrossRef]
41. Owens, F.; Goetsch, A. Ruminal fermentation. In *The Ruminant Animal: Digestive Physiology and Nutrition*; Church, D.C., Ed.; Waveland Press: Long Grove, IL, USA, 1993; pp. 145–171.
42. Dewhurst, R.J.; Davies, D.R.; Merry, R.J. Microbial protein supply from the rumen. *Anim. Feed Sci. Technol.* **2000**, *85*, 1–21. [CrossRef]
43. Poppi, D.P.; McLennan, S.R. Protein and energy utilization by ruminants at pasture. *J. Anim. Sci.* **1995**, *73*, 278–290. [CrossRef]
44. Leng, R.A. Factors affecting the utilization of "poor-quality" forages by ruminants particularly under tropical conditions. *Nutr. Res. Rev.* **1990**, *3*, 277–303. [CrossRef]
45. Nagaraja, T.G.; Newbold, C.J.; van Nevel, C.J.; Demeyer, D.I. Manipulation of ruminal fermentation. In *The Rumen Microbial Ecosystem No. 1*; Hobson, P.N., Stewart, C.S., Eds.; Blackie Academic & Professional: London, UK, 1997; pp. 523–632.
46. Bourges, H.; Sotomayor, A.; Mendoza, E.; Chavez, A. Utilization of the alga Spirulina as a protein source. *Nutr. Rep. Int.* **1971**, *4*, 31–43.
47. Wild, K.J.; Steingaß, H.; Rodehutscord, M. Variability of in vitro ruminal fermentation and nutritional value of cell-disrupted and nondisrupted microalgae for ruminants. *GCB Bioenergy* **2019**, *11*, 345–359. [CrossRef]
48. Hippenstiel, F.; Kivitz, A.; Benninghoff, J.; Südekum, K.H. Estimation of intestinal protein digestibility of protein supplements for ruminants using a three-step enzymatic in vitro procedure. *Arch. Anim. Nutr.* **2015**, *69*, 310–318. [CrossRef] [PubMed]
49. Bowen, M.K.; Poppi, D.P.; McLennan, S.R. Efficiency of rumen microbial protein synthesis in cattle grazing tropical pastures as estimated by a novel technique. *Anim. Prod. Sci.* **2017**, *57*, 1702–1712. [CrossRef]
50. Commonwealth Scientific and Industrial Research Organization. *Nutrient Requirements of Domesticated Ruminants*; CSIRO Publications: Collinwood, VIC, Australia, 2007.
51. Koletzko, B.; Schmidt, E.; Bremer, H.J.; Haug, M.; Harzer, G. Effects of dietary long-chain polyunsaturated fatty acids on the essential fatty acid status of premature infants. *Eur. J. Pediatr.* **1989**, *148*, 669–675. [CrossRef]
52. Hoffman, D.R.; Birch, E.E.; Birch, D.G.; Uauy, R.D. Effects of supplementation with omega 3 long-chain polyunsaturated fatty acids on retinal and cortical development in premature infants. *Am. J. Clin. Nutr.* **1993**, *57*, 807S–812S. [CrossRef]
53. SanGiovanni, J.P.; Parra-Cabrera, S.; Colditz, G.A.; Berkey, C.S.; Dwyer, J.T. Meta-analysis of dietary essential fatty acids and long-chain polyunsaturated fatty acids as they relate to visual resolution acuity in healthy preterm infants. *Pediatrics* **2000**, *105*, 1292–1298. [CrossRef]
54. Scollan, N.D.; Dhanoa, M.S.; Choi, N.J.; Maeng, W.J.; Enser, M. Biohydrogenation and digestion of long chain fatty acids in steers fed on different sources of lipid. *J. Agric. Sci.* **2001**, *136*, 345–355. [CrossRef]
55. Bauman, D.E.; Harvatine, K.J.; Lock, A.L. Nutrigenomics, rumen-derived bioactive fatty acids, and the regulation of milk fat synthesis. *Annu. Rev. Nutr.* **2011**, *31*, 299–319. [CrossRef]
56. Jiang, Y.; Fan, K.-W.; Wong, R.T.; Chen, F. Fatty acid composition and squalene content of the marine microalga Schizochytrium mangrovei. *J. Agric. Food Chem.* **2004**, *52*, 1196–2000. [CrossRef]
57. Hauvermale, A.; Kunera, J.; Rosenzweiga, B.; Guerrab, D.; Diltza, S.; Metza, J.G. Fatty acid production in Schizochytrium sp.: Involvement of a polyunsaturated fatty acid synthase and a type I fatty acid synthase. *Lipids* **2006**, *41*, 739–747. [CrossRef] [PubMed]
58. Van Pelt, C.K.; Huang, M.-C.; Tschanz, C.L.; Brenna, J.T. An octaene fatty acid, 4,7,10,13,16,19,22,25- octacosaoctaenoic acid (28:8n;–3), found in marine oils. *J. Lipid Res.* **1999**, *40*, 1501–1505. [CrossRef] [PubMed]
59. Or-Rashid, M.M.; Kramer, J.K.G.; Wood, M.A.; McBride, B.W. Supplemental algal meal alters the ruminal trans-18:1 fatty acid and conjugated linoleic acid composition in cattle. *J. Anim. Sci.* **2008**, *86*, 187–196. [CrossRef] [PubMed]
60. Ashes, J.R.; Siebert, B.D.; Gulati, S.K.; Cuthbertson, A.Z.; Scott, T.W. Incorporation of n-3 fatty acids of fish oil into tissue and serum lipids of ruminants. *Lipids* **1992**, *27*, 629–631. [CrossRef] [PubMed]
61. Polan, C.E.; McNeill, J.J.; Tove, S.B. Biohydrogenation of unsaturated fatty acids by rumen bacteria. *J. Bacteriol.* **1964**, *88*, 1056–1064. [CrossRef]

62. Sehat, N.; Rickert, R.; Mossoba, M.; Kramer, J.; Yurawecz, M.; Roach, J.; Adlof, R.; Morehouse, K.; Fritsche, J.; Eulitz, K.; et al. Improved separation of conjugated fatty acid methyl esters by silver ion-high-performance liquid chromatography. *Lipids* **1999**, *34*, 407–413. [CrossRef]
63. Smith, S.B.; Kawachi, H.; Choi, C.B.; Choi, C.W.; Wu, G.; Sawyer, J.E. Cellular regulation of bovine intramuscular adipose tissue development and composition. *J. Anim. Sci.* **2008**, *87*, 1–38. [CrossRef]
64. Mattos, R.; Staples, C.R.; Thatcher, W.W. Effects of dietary fatty acids on reproduction in ruminants. *Rev. Reprod.* **2000**, *5*, 38–45. [CrossRef]
65. Staples, C.R.; Burke, J.M.; Thatcher, W.W. Influence of supplemental fats on reproductive tissues and performance of lactating cows. *J. Dairy Sci.* **1998**, *81*, 856–871. [CrossRef]
66. Costa, D.F.A.; Quigley, S.P.; Isherwood, P.; McLennan, S.R.; Sun, X.Q.; Gibbs, S.J.; Poppi, D.P. Small differences in biohydrogenation resulted from the similar retention times of fluid in the rumen of cattle grazing wet season C3 and C4 forage species. *Anim. Feed Sci. Technol.* **2019**, *253*, 101–112. [CrossRef]
67. Costa, D.F.A.; Quigley, S.P.; Isherwood, P.; McLennan, S.R.; Sun, X.Q.; Gibbs, S.J.; Poppi, D.P. The inclusion of low quantities of lipids in the diet of ruminants fed low quality forages has little effect on rumen function. *Anim. Feed Sci. Technol.* **2017**, *234*, 20–28. [CrossRef]
68. Costa, D.F.A. Cattle responses to small inclusions of lipids in the diet. *Appl. Food Sci. J.* **2018**, *2*, 21.
69. McCauley, J.; Labeeuw, L.; Jaramillo-Madrid, A.; Nguyen, L.; Nghiem, L.; Chaves, A.; Ralph, P. Management of Enteric Methanogenesis in Ruminants by Algal-Derived Feed Additives. *Curr. Pollut. Rep.* **2020**, *6*, 188–205. [CrossRef]
70. Almeida, A.K.; Hegarty, R.S.; Cowie, A. Meta-analysis quantifying the potential of dietary additives and rumen modifiers for methane mitigation in ruminant production systems. *Anim. Nutr.* **2021**, *7*, 1219–1230. [CrossRef]
71. Lean, I.J.; Golder, H.M.; Grant, T.M.D.; Moate, P.J. A meta-analysis of effects of feeding seaweed on beef and dairy cattle performance and methane yield. *PLoS ONE* **2021**, *16*, e0249053. [CrossRef]
72. Glasson, C.R.K.; Kinley, R.D.; de Nys, R.; King, N.; Adams, S.L.; Packer, M.A.; Svenson, J.; Eason, C.T.; Magnusson, M. Benefits and risks of including the bromoform containing seaweed *Asparagopsis* in feed for the reduction of methane production from ruminants. *Algal Res.* **2022**, *64*, 102673. [CrossRef]
73. Klieve, A.; Harper, K.; Martinez, E.; Ouwerkerk, D. *Increasing Productivity and Reducing Methane Emissions by Supplementing Feed with Dietary Lipids*; Meat & Livestock Australia: Sydney, NSW, Australia, 2012.
74. Sucu, E. Effects of Microalgae Species on Rumen Fermentation Pattern and Methane Production. *Ann. Anim. Sci.* **2020**, *20*, 207–218. [CrossRef]
75. Kiani, A.; Wolf, C.; Giller, K.; Eggerschwiler, L.; Kreuzer, M.; Schwarm, A. In vitro ruminal fermentation and methane inhibitory effect of three species of microalgae. *Can. J. Anim. Sci.* **2020**, *100*, 485–493. [CrossRef]
76. Meehan, D.J.; Cabrita, A.R.J.; Silva, J.L.; Fonseca, A.J.M.; Maia, M.R.G. Effects of *Chlorella vulgaris*, *Nannochloropsis oceanica* and *Tetraselmis* sp. supplementation levels on in vitro rumen fermentation. *Algal Res.* **2021**, *56*, 102284. [CrossRef]
77. Roselet, F.; Vandamme, D.; Muylaert, K.; Abreu, P.C. Harvesting of Microalgae for Biomass Production. In *Microalgae Biotechnology for Development of Biofuel and Wastewater Treatment*; Alam, M.A., Wang, Z., Eds.; Springer Singapore: Singapore, 2019; pp. 211–243.
78. Richmond, A. *Handbook of Microalgal Culture: Biotechnology and Applied Phycology*; Blackwell Publishing Company: Oxford, UK, 2004.
79. Zullaikah, S.; Utomo, A.T.; Yasmin, M.; Ong, L.K.; Ju, Y.H. Ecofuel conversion technology of inedible lipid feedstocks to renewable fuel. In *Advances in Eco-Fuels for a Sustainable Environment*; Azad, K., Ed.; Woodhead Publishing: Sawston, UK, 2019; pp. 237–276.
80. Radmann, E.M.; Reinehr, C.O.; Costa, J.A.V. Optimization of the repeated batch cultivation of microalga *Spirulina platensis* in open raceway ponds. *Aquaculture* **2007**, *265*, 118–126. [CrossRef]
81. Tiburcio, P.C.; Galvez, F.C.F.; Cruz, L.J.; Gavino, V.C. Determination of shelf life of *Spirulina platensis* (MI2) grown in the Philippines. *J. Appl. Phycol.* **2007**, *19*, 727–731. [CrossRef]
82. Caroppo, C. Using Satellite Remote Sensing of Harmful Algal Blooms (HABs) in a Coastal European Site. *Phycologia* **2017**, *56*, 28.
83. Omidi, A.; Pflugmacher, S.; Kaplan, A.; Kim, Y.J.; Esterhuizen, M. Reviewing Interspecies Interactions as a Driving Force Affecting the Community Structure in Lakes via Cyanotoxins. *Microorganisms* **2021**, *9*, 1583. [CrossRef] [PubMed]
84. Costa, D.F.A.; Poppi, D.P.; McLennan, S.R. Beef cattle production in northern Australia—Management and supplementation strategies. In Proceedings of the 7th International Congress on Beef Cattle, Sao Pedro, SP, Brazil, 19–21 December 2012; p. 19.
85. Williams, T.; Wilson, C.; Wynn, P.; Costa, D. Opportunities for precision livestock management in the face of climate change: A focus on extensive systems. *Anim. Front.* **2021**, *11*, 63–68. [CrossRef] [PubMed]
86. Du, T.; Bogush, A.; Edwards, P.; Stanley, P.; Lombardi, A.T.; Campos, L.C. Bioaccumulation of metals by algae from acid mine drainage: A case study of Frongoch Mine (UK). *Environ. Sci. Pollut. Res.* **2022**, *29*, 32261–32270. [CrossRef]
87. Baweja, P.; Kumar, S.; Kumar, G. Organic Fertilizer from Algae: A Novel Approach Towards Sustainable Agriculture. In *Biofertilizers for Sustainable Agriculture and Environment*; Giri, B., Prasad, R., Wu, Q.-S., Varma, A., Eds.; Springer International Publishing: Cham, Switzerland, 2019; pp. 353–370.
88. Magnusson, M.; Vucko, M.J.; Neoh, T.L.; de Nys, R. Using oil immersion to deliver a naturally-derived, stable bromoform product from the red seaweed *Asparagopsis taxiformis*. *Algal Res.* **2020**, *51*, 102065. [CrossRef]
89. Rinehart, K.; Namikoshi, M.; Choi, B. Structure and biosynthesis of toxins from blue-green algae (cyanobacteria). *J. Appl. Phycol.* **1994**, *6*, 159–176. [CrossRef]

90. Hammond, B.G.; Mayhew, D.A.; Holson, J.F.; Nemec, M.D.; Mast, R.W.; Sander, W.J. Safety assessment of DHA-rich microalgae from *Schizochytrium* sp.: II. Developmental toxicity evaluation in rats and rabbits. *Regul. Toxicol. Pharmacol.* **2001**, *33*, 205–217. [CrossRef]
91. Mabeau, S.; Fleurence, J. Seaweed in food products: Biochemical and nutritional aspects. *Trends Food Sci. Technol.* **1993**, *4*, 103–107. [CrossRef]
92. Agricultural and Food Research Council. *Energy and Protein Requirements of Ruminants*; CABI International: Wallingford, UK, 1993.
93. Muizelaar, W.; Groot, M.; van Duinkerken, G.; Peters, R.; Dijkstra, J. Safety and transfer study: Transfer of bromoform present in *Asparagopsis taxiformis* to milk and urine of lactating dairy cows. *Foods* **2021**, *10*, 584. [CrossRef]
94. Codd, G.A.; Morrison, L.F.; Metcalf, J.S. Cyanobacterial toxins: Risk management for health protection. *Toxicol. Appl. Pharmacol.* **2005**, *203*, 264–272. [CrossRef]
95. Hawkins, P.R.; Chandrasena, N.R.; Jones, G.J.; Humpage, A.R.; Falconer, I.R. Isolation and toxicity of *Cylindrospermopsis raciborskii* from an ornamental lake. *Toxicon* **1997**, *35*, 341–346. [CrossRef] [PubMed]
96. Lawton, L.A.; Codd, G.A. Cyanobacterial (Blue-Green Algal) Toxins and their Significance in UK and European Waters. *Water Environ. J.* **1991**, *5*, 460–465. [CrossRef]
97. Borowitzka, M.A. Commercial production of microalgae: Ponds, tanks, tubes and fermenters. *J. Biotechnol.* **1999**, *70*, 313–321. [CrossRef]
98. Gieseck, D.; Tiemeyer, W. Availability and metabolism of purines of single-cell proteins in monogastric animals. *Proc. Nutr. Soc.* **1982**, *41*, 319–327. [CrossRef] [PubMed]
99. Smith, R.H.; McAllan, A.B. Nucleic acid metabolism in the ruminant. *Br. J. Nutr.* **1970**, *24*, 545–556. [CrossRef] [PubMed]

MDPI
St. Alban-Anlage 66
4052 Basel
Switzerland
Tel. +41 61 683 77 34
Fax +41 61 302 89 18
www.mdpi.com

Microorganisms Editorial Office
E-mail: microorganisms@mdpi.com
www.mdpi.com/journal/microorganisms

www.ingramcontent.com/pod-product-compliance
Lightning Source LLC
Chambersburg PA
CBHW061510210926
51850CB00008B/1872
9783036569956